# Electrochemical
# Systems

# Electrochemical

# Systems

John S. Newman
*University of California*

Prentice-Hall, Inc.
Englewood Cliffs, N.J.

**142935**

*Library of Congress Cataloging in Publication Data*

NEWMAN, JOHN
  Electrochemical Systems
  (Prentice-Hall international series in the
physical and chemical engineering sciences
  Includes bibliographical references
  1. Electrochemistry, Industrial.  I. Title.
TP255.N48   660.2'9'7   72-10042
ISBN   0–13–248922–8

© 1973 by Prentice-Hall, Inc.
Englewood Cliffs, N.J.

10  9  8  7  6  5  4  3  2  1

Printed in the United States of America

PRENTICE-HALL INTERNATIONAL, INC. *London*
PRENTICE-HALL OF AUSTRALIA, PTY. LTD. *Sydney*
PRENTICE-HALL OF CANADA LTD., *Toronto*
PRENTICE-HALL OF INDIA PRIVATE LIMITED, *New Delhi*
PRENTICE-HALL OF JAPAN, INC., *Tokyo*

# Contents

v

# Preface

Electrochemistry is involved to a significant extent in the present-day industrial economy. Examples are found in primary and secondary batteries and fuel cells; in the production of chlorine, caustic soda, aluminum, and other chemicals; in electroplating, electromachining, and electrorefining; and in corrosion. In addition, electrolytic solutions are encountered in desalting water and in biology. The decreasing relative cost of electric power has stimulated a growing rôle for electrochemistry. The electrochemical industry in the United States amounts to 1.6 percent of all U.S. manufacturing and is about one third as large as the industrial chemicals industry.[1]

The goal of this book is to treat the behavior of electrochemical systems from a practical point of view. The approach is therefore macroscopic rather than microscopic or molecular. An encyclopedic treatment of many specific systems is, however, not attempted. Instead, the emphasis is placed on fundamentals, so as to provide a basis for the design of new systems or processes as they become economically important.

Thermodynamics, electrode kinetics, and transport phenomena are the three fundamental areas which underlie the treatment, and the attempt is made to illuminate these in the first three parts of the book. These areas are interrelated to a considerable extent, and consequently the choice of the proper sequence of material is a problem. In this circumstance, we have pursued each

[1]G. M. Wenglowski, "An Economic Study of the Electrochemical Industry in the United States," J. O'M. Bockris, ed., *Modern Aspects of Electrochemistry*, no. 4 (London: Butterworths, 1966), pp. 251–306.

subject in turn, notwithstanding the necessity of calling upon material which is developed in detail only at a later point. For example, the open-circuit potentials of electrochemical cells belong, logically and historically, with equilibrium thermodynamics, but a complete discussion requires the consideration of the effect of irreversible diffusion processes.

The fascination of electrochemical systems comes in great measure from the complex phenomena which can occur and the diverse disciplines which find application. Consequences of this complexity are the continual rediscovery of old ideas, the persistence of misconceptions among the uninitiated, and the development of involved programs to answer unanswerable or poorly conceived questions. We have tried, then, to follow a straightforward course. Although this tends to be unimaginative, it *does* provide a basis for effective instruction.

The treatment of these three fundamental aspects is followed by a fourth part, on applications, in which thermodynamics, electrode kinetics, and transport phenomena may all enter into the determination of the behavior of electrochemical systems. These four main parts are preceded by an introductory chapter in which are discussed, mostly in a qualitative fashion, some of the pertinent factors which will come into play later in the book. These concepts are illustrated with rotating cylinders, a system which is moderately simple from the point of view of the distribution of current.

The book is directed toward seniors and graduate students in science and engineering and toward practitioners engaged in the development of electrochemical systems. A background in calculus and classical physical chemistry is assumed.

William H. Smyrl, currently of Sandia Laboratories, prepared the first draft of chapter 2, and Wa-She Wong, now at the General Motors Science Center, wrote the first draft of chapter 5. The author acknowledges with gratitude the support of his research endeavors by the United States Atomic Energy Commission, through the Inorganic Materials Research Division of the Lawrence Berkeley Laboratory.

# *Introduction* 1

Since the fundamental aspects of electrochemistry are to be developed, in order, in parts A, B, and C, this chapter has been designed to introduce important concepts related to the flow of fluids, mass transfer, interfacial phenomena, and electrochemical thermodynamics. For accomplishing this purpose, it seems appropriate to begin with a superficial consideration of the behavior of a particular electrochemical system. In this way the reader can see how these factors act, and interact with each other, to determine system behavior. He can then proceed with an overall view of the ultimate utility and application of the detailed material as it is presented subsequently. Since essentially all the material in chapter 1 will be repeated later with a more thorough development, reference to original work and collateral reading will be postponed.

## 1. Thermodynamics, electrode kinetics, and transport processes

The analysis of electrochemical systems draws primarily on three fundamental areas of electrochemistry.

*Thermodynamics* provides the framework for describing the properties of electrolytic solutions and their dependence on composition, temperature, and pressure. This is a macroscopic science and hence provides an appropriate basis for our studies, since the system behavior need not be correlated with microscopic or molecular concepts. Thermodynamics also provides a framework for describing reaction equilibria and thermal effects, which manifest themselves most directly in equilibrium cell potentials. Furthermore, the driving forces for irreversible processes are conveniently expressed in thermodynamic terms.

Departures from equilibrium conditions are inherent in the application of electrochemical systems. *Electrode kinetics* concerns the nonequilibrium driving force, or *surface overpotential*, necessary to make heterogeneous electrode reactions proceed at appreciable rates. Here again, we seek to express relationships among macroscopically measurable quantities as they will affect system behavior.

Of equal importance are irreversibilities associated with *transport* in electrolytic solutions. These are responsible for ohmic losses and heating in the solutions, and for limited rates of transfer of reactants to electrodes and products away from electrodes.

## 2. Rotating cylinders

To illustrate the applications of these fundamental areas of thermodynamics, electrode kinetics, and transport phenomena, chapter 1 will consider their use in the analysis of a simple electrochemical system; namely, two concentric, cylindrical electrodes with an electrolytic solution in the annulus between the two, as shown in figure 2-1. The application of an electric poten-

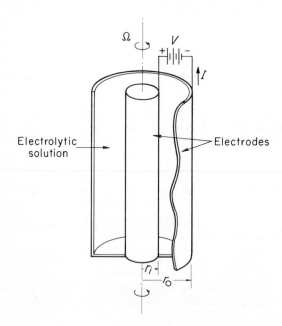

**Figure 2-1.** System of cylindrical electrodes, the inner of which can rotate.

tial difference between these electrodes leads to the passage of current and the occurrence of electrode reactions. We include the possibility of rotating one or both of the electrodes about the axis to stir the solution.

This electrode system is not utilized extensively in any commercial application nor as a routine research tool, although the latter possibility has perhaps been overlooked. It is used here because it can illustrate a variety of complex behavior, and, at the same time, it is simple enough to permit some definite descriptions of this behavior.

## 3. Electrolytic conduction

Suppose that the solution placed between the electrodes is an aqueous solution of cupric sulfate and that the electrodes are copper. The process consists of the dissolution of copper at the positive electrode, the anode;

$$Cu \longrightarrow Cu^{++} + 2e^-, \tag{3-1}$$

the passage of cupric ions through the solution; and the deposition of copper at the negative electrode, the cathode.

In the solution, cupric sulfate will dissociate into charged cupric ions $Cu^{++}$ and sulfate ions $SO_4^=$, which are driven through the solution by the electric field; the cupric ions are driven toward the negative electrode, and the sulfate ions toward the positive electrode. The velocity of migration of an ion is proportional to the charge on the ion and the electric field, the negative of the gradient of electric potential

$$v_i = -z_i u_i F \frac{d\Phi}{dr}. \tag{3-2}$$

Here $\Phi$ is the potential; $r$ is the radial distance from the axis; $z_i$ is the charge number of the ion; and $F$ is Faraday's constant, equal to 96,487 C/equiv. Thus $z_i F$ is the charge per gram ion. The proportionality factor $u_i$ is called the *mobility* of the ion and has units of $cm^2$-mole/J-sec.

The current density $i$ is the sum of the currents carried by the two kinds of ions and has the units of $A/cm^2$. It is obtained by multiplying the ionic velocities by the concentrations and the charges and adding:

$$i = z_+ c_+ F v_+ + z_- c_- F v_-. \tag{3-3}$$

Since the solution is electrically neutral, that is, the charges on the cations are balanced by the charges on the anions, the ionic concentrations are related by

$$z_+ c_+ + z_- c_- = 0. \tag{3-4}$$

In this case $z_+ = 2$, $z_- = -2$, and the ionic concentrations $c_+$ or $c_-$ can be identified with the stoichiometric concentration of cupric sulfate.

Equations 2, 3, and 4 can be combined to yield

$$i = z_+ c_+ F(v_+ - v_-) = -z_+ c_+ F^2 (z_+ u_+ - z_- u_-) \frac{d\Phi}{dr}. \tag{3-5}$$

The coefficient in this relationship between the current density and the gradient of the potential is a transport property of the solution known as the conductivity $\kappa$ with units of ohm$^{-1}$-cm$^{-1}$:

$$\kappa = z_{-}c_{+}F^{2}(z_{+}u_{+} - z_{-}u_{-}). \tag{3-6}$$

Consequently, equation 5 can be written as Ohm's law:

$$i = -\kappa \frac{d\Phi}{dr}. \tag{3-7}$$

This relationship is valid only if the concentration is uniform, as we shall see presently.

Having in hand this expression for the current density, we should like to determine the distribution of potential in the solution and the overall electric resistance of the system. The total current $I$ passing from one electrode to the other is obtained by multiplying the current density by the area at a section in the solution:

$$I = 2\pi\, rHi = -2\pi\, rH\kappa \frac{d\Phi}{dr}, \tag{3-8}$$

where $H$ is the depth of solution. Since $I$ is a constant independent of position, this equation can be integrated to yield the potential distribution:

$$\Phi(r) - \Phi(r_i) = -\frac{I}{2\pi\, H\kappa} \ln \frac{r}{r_i}, \tag{3-9}$$

where $r_i$ is the radius of the inner electrode. Correspondingly, the relationship between the current $I$ and the potential difference $V$ applied between the electrodes is

$$V = \Phi(r_i) - \Phi(r_0) = \frac{I}{2\pi\, H\kappa} \ln \frac{r_0}{r_i} = IR, \tag{3-10}$$

where we can identify the resistance of this system as

$$R = \frac{\ln \dfrac{r_0}{r_i}}{2\pi\, H\kappa}. \tag{3-11}$$

The conductivity of an aqueous cupric sulfate solution is relatively low, say $\kappa = 0.00872$ ohm$^{-1}$-cm$^{-1}$ for a 0.1 $M$ solution. Then for $H = 10$ cm, $r_0 = 3$ cm, and $r_i = 2$ cm, we calculate a resistance of $R = 0.74$ ohm.

This analysis illustrates the application of Ohm's law and conservation of current in a cylindrical annulus where the area available for the passage of current varies with radial position. The potential distribution is sketched in figure 3-1, which shows that the electric field becomes greater near the inner electrode.

As has been said, the cations migrate toward the cathode and the anions toward the anode, and both contribute to the current density, as indicated in equation 3. The fraction of the current carried by an ion (in the absence of concentration variations) is a transport property of the solution known

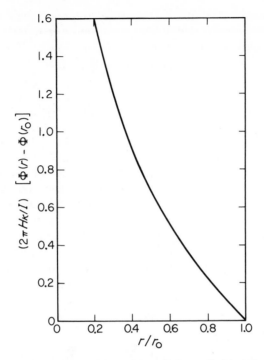

**Figure 3-1.** Distribution of potential applied to cylindrical electrodes.

as the *transference number*. For a solution containing only two species of ions, we have, from equations 2 and 3,

$$t_+ = \frac{z_+ c_+ F v_+}{i} = \frac{z_+ u_+}{z_+ u_+ - z_- u_-}. \tag{3-12}$$

The value of the transference number of cupric ions in 0.1 $M$ cupric sulfate solution is about 0.363.

A paradox appears at this point. Since we have considered the solution to be of uniform concentration, both cupric ions and sulfate ions contribute to the conduction mechanism in the solution. However, only the cupric ions enter into the electrode reaction and pass through the electrode-solution interface. Thus, if the sulfate ions are carrying part of the current in the solution but are not being passed through the electrode boundary, they must accumulate near the anode and become depleted near the cathode.

This brings up the consideration of concentration variations and diffusion; these subjects will be discussed later. Nevertheless, the material discussed in the present section does have applicability just after current begins to pass and before concentration differences can develop, and also if there is sufficient stirring to ensure a uniform concentration. Furthermore, the two transport properties introduced, the conductivity and the transference number, contain information needed to treat solutions with concentration gradients. Thus, they are used in such situations even though their physical interpretation must be modified, because Ohm's law no longer applies and transport mechanisms other than migration are present.

## 4. Fluid flow

We have included the possibility of rotating one of our electrodes. Flow patterns of the electrolytic solution produced by such rotation are quite important in determining the behavior of electrochemical systems, since they provide a mechanical mixing that helps to reduce concentration variations and contributes materially to the transport of reactants to electrode surfaces. In the rotating-cylinder system, these flow patterns are sufficiently complicated to warrant separate discussion.

Very low rotation speeds lead to simple flow of the fluid in concentric circles as, shown in figure 4-1. Such a flow is not of much practical interest because the fluid velocity is perpendicular to the direction of mass transfer. The flow simply carries the material in circles and does not lead to any enhancement of the rate of mass transfer.

This simple flow pattern becomes unstable at higher rotation speeds, particularly if the inner electrode rotates, as we shall suppose here. The flow then has a cellular motion superimposed upon the flow around the inner

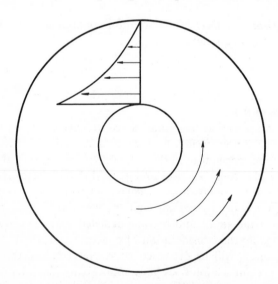

**Figure 4-1.** Velocity distribution for concentric streamlines between rotating cylinders.

cylinder. These so-called Taylor vortices are shown in figures 4-2 and 4-3. Now there is a component of the velocity in the direction from one cylinder to the other, and rates of mass transfer can be enhanced. However, a deposit of copper made under these conditions will be irregular, corresponding to the cellular flow pattern.

At still higher rotation speeds, the flow becomes turbulent and is characterized by rapid and random fluctuations of velocity and pressure. These include a fluctuating velocity component in the direction from one cylinder to the other. Hence, the rate of mass transfer can be enhanced considerably, in a uniform manner over the surface of the cylinder. At a solid surface, the fluid velocity is equal to that of the solid, since the fluid cannot flow through the surface and since frictional effects do not allow a discontinuity in the tangential velocity. For this reason, the velocity fluctuations die away as a solid surface is approached.

For the rotating cylinders, we thus encounter three flow regimes, which we might not have anticipated. The simple laminar flow with concentric streamlines requires the least torque to rotate the inner cylinder and leads

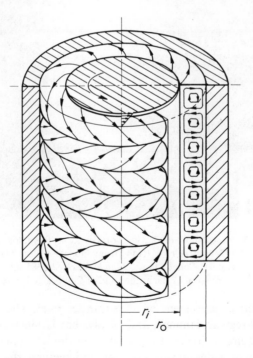

**Figure 4-2.** Sketch of Taylor vortices (after H. Schlichting, *Boundary-Layer Theory*, 1968, p. 501, with permission of McGraw-Hill Book Company).

**Figure 4-3.** Photograph of Taylor vortices at a Reynolds number of
143 with $r_0/r_i = 1.144$ (from Donald Coles, "Transition in
circular Couette flow," *J. Fluid Mech.*, *21*, 385–425 (1965), with
permission of the author and of Cambridge University Press).

to no particular enhancement of mass-transfer rates. The flow with Taylor
vortices is still regular, laminar, and steady, but is much more complex to
analyze. It requires a higher torque and contributes to mass-transfer rates.
Turbulent flow provides effective stirring and requires the highest torques.
Turbulent flow cannot be analyzed satisfactorily from fundamental principles;
but many geometric arrangements, including rotating cylinders, have been
studied empirically so that one can predict with confidence the effect on mass
transfer.

The flow between rotating cylinders is characterized by two dimensionless ratios: the ratio of the gap distance $r_0 - r_i$ to the inner radius $r_i$; and the Reynolds number $Re = (r_0 - r_i)r_i\Omega/\nu$, where $\Omega$ is the rotation speed (radian/sec) and $\nu$ is the kinematic viscosity of the fluid (cm²/sec). (If the outer cylinder also rotates, a third dimensionless ratio, say, the ratio of the rotation speeds of the inner and outer cylinders, is also required.) For small values of $(r_0 - r_i)/r_i$, the simple laminar flow is stable for a Taylor number

$$Ta = Re^2 \frac{r_0 - r_i}{r_i} = \frac{r_i\Omega^2}{\nu^2}(r_0 - r_i)^3 \qquad (4\text{-}1)$$

less than 1708. At higher values of the Taylor number, one encounters Taylor vortices or turbulent flow. For example, for $(r_0 - r_i)/r_i = 0.19$, turbulent flow prevails for Reynolds numbers greater than 3960 or Taylor numbers greater than about $3 \times 10^6$.

These conditions apply when only the inner cylinder is rotated. For rotation of the outer electrode, the simple laminar flow is considerably more stable, although turbulent flow can still be achieved. The criteria determining the prevailing flow regime are, of course, more complex when one includes the possibility of rotating both electrodes, possibly in opposite directions. It is usually desirable to rotate the inner electrode in order to achieve turbulent flow and to operate under conditions which have been well characterized.

## 5.  The diffusion layer

We return now to the treatment of concentration variations, touched on in section 3. The flux $\mathbf{N}_i$ of a species $i$ is equal to its velocity multiplied by its concentration, and represents the number of moles passing per unit time through a unit area oriented perpendicular to the velocity. In dilute solutions, equation 3-2 can be generalized to express the flux of an ionic species as

$$\mathbf{N}_i = c_i\mathbf{v}_i = -z_iu_iFc_i\,\nabla\Phi - D_i\,\nabla c_i + c_i\mathbf{v}. \qquad (5\text{-}1)$$

This equation is written in vector notation; its component in the radial direction is

$$N_{ir} = -z_iu_iFc_i\frac{\partial\Phi}{\partial r} - D_i\frac{\partial c_i}{\partial r} + c_iv_r. \qquad (5\text{-}2)$$

The first term on the right represents transport by migration, just as in equation 3-2. The second term represents transport by diffusion and is proportional to the gradient of concentration, the proportionality factor being called the ionic diffusion coefficient, with units of cm²/sec. The last term represents transport by convection, with the fluid velocity $\mathbf{v}$.

Figure 5-1 is a sketch of the concentration profile of cupric sulfate in the solution between the electrodes. Consider first the dashed curve which would apply when there is no radial component of the velocity, for example, in slow, laminar flow with concentric streamlines. In the steady state, there is no transport of sulfate ions, since only the cupric ions react at the electrodes. The sulfate ions are driven by migration toward the positive electrode, that is, toward the left in figure 5-1. Since there is no net movement of these ions, this tendency for migration is exactly compensated by a tendency for diffusion toward the right, toward a region of lower concentration. Thus, the precise form of the concentration profile is dictated by this requirement for a balance between diffusion and migration of sulfate ions.

The cupric ions, on the other hand, are driven toward the right by migration. The diffusion tendency is also toward the right, thus aiding migration; and cupric ions can be said to carry all the current from the anode to the cathode.

The solid curve in figure 5-1 indicates the effect of stirring the solution by rotation of the inner cylinder to create turbulent flow. The convective transport tends to eliminate concentration variations in the middle of the annular space. However, turbulent fluctuations are damped near the solid electrodes, and in these regions migration and diffusion again become the dominant modes of transport. The electrode reactions still lead to depletion of the solution near the cathode and an enhancement of the concentration near the anode.

For high rates of stirring, we are led to the concept that concentration variations are confined to thin *boundary layers* near the electrodes. This concept has important consequences in simplifying the quantitative treatment of transport in electrochemical systems. As the rotation speed is increased, these diffusion layers become thinner.

Figure 5-1 shows the concentration going to zero at the cathode surface. This is, of course, a limiting condition. At zero current, the concentration will be uniform and equal to its average value. As the current is increased, departures from the average concentration will be magnified in the vertical direction, but the shape of the curve will be similar to one of those shown in figure 5-1. At a current corresponding to the *limiting current* for copper deposition, the concentration will drop to zero at the cathode. A higher current can be passed only if another electrode process, evolution of hydrogen in this case, also begins to occur.

At the surfaces of the electrodes, the velocity in the radial direction is zero, and the flux of sulfate ions is zero. Consequently, if equation 2 is written for both the positive and the negative ions, the potential derivative can be eliminated between these two equations with the aid of the electroneutrality equation 3-4. The current density can then be related to the concentration derivative at the electrode surface by means of equation 3-3, with the result

$$i = -\frac{z_+ FD}{1 - t_+} \frac{\partial c_+}{\partial r} \quad \text{at an electrode,} \tag{5-3}$$

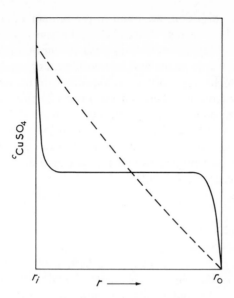

**Figure 5-1.** Concentration profile in the annular space between the electrodes. The dashed curve refers to the absence of a radial component of velocity. The solid curve refers to the presence of a turbulent flow regime.

where
$$t_+ = \frac{z_+ u_+}{z_+ u_+ - z_- u_-}$$ (5-4)

(see equation 3-12) and

$$D = \frac{z_+ u_+ D_- - z_- u_- D_+}{z_+ u_+ - z_- u_-}.$$ (5-5)

This equation applies at currents up to the limiting current and demonstrates that the concentration gradients become steeper as the current is increased. Hence, the way that figure 5-1 is drawn indicates that the limiting current will be much higher when turbulent flow is present than when there is simple flow with concentric streamlines. In the latter case, we can estimate that the limiting current density at the cathode is 0.37 mA/cm² for a 0.1 $M$ cupric sulfate solution with $r_0 = 3$ cm and $r_i = 2$ cm. If we now rotate the inner cylinder at 900 rpm, corresponding to a Reynolds number of 20,000, the limiting current density should be increased to about 79 mA/cm². Thus, stirring of the solution will have a large effect on the permissible currents that can be passed.

We might note that the transference number $t_+$ finds application in equation 3 in a region of nonuniform concentration. It is still a fundamental transport property related to the mobilities by equation 4, but it no longer represents the fraction of the current carried by an ionic species. Because of the contribution of diffusion, we found in the present situation that the cupric ion carries all the current. Nevertheless, the transference number retains its numerical value, say, $t_+ = 0.363$ in a 0.1 $M$ cupric sulfate solution. The coefficient $D$ turns out to be the diffusion coefficient of cupric sulfate, governing the rate at which cupric sulfate would diffuse in the absence of current. It represents a compromise between the ionic diffusion coefficients of the cupric ions and the sulfate ions. For example, if $D_+ = 0.713 \times 10^{-5}$ cm²/sec and $D_- = 1.065 \times 10^{-5}$ cm²/sec, then $D = 0.854 \times 10^{-5}$ cm²/sec.* The conductivity $\kappa$, the transference number $t_+$, and the diffusion coefficient $D$ are the three transport properties that are measured directly for mass transport in a binary electrolytic solution.

## 6. Concentration cells

We need to examine further the consequences of the concentration variations between the cylinders. The potential effect treated in this section enters into the determination of the *concentration overpotential* in section 7.

Figure 6-1 shows a simple electrochemical cell consisting of two copper electrodes partially immersed in two cupric sulfate solutions of different concentrations. The two solutions are separated by a porous glass disk that prevents rapid mixing of the solutions while allowing the passage of current and slow diffusion between the solutions.

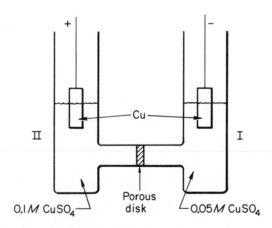

**Figure 6-1.** Concentration cell.

*These represent values of $D_+$, $D_-$, and $D$ at infinite dilution (see table 75-1). For a 0.1 $M$ copper sulfate solution, $D$ is about $0.558 \times 10^{-5}$ cm²/sec.

Because of the concentration difference, there is a tendency for cupric ions to discharge from the 0.1 $M$ solution and for copper to dissolve into the 0.05 $M$ solution. This manifests itself in a potential difference between the electrodes in order to prevent the flow of current, the electrode in the more concentrated solution being positive relative to the other electrode. If these electrodes were connected through an external resistor, current would flow through the resistor from the positive to the negative electrode and through the solution from the negative to the positive. This will speed the process of reducing the concentration difference between the solutions, which process is already occurring by diffusion through the porous disk.

In the absence of the flow of current, the potential difference between the electrodes can be expressed approximately as

$$U = \Phi_I - \Phi_{II} = (1 - t_+)\frac{RT}{F}\ln\frac{c_I}{c_{II}}, \tag{6-1}$$

where $t_+$ is the transference number of cupric ions, assumed to be constant in the region of varying concentration; $R$ is the universal gas constant; and where activity-coefficient corrections have been ignored.

## 7. Concentration overpotential

As indicated in the preceding section, a concentration variation leads to potential differences between electrodes introduced into a solution. If we take into account the presence of concentration variations near the electrodes in an electrochemical cell (see figure 5-1), we find that the relationship between current and applied potential is not as simple as that expressed in equation 3-10, which considered only the ohmic potential drop in a solution of uniform concentration.

For the purpose of assessing potential variations in a solution, it is convenient to conceive of placing *reference electrodes* into the solution at appropriate locations. In the present case, we mean small copper electrodes, through which only a negligible current is passed and which do not alter conditions significantly from those prevailing in their absence. We then imagine that we can use these idealized electrodes to probe potential variations in the solution.

In the absence of concentration variations, two such electrodes will measure the ohmic potential drop between two points:

$$\Delta\Phi = \Phi_1 - \Phi_2 = \frac{1}{\kappa}\int_1^2 \mathbf{i}\cdot\mathbf{d\ell}. \tag{7-1}$$

For example, equation 3-9 represents such a potential difference between a reference electrode at a position $r$ and another placed adjacent to the inner electrode at $r = r_i$. Thus, the reference electrodes can be placed in a solution carrying current even though the reference electrodes themselves pass no current.

In the presence of concentration variations, it is necessary to include the potential difference $U$ that would exist in the absence of current flow

$$\Delta\Phi = \Phi_1 - \Phi_2 = U + \int_1^2 \frac{\mathbf{i} \cdot \mathbf{d\ell}}{\kappa}. \qquad (7\text{-}2)$$

The conductivity $\kappa$ must now be included under the integral sign since it is no longer constant. $U$ can be expressed approximately by equation 6-1, with the result

$$\Delta\Phi = \Phi_1 - \Phi_2 = (1 - t_+)\frac{RT}{F}\ln\frac{c_1}{c_2} + \int_1^2 \frac{\mathbf{i} \cdot \mathbf{d\ell}}{\kappa}. \qquad (7\text{-}3)$$

To decompose the potential applied to the rotating-cylinder cell, we need to introduce three reference electrodes, one adjacent to each of the cylindrical electrodes and one in the middle, as indicated in figure 7-1. It is then customary to speak of the potential difference $\Phi_1 - \Phi_2$ as composed of an

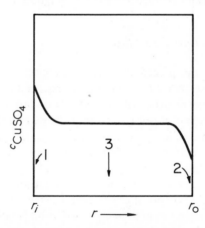

**Figure 7-1.** Placement of reference electrodes (1, 2, and 3) in the solution between cylindrical electrodes. The indicated concentration profile corresponds to turbulent flow at a current somewhat below the limiting current.

ohmic portion and a *concentration overpotential*. Such a decomposition is apparent in equation 2 or 3. The ohmic portion is proportional to the current and will disappear immediately if the current is interrupted. This provides a means for distinguishing between the two. The total potential difference between two reference electrodes is measured with the current flowing. The contribution of concentration variations is that value measured just after the current is interrupted but before the concentration distribution can change

by diffusion or convection. The difference between these two measurements
is the ohmic contribution. (It should be pointed out that in most geometries
interruption of the external current does not automatically ensure that the
local current density is everywhere equal to zero. For this reason, measurement of ohmic potential drops by interruption methods should perhaps not
be encouraged.)

Frequently one tries to decompose further the concentration overpotential into contributions related to the concentration variations near the cathode
and those near the anode. For this purpose, we use the third reference electrode, located in the bulk of the solution. The concentration overpotential
at the anode is then the value of $\Phi_1 - \Phi_3$, and the cathodic overpotential
is that of $\Phi_2 - \Phi_3$, just after the current is interrupted. This decomposition
of the concentration overpotential depends on the concept of thin diffusion
layers near the electrodes and the existence of a bulk solution where the
concentration does not vary significantly. Then the anodic and cathodic
overpotentials are independent of the precise location of the third reference
electrode, since it is in a region of uniform concentration and there is no current flow.

Electrode potentials and overpotentials for working electrodes are
commonly measured relative to the solution. This explains the convention
of defining the concentration overpotential at a working electrode as the
potential of a reference electrode near its surface relative to one in the bulk
solution (after interruption of the current). Consequently, the overpotential
at an anode will generally be positive, and that at a cathode will be negative.
In most geometries, the concentration will vary along an electrode surface,
and, as a result, the concentration overpotential will also depend on position
along that surface.

Another decomposition of potentials is conceivable and is frequently
useful as an aid in calculations. Let $\Delta\Phi$ be the potential difference between
a reference electrode adjacent to the surface and one in the bulk, and let
$\Delta\Phi_{\text{ohm}}$ be that which would be measured between the same electrodes if the
current distribution were unchanged but there were no concentration differences between the electrode surface and the bulk solution. The difference
between the two is a concentration overpotential which can clearly be associated with concentration changes at the electrode and will be denoted by $\eta_c$

$$\eta_c = \Delta\Phi - \Delta\Phi_{\text{ohm}}. \tag{7-4}$$

(The symbol $\eta$ is frequently used for overpotentials.)

Based on the concept of thin diffusion layers, $\eta_c$ can be written as

$$\eta_c = U + \int_0^b \left(\frac{1}{\kappa} - \frac{1}{\kappa_b}\right) i_y \, dy, \tag{7-5}$$

where $y$ is the perpendicular distance from the electrode and $b$ denotes the bulk solution. Since $\Delta\Phi$ is given by equation 2, subtraction of $i_y/\kappa_b$ corresponds to subtracting the ohmic contribution that would exist in the absence of concentration variations. This definition of concentration overpotential differs from the preceding one by the integral in equation 5, which is the difference in the ohmic contribution with and without concentration variations. This term can thus be logically associated with concentration variations near electrodes.

In the present case, with the conductivity related to the concentration by equation 3-6 and with an assumed linear variation of concentration in the diffusion layer, the concentration overpotential can be expressed as

$$\eta_c = \frac{RT}{F}\left[\ln\frac{c_0}{c_b} + t_+\left(1 - \frac{c_0}{c_b}\right)\right], \tag{7-6}$$

where $c_0$ refers to the concentration immediately adjacent to the surface of the working electrode.

One of the advantages of this modified definition of the concentration overpotential stems from the fact that the potential difference $\Phi_1 - \Phi_2$ can now be expressed as

$$\Phi_1 - \Phi_2 = \Delta\Phi_{\mathrm{ohm}} + \eta_c(\text{anode}) - \eta_c(\text{cathode}), \tag{7-7}$$

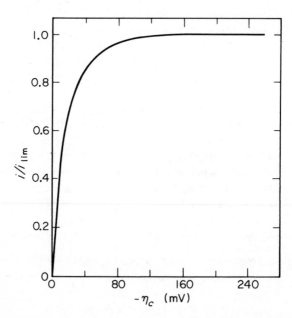

**Figure 7-2.** Concentration overpotentials at a cathode in 0.1 $M$ CuSO$_4$.

where in this case $\Delta\Phi_{ohm}$ can be calculated without regard for the concentration variations near electrodes and is, in fact, given by equation 3-10. The difference between these two definitions is probably not important except for solutions of a single electrolyte, such as the cupric sulfate solutions being considered here. In detailed calculations for many electrochemical systems, it is desirable to calculate the potential distribution in the bulk solution somewhat separated from the details of the calculations of concentration variations near electrodes. The principal disadvantage of the second decomposition of potentials lies in the fact that it is different from the decomposition of potential differences by interrupter methods.

Figure 7-2 shows the concentration overpotential at a cathode for currents up to the limiting current.

## 8. Surface overpotential

An additional contribution to the overall cell potential is the driving force required to make the electrode reactions proceed at appreciable rates. The *surface overpotential* is defined as the potential of the working electrode relative to a reference electrode of the same kind placed in the solution adjacent to the surface of the working electrode. This reference electrode is thus one of those used to define the concentration overpotential. For example, at the anode in figure 7-1, the surface overpotential is

$$\eta_s = \Phi(\text{anode}) - \Phi_1. \tag{8-1}$$

The rate of the electrode reactions is the rate of dissolution or deposition of copper, and this can be conveniently measured by the current density at the electrode. By convention, the current density is positive when it flows from the electrode into the solution. Thus, current densities are positive at anodes and negative at cathodes.

The current density depends on the driving force and thus is related to the surface overpotential and the composition of the solution at the interface, as well as the temperature. For example, the current density can frequently be expressed as

$$i = i_0 \left[ \exp\left(\frac{\alpha_a F}{RT}\eta_s\right) - \exp\left(-\frac{\alpha_c F}{RT}\eta_s\right) \right]. \tag{8-2}$$

This is similar to the expression of the rate of a nonelectrochemical, heterogeneous reaction. The first exponential term can be regarded as representing the rate of the anodic process, and the second term, that of the cathodic process. These are governed by activation energies which depend on the surface overpotential.

When $\eta_s = 0$, the anodic and cathodic currents are equal in magnitude to each other and to $i_0$. This is an important kinetic parameter and is termed the *exchange current density*. Different reactions may have exchange current densities that differ by many orders of magnitude. Furthermore, the exchange current density depends strongly on the composition at the interface and on the temperature. $\alpha_a$ and $\alpha_c$ are two additional kinetic parameters and may be termed *apparent transfer coefficients*. They usually have values between 0.2 and 2.

Equation 2 serves to show explicitly a typical dependence of reaction rate on surface overpotential. The dependence on composition and temperature is not shown explicitly, since $i_0$ depends on these factors in an unspecified manner. Many reactions, such as those involving oxide formation, are complicated and do not follow equation 2. Even the relatively simple copper electrode is complicated. At high anodic rates, appreciable amounts of cuprous ions are formed. These subsequently disproportionate in the solution to yield cupric ions and copper, which precipitates,

$$2\,Cu^+ \longrightarrow Cu^{++} + Cu. \tag{8-3}$$

It should also be noted that the value of $i_0$ can depend on the composition and preparation of the electrode and on the presence of impurities.

With these words of caution, we introduce equation 2 to describe electrochemical kinetics in general or in the absence of detailed information. It should be emphasized that the goal here is to relate the reaction rate or current density to conditions prevailing at the interface itself. Thus, we define the surface overpotential in terms of a reference electrode well within the diffusion layer, adjacent to the electrode surface; and we seek to relate the kinetic parameters to the composition adjacent to the electrode, not to the composition of the bulk solution or to concentration gradients near the surface.

The exchange current density is a measure of the freedom from kinetic limitations. A reaction with a large value of $i_0$ is frequently said to be *fast* or *reversible*. For large values of $i_0$, a given current density can be obtained with small surface overpotentials.

Figures 8-1 and 8-2 illustrate the behavior of equation 2. One can see that higher values of $i_0$ give higher values of the current density at a given surface overpotential. Figure 2 is a so-called Tafel plot, used because, at high surface overpotentials, one of the terms in equation 2 becomes negligible, and a straight line is obtained on a semi-logarithmic plot. Thus:

$$i = i_0 \exp\left(\frac{\alpha_a F}{RT}\eta_s\right) \quad \text{or} \quad \eta_s = 2.303\frac{RT}{\alpha_a F} \log\left(\frac{i}{i_0}\right)$$

$$\text{for} \quad \alpha_a F\eta_s \gg RT; \tag{8-4}$$

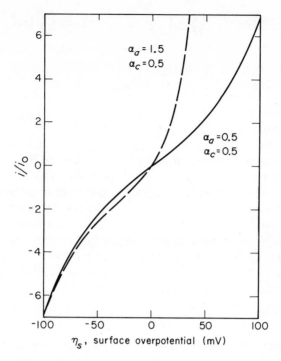

**Figure 8-1.** Dependence of current density on surface overpotential at 25°C.

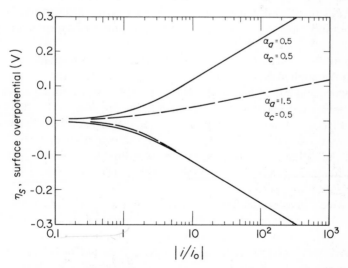

**Figure 8-2.** Tafel plot of relationship between current density and surface overpotential at 25°C.

$$i = -i_0 \exp\left(-\frac{\alpha_c F}{RT}\eta_s\right) \quad \text{or} \quad \eta_s = -2.303\frac{RT}{\alpha_c F}\log\left|\frac{i}{i_0}\right|$$

$$\text{for} \quad \alpha_c F \eta_s \ll -RT. \tag{8-5}$$

The *Tafel slope*, either $2.303\ RT/\alpha_a F$ or $2.303\ RT/\alpha_c F$, is thus seen to be inversely proportional to the apparent transfer coefficients.

## 9. The cell potential

We have now introduced enough of the elements to be in a position to discuss the composition of the overall cell potential. This is due first of all to the ohmic potential drop in the solution, discussed in section 3. In addition, there is a potential loss associated with the concentration variations in the solution near electrodes, which we have termed the concentration overpotential. Finally, there is the surface overpotential due to the limited rates of the electrode reaction. The sum of these is the cell potential. Its decomposition into these parts is facilitated by the definitions of $\eta_c$ and $\eta_s$, involving reference electrodes in various parts of the solution.

$$V = \Phi(\text{anode}) - \Phi(\text{cathode})$$
$$= \Phi(\text{anode}) - \Phi_1 + (\Phi_1 - \Phi_2) - \Phi(\text{cathode}) + \Phi_2$$
$$= \eta_s(\text{anode}) + \eta_c(\text{anode}) + \Delta\Phi_{\text{ohm}} - \eta_c(\text{cathode}) - \eta_s(\text{cathode}). \tag{9-1}$$

$\Delta\Phi_{\text{ohm}}$ is given by equation 3-10, $\eta_c$ by equation 7-5 or approximately by equation 7-6, and $\eta_s$ by equation 8-2. The terms $\eta_c$ and $\eta_s$ for the cathode enter with negative signs because of the conventions that have been adopted. Since they are generally negative, they make a positive contribution to the cell potential. Thus, none of these terms, ohmic drop or overpotentials, represents a source of energy.

Figure 9-1 indicates what we should expect for the response of the current $I$ to the applied potential $V$. At low currents, most of the applied potential is consumed by ohmic losses. However, there is a limiting current for copper deposition, estimated in section 5 to be $2\pi r_0 H i_{\text{lim}} = 1.5$ A when the inner electrode is rotated at 900 rpm. The anodic overpotentials are not shown in the figure. At sufficiently large voltages, the evolution of hydrogen will begin at the cathode, as indicated by a broken line on the figure. Any additional stirring due to hydrogen bubbles will, at the same time, increase the limiting rate of copper deposition.

In a cell involving different reactions at the anode and cathode (for example, batteries or cells for production of chlorine and sodium hydroxide), an additional potential related to the overall reaction would be involved. For the chosen example of deposition and dissolution of copper, no such term is necessary in equation 1.

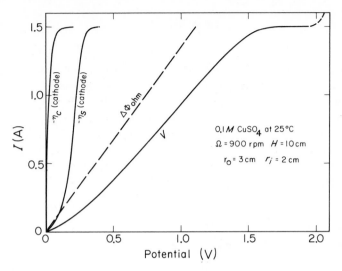

**Figure 9-1.** Current-potential relations for rotating cylinders, the inner of which rotates at 900 rpm. The overpotentials for the anode are not shown.

## 10. Supporting electrolyte

We have given rather detailed discussions of the phenomena involved in the passage of current through a solution containing cupric sulfate. In practice, it is common to add an inert electrolyte, which does not participate in the electrode reactions, principally to increase the conductivity and thereby decrease the ohmic potential drop. For example, a solution 0.1 $M$ in $CuSO_4$ and 1.53 $M$ in $H_2SO_4$ will have a conductivity of about 0.548 ohm$^{-1}$-cm$^{-1}$. Equation 3-11 will then yield a resistance of only 0.0118 ohm, in contrast to a value of 0.74 ohm for a 0.1 $M$ $CuSO_4$ solution containing no sulfuric acid.

A second effect of the addition of sulfuric acid would be the reduction of the contribution of migration to the flux of cupric ions. Since the electric field would be reduced substantially, migration becomes unimportant compared to diffusion and convection. Consequently, the limiting current for copper deposition at the cathode will also be smaller. With due allowance for the variation of viscosity and diffusion coefficients on adding sulfuric acid, we now estimate the limiting current density to be 48 mA/cm² at 900 rpm in contrast to the value of 79 mA/cm² estimated in section 5.

In the presence of an excess of sulfuric acid, we should express the concentration overpotential in the approximate form

$$\eta_c = \frac{RT}{2F} \ln \frac{c_0}{c_b},$$    (10-1)

instead of using equation 7-6. Here we can neglect the variation of conductivity in the diffusion layer; furthermore, the neglect of activity-coefficient corrections is a better approximation here than in the case of a solution of cupric sulfate alone. Finally, the addition of sulfuric acid can be expected to modify the parameters used to describe the electrode kinetics.

Figure 10-1 is a sketch of the concentration profile of cupric ions in the diffusion layer near the cathode. In the absence of added sulfuric acid, both migration and diffusion will tend to drive cations toward the electrode, since the electric field is in this direction and the concentration decreases toward

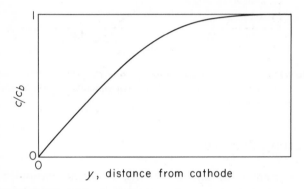

**Figure 10-1.** Concentration profile of cupric ions in the diffusion layer near the cathode. The scale of distance will depend on the rotation speed since the diffusion layer becomes thinner at high rates of stirring.

the electrode. Far from the electrode, convective transport dominates, and the concentration becomes uniform. Because of the depletion of the solution, the electric field is higher near the electrode, and migration is important even though the concentration decreases. (The migration flux is proportional to the electric field times the concentration, see equation 5-2.) Equation 5-3 includes the contribution of migration in a solution of a single electrolyte. Addition of sulfuric acid decreases the electric field and effectively eliminates the contribution of migration. This reduces the limiting current or the rate of transport of cupric ions under a given convective condition.

Figure 10-2 indicates the expected relationship of current to applied potential in the presence of sulfuric acid. The ohmic potential drop is reduced relative to figure 9-1, and the concentration and surface overpotentials now constitute a larger fraction of the applied potential. The magnitude of the limiting current is also reduced, as discussed above.

The conductivity of the solution could also have been increased by adding more cupric sulfate. However, in refining a precious metal, it is desirable to maintain the inventory in the system at a low level. Furthermore, the solubility of cupric sulfate is only 1.4 $M$. To avoid supersaturation at the anode, we might set an upper limit of 0.7 $M$, at which concentration the conductivity is still only 0.037 ohm$^{-1}$-cm$^{-1}$. An excess of supporting electrolyte is usually used in electroanalytical chemistry and in studies of electrode kinetics or of mass transfer, not only because the potential variations in the solution are kept small but also because activity coefficients, transport properties, and even the properties of the interface change little with small changes of the reactant concentration.

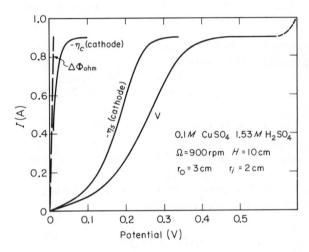

**Figure 10-2.** Current-potential relations with sulfuric acid present.

## 11. Free convection

If we now stop the rotation of the inner electrode, we find that we have not eliminated convection, nor is the concentration profile the same as that for simple concentric streamlines (see figure 5-1). Instead we find that the concentration differences that exist near the electrodes produce density differences, which in turn result in fluid motion near the electrode surfaces. This is known as *free* or *natural convection* and is a very important mode of stirring in many electrochemical systems.

The less dense fluid near the cathode will flow up, and the denser fluid near the anode will flow down. The resulting pattern of streamlines is shown in figure 11-1. A limiting current, corresponding to a zero concentration of

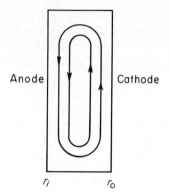

Anode                              Cathode

$r_i$                    $r_o$

**Figure 11-1.** Streamlines for free convection in the annular space between two cylindrical electrodes.

cupric ions along the cathode surface, will still manifest itself in this system. The corresponding current distribution on the cathode will now be nonuniform, tending to be high near the bottom and decreasing farther up the cathode as the solution becomes depleted while flowing along the electrode surface.

In the present case (0.1 $M$ CuSO$_4$, $H = 10$ cm), the average limiting current density can be estimated to be about 9.1 mA/cm$^2$. This can be compared with estimates of section 5: 0.37 mA/cm$^2$ in the absence of a radial component of the velocity and 79 mA/cm$^2$ for a rotation speed of 900 rpm. Thus, free convection by itself is comparable to moderate forced convection achieved by rotating the inner electrode.

## NOTATION

| | |
|---|---|
| $c_i$ | concentration of species $i$, mole/cm$^3$ |
| $c_b$ | concentration in bulk solution, mole/cm$^3$ |
| $c_0$ | concentration at electrode surface, mole/cm$^3$ |
| $D$ | diffusion coefficient of electrolyte, cm$^2$/sec |
| $D_i$ | diffusion coefficient of species $i$, cm$^2$/sec |
| $F$ | Faraday's constant, 96,487 C/equiv |
| $H$ | depth of solution, cm |
| $i$ | current density, A/cm$^2$ |
| $i_0$ | exchange current density, A/cm$^2$ |
| $I$ | total current, A |
| $\ell$ | distance, cm |
| $\mathbf{N}_i$ | flux of species $i$, mole/cm$^2$-sec |
| $r$ | radial position, cm |
| $r_i$ | radius of inner electrode, cm |

| | |
|---|---|
| $r_0$ | radius of outer electrode, cm |
| $R$ | resistance, ohm |
| $R$ | universal gas constant, 8.3143 J/mole-deg |
| $Re$ | Reynolds number |
| $t_+$ | cation transference number |
| $T$ | absolute temperature, deg K |
| $Ta$ | Taylor number |
| $u_i$ | mobility of species $i$, cm$^2$-mole/J-sec |
| $U$ | cell potential at open circuit, V |
| $\mathbf{v}$ | fluid velocity, cm/sec |
| $\mathbf{v}_i$ | average velocity of species $i$, cm/sec |
| $V$ | cell potential, V |
| $y$ | distance from electrode, cm |
| $z_i$ | charge number of species $i$ |
| $\alpha_a, \alpha_c$ | anodic and cathodic transfer coefficients |
| $\eta_c$ | concentration overpotential, V |
| $\eta_s$ | surface overpotential, V |
| $\kappa$ | conductivity, mho/cm |
| $\nu$ | kinematic viscosity, cm$^2$/sec |
| $\Phi$ | potential, V |
| $\Delta\Phi_{ohm}$ | ohmic potential drop, V |
| $\Omega$ | rotation speed, radian/sec |

*Part* **A**

# *Thermodynamics of Electrochemical Cells*

For a discussion of the thermodynamics of electrochemical cells, we first need to introduce free energies, chemical potentials, and activity coefficients. If we restrict ourselves to electrodes in equilibrium with the solution adjacent to them, then the cell potential can be obtained by expressing these phase equilibria in terms of the electrochemical potentials of species present in the electrodes and in the solutions. The condition of phase equilibrium precludes the passage of anything but an infinitesimal current; it also precludes the possibility of the occurrence of spontaneous reactions which require no net current. Under certain conditions it is possible, however, to have more than one reaction simultaneously in equilibrium.

In all but the simplest cells, the expression of the phase equilibria does not lead to an immediately useful result. The solutions adjacent to the two electrodes of a cell usually have different compositions, and in order to preclude spontaneous reactions at the electrodes, it is necessary to prevent the reactants for one electrode from reaching the other. Thus, somewhere between the electrodes there must be a region of nonuniform composition; and in this region, which is referred to as a *liquid junction*, spontaneous diffusion occurs. To treat the potentials of any but the simplest cells therefore requires some consideration of the irreversible process of diffusion. The necessary results are carried over from part C on transport processes in electrolytic solutions.

The treatment of cell potentials follows a certain pattern. The description of phase equilibria allows the cell potential to be expressed in terms of the electrochemical potentials of species in the solutions adjacent to the electrodes. To obtain useful results, these electrochemical potentials must be related to each other, usually by a consideration of the transport process in the junction region.

The concept of the potential is important in electrochemistry, but an understanding of the concept is made difficult by the background which most of us acquire in classical electrostatics. A discussion of the potential and its use in electrochemistry is therefore in order. Many equations in common use can be clarified by this study, and, at the same time, the assumptions inherent in their derivation can be exposed.

The remainder of part A deals with practical and theoretical aspects of electrochemical thermodynamics—in particular, with activity coefficients and reference electrodes.

# Thermodynamics in Terms of Electrochemical Potentials

**2**

## 12. Phase equilibrium

An electrochemical cell necessarily consists of several phases. These phases may be two electrode metals and an electrolytic solution (three phases), but additional phases, such as a solid salt or a gas, are included in most cells of practical interest. Equilibria between these individual phases (for example, electrode metal $\beta$ in equilibrium with the solution $\delta$) characterize an electrochemical cell used for thermodynamic measurements.

The system below is illustrative of the type of cell commonly called a *cell without transference*:

| metal | metal | electrolytic solution | solid salt | metal | metal | |
|-------|-------|-----------------------|------------|-------|-------|---|
| $\alpha$ | $\beta$ | $\delta$ | $\epsilon$ | $\phi$ | $\alpha'$ | (12-1) |

The vertical lines denote phase separation. The several phase equilibria which may be attained are:

phase $\alpha$ in equilibrium with phase $\beta$

phase $\beta$ in equilibrium with phase $\delta$

phase $\delta$ in equilibrium with phase $\epsilon$
phase $\epsilon$ in equilibrium with phase $\phi$
phase $\phi$ in equilibrium with phase $\alpha'$.

Phases $\alpha$ and $\alpha'$ are composed of the same metal, but are not necessarily in equilibrium since they may not be at the same electrical potential.

If two phases are in equilibrium and if a neutral species $A$ is present in each phase, then the chemical potential of $A$ is the same in the two phases; that is,

$$\mu_A^\delta = \mu_A^\epsilon. \tag{12-2}$$

Here the subscript $A$ refers to the species, and the superscript refers to the phase. Similar equations hold for other equilibrated species. Indeed, this equation must be obeyed for each species which exists in both phases before the phases may be said to be in equilibrium.

When two phases are at the same temperature but are not in equilibrium, upon contact there will be transport of material across the phase boundary until the condition described by equation 2 is attained. Thus, each of the phases in equilibrium may be considered to be *open* with respect to those species which can be transported between the phases. It should be pointed out that phase equilibrium as used here does not require that all species be present in each phase or that each species be at the same chemical potential in each phase. For example, if phase $\alpha$ is platinum and phase $\beta$ is a potassium amalgam, platinum is not present in phase $\beta$, and neither potassium nor mercury is present in phase $\alpha$. It is electrons which are equilibrated between these phases. Similarly, the electrons are presumed to be absent in the solution phase $\delta$. If all species were present in all phases and were equilibrated, then phases $\alpha$ and $\alpha'$ would be in equilibrium, and there would be no electrical potential difference between them.

To repeat, phase equilibrium is taken to be the thermodynamic state in which equation 2 applies to those species which are present in both phases.

In electrochemical systems, there are also equilibria which involve ionic or charged species. Let one mole of the neutral species $A$ dissociate into $v_+^A$ gram ions of a positively charged species and $v_-^A$ gram ions of a negatively charged species. The chemical potential of $A$ can then be expressed as

$$\mu_A^\alpha = v_+^A \mu_+^\alpha + v_-^A \mu_-^\alpha, \tag{12-3}$$

where $\mu_+$ and $\mu_-$ are the electrochemical potentials of the charged species and depend on the temperature, pressure, chemical composition, and electrical state of the phase. We shall return to these terms in the next section. Since species $A$ is electrically neutral, the coefficients $v_i^A$ are subject to the restriction

$$\sum_i z_i v_i^A = 0, \tag{12-4}$$

where $z_i$ is the charge number of species $i$, and the superscript $A$ refers to the particular neutral species. For example, potassium chloride is made up of potassium ions and chloride ions, for which $v_+ = v_- = z_+ = -z_- = 1$. Copper metal can be regarded as composed of cupric ions and electrons, for which $v_+ = 1$, $v_- = 2$, $z_+ = 2$, and $z_- = -1$.

Equation 2 expresses the condition of phase equilibrium involving neutral species. For the charged species $i$, the corresponding condition of phase equilibrium between the two phases $\alpha$ and $\beta$ is

$$\mu_i^\alpha = \mu_i^\beta. \tag{12-5}$$

If there are several ionic species present in each phase, equation 5 must be obeyed for each ionic species present in both phases before phase equilibrium is attained.

Each phase individually will also be electrically neutral even though all the ionic species are not present in each phase. Thus, the composition of any phase is determined by specifying the concentrations of all but one of the charged species, the concentration of the remaining species then being given by this condition of electrical neutrality.

Occasionally it is desirable and convenient to express the condition of equilibrium for an electrode reaction all at once. For the general electrode reaction

$$\sum_i s_i M_i^{z_i} \longrightarrow ne^-, \tag{12-6}$$

this condition is

$$\sum_i s_i \mu_i = n\mu_{e^-}. \tag{12-7}$$

Here $s_i$ is the stoichiometric coefficient of species $i$ in the electrode reaction and $M_i$ is a symbol for the chemical formula of species $i$. Superscripts for the appropriate phases in which the species exist should be added.

## 13. Chemical potential and electrochemical potential

The chemical potential of a neutral species is a function of the temperature, pressure, and chemical composition of the phase in which it exists. The chemical potential is defined as

$$\mu_A = \left(\frac{\partial G}{\partial n_A}\right)_{\substack{T, p, n_B \\ B \neq A}} = \left(\frac{\partial A}{\partial n_A}\right)_{\substack{T, V, n_B \\ B \neq A}} = \left(\frac{\partial U}{\partial n_A}\right)_{\substack{S, V, n_B \\ B \neq A}} = \left(\frac{\partial H}{\partial n_A}\right)_{\substack{S, p, n_B \\ B \neq A}}, \tag{13-1}$$

where $G$ is the Gibbs free energy, $A$ is the Helmholtz free energy, $U$ is the internal energy, $H$ is the enthalpy, $S$ is the entropy, $V$ is the volume, $T$ is the temperature, and $p$ is the pressure. In making measurements, one always determines a difference in the chemical potential between different thermodynamic states and never the absolute value in a particular state. However, in tabulating data, it is convenient to assign a value to each thermodynamic state. One can do this by arbitrarily assigning the value of the chemical potential in some state and determining the value in other states by comparison to this reference state. For example, the chemical potentials of pure elements at 0°C and one atmosphere pressure can be taken to be zero. Once the reference state is clearly specified and the values of the chemical potential in other states are tabulated, one can easily reproduce the experimental results. This will be mentioned again in the treatment of data from electrochemical cells.

The electrochemical potential of an ion was introduced by Guggenheim,[1] the difference between its values in two phases being defined as the work of transferring reversibly, at constant temperature and constant volume, one gram ion from one phase to the other.* It is a function of temperature, pressure, chemical composition, and electrical state of the phase. It is still necessary to determine how well defined these independent variables are. Consider the following cases where transference of ions may be involved:

1. Constant temperature and pressure, identical chemical composition of phases $\alpha$ and $\beta$. The only differences between the two phases will be electrical in nature.

(a) For the transfer of one gram ion of species $i$ from phase $\beta$ to phase $\alpha$, the work of transference is

$$w = \mu_i^\alpha - \mu_i^\beta = z_i F(\Phi^\alpha - \Phi^\beta), \qquad (13\text{-}2)$$

where, in the second equation, the difference in electrical state of the two phases can be characterized by the difference in electrical potential of the two phases, as defined by equation 2.

(b) For the transfer of $v_1$ gram ions of species 1 and $v_2$ gram ions of species 2 such that

$$\sum_i z_i v_i = 0, \qquad (13\text{-}3)$$

the work of transference is zero. Such electrically neutral combinations of ions do not depend on the electrical state of the phase, and we can utilize this fact to examine the potential difference defined above. Since the total work of transference will be zero for neutral combinations such that equation 3 holds, we have

$$w = 0 = v_1(\mu_1^\alpha - \mu_1^\beta) + v_2(\mu_2^\alpha - \mu_2^\beta). \qquad (13\text{-}4)$$

If we take equation 2 to apply to the ionic species 1, we can combine equations 2, 3, and 4 to express the electrochemical potential difference of the ionic species 2 as

$$\mu_2^\alpha - \mu_2^\beta = -\frac{v_1}{v_2}(\mu_1^\alpha - \mu_1^\beta) = -\frac{z_1 v_1}{v_2} F(\Phi^\alpha - \Phi^\beta) = z_2 F(\Phi^\alpha - \Phi^\beta). \qquad (13\text{-}5)$$

---

[1] E. A. Guggenheim, "The Conceptions of Electrical Potential Difference Between Two Phases and the Individual Activities of Ions," *The Journal of Physical Chemistry*, **33** (1929), 842–849.

*This is analogous to the definition of the chemical potential of a neutral species. For condensed phases, a distinction between a constant volume process and a constant pressure process is of little practical significance.

Therefore, the electric potential difference $\Phi^\alpha - \Phi^\beta$ defined by equation 2 does not depend on which charged species, 1 or 2, is used in equation 2. In this sense the electric potential difference is well defined and coincides with our usual concept of a potential difference.

2. Constant temperature and pressure, different chemical composition of the two phases. For transfer of neutral combinations such that equation 3 holds, there is no dependence on the electrical state of either phase. Thus, the work of transfer will depend only on the differences in chemical composition. The work of transfer of a charged species will still be given by

$$w = \mu_i^\alpha - \mu_i^\beta, \tag{13-6}$$

but this can no longer be expressed simply in terms of differences of electrical potential because the chemical environment of the transferred species will be different in the two phases.

It should be noted that no quantitative characterization or measure of the difference of electrical state of two phases has yet been given when the phases are of different chemical composition. It is possible (and even expedient for some purposes of computation) to define such an electrical variable, but this involves an unavoidable element of arbitrariness and is not essential to a treatment of the thermodynamic phenomena involved. Several possible methods of doing this are discussed in chapter 3. Our usual concept of electric potentials has an electrostatic rather than a thermodynamic basis, and the use of electrochemical potentials is more appropriate here.

Of interest here is the state of a phase and the question of whether two phases are in the same state. If two phases have different compositions, the question of whether they are in the same electrical state is not thermodynamically significant. On the other hand, if the two phases are chemically identical, it is convenient to describe their electrical states quantitatively, in a way which coincides with our usual concept of a potential.

## 14.  The definition of some thermodynamic functions

In this section the absolute activity, activity coefficient, mean activity coefficient, and osmotic coefficient are introduced. The last two are useful for the tabulation of the composition dependence of the thermodynamic properties of electrolytic solutions but may appear cumbersome in the theoretical treatment of cell potentials.

The absolute activity $\lambda_i$ of an ionic or a neutral species, used extensively by Guggenheim,[2] is defined by

$$\mu_i = RT \ln \lambda_i. \tag{14-1}$$

---

[2] E. A. Guggenheim, *Thermodynamics* (Amsterdam: North-Holland Publishing Company, 1959).

It has the advantage of being zero when the species is absent, whereas the chemical potential is equal to minus infinity in such a case. Furthermore, $\lambda_i$ is dimensionless. It also has the advantage that it can be manipulated like conventional activities but is independent of any secondary reference states that might be adopted for a particular solution or solvent at a particular temperature and pressure.

For solute species in a solution, $\lambda_i$ is further broken down as follows:

$$\lambda_i = m_i \gamma_i \lambda_i^\theta, \tag{14-2}$$

where $m_i$ is the molality or moles of solute per unit mass of solvent (usually expressed in gram moles or gram ions per kilogram of solvent), $\gamma_i$ is the activity coefficient of species $i$, and $\lambda_i^\theta$ is a proportionality constant, independent of composition and electrical state, but charcteristic of the solute species and the solvent and dependent on temperature and pressure. For condensed phases, the pressure dependence is frequently ignored.

Other concentration scales can be used, but the activity coefficient and the constant are changed so that $\lambda_i$ is independent of the concentration scale used. Another concentration scale in common use is molarity or moles per unit volume of solution (usually expressed in moles per liter denoted $M$), and $\lambda_i$ is related to this scale by

$$\lambda_i = c_i f_i a_i^\theta, \tag{14-3}$$

where $c_i$ is the molarity of species $i$, $f_i$ is the activity coefficient, and $a_i^\theta$ is a proportionality constant (analogous to $\lambda_i^\theta$). The molality is related to the molarity according to

$$m_i = \frac{c_i}{\rho - \sum_{j \neq 0} c_j M_j} = \frac{c_i}{c_0 M_0}, \tag{14-4}$$

where $\rho$ is the density of the solution (g/cm³), $M_j$ is the molecular weight of species $j$ (g/g-mole or g/g-ion), and where the sum does not include the solvent, denoted by the subscript 0.*

The molality is perhaps popular among experimentalists in the physical chemistry of solutions because it can be calculated directly from the masses of the components in the solution, without a separate determination of the density. The concentration on a molar scale is more directly useful in the analysis of transport processes in solutions. Furthermore, the molality is particularly inconvenient if the range of concentrations includes the molten salt, since the molality is then infinite. A mole fraction scale can be used, but then a decision has to be made on how to treat a dissociated electrolyte. The mass fraction has the advantage of depending only on the masses of the components and is also independent of the scale of atomic weights, which has been known to change even in recent years. However, a mass fraction scale

*Consistent units for equation 4 would be $m_i$ in g-mole/g, $c_i$ in g-mole/cm³, and $\rho$ in g/cm³.

does not allow a simple account of the colligative properties of solutions (freezing point depression, boiling point elevation, and vapor pressure lowering) nor of the properties of dilute solutions of electrolytes. Of these several scales, the molar concentration is the only one which changes with temperature when a particular solution is heated.

The secondary reference states necessary to specify $\lambda_i^\theta$ or $a_i^\theta$ are that certain combinations of activity coefficients should approach unity in infinitely dilute solutions; namely,

$$\prod_i (\gamma_i)^{\nu_i} \longrightarrow 1 \quad \text{as} \quad \sum_{i \neq 0} m_i \longrightarrow 0 \tag{14-5}$$

and
$$\prod_i (f_i)^{\nu_i} \longrightarrow 1 \quad \text{as} \quad \sum_{i \neq 0} c_i \longrightarrow 0 \tag{14-6}$$

for all such combinations of $\gamma_i$ and $f_i$ where the $\nu_i$'s satisfy equation 13-3. In particular, the activity coefficient of any neutral, undissociated species approaches unity as the concentrations of all solutes approach zero. If we take the activity coefficients to be dimensionless, then $\lambda_i^\theta$ and $a_i^\theta$ have the reciprocals of the dimensions of $m_i$ and $c_i$. In view of the definitions 5 and 6 of the secondary reference states, $\lambda_i^\theta$ and $a_i^\theta$ are then related by

$$\lambda_i^\theta = \rho_0 a_i^\theta, \tag{14-7}$$

where $\rho_0$ is the density of the pure solvent (g/cm³).

For an ionic species, $\lambda_i$ depends on the electrical state of the phase. Since $\lambda_i^\theta$ and $m_i$ are taken to be independent of the electrical state, we conclude that $\gamma_i$ is dependent upon this state. A similar statement applies to the activity coefficient $f_i$. In contrast, Guggenheim takes $\gamma_i$ to be independent of electrical state and $\lambda_i^\theta$ to be dependent upon it. This leaves us with the unsatisfactory situation that $\gamma_i$ should depend on composition at constant electrical state. However, a constant electrical state has not yet been defined for solutions of different composition.

To illustrate further the nature of these activity coefficients, consider a solution of a single electrolyte $A$ which dissociates into $\nu_+$ cations of charge number $z_+$ and $\nu_-$ anions of charge number $z_-$. (Since only a single electrolyte is involved, the superscript $A$ on $\nu_+$ and $\nu_-$ is omitted.) Then the stoichiometric concentration of the electrolyte can be represented as

$$m = \frac{m_+}{\nu_+} = \frac{m_-}{\nu_-} \quad \text{or} \quad c = \frac{c_+}{\nu_+} = \frac{c_-}{\nu_-}. \tag{14-8}$$

The chemical potential of $A$ can then be expressed by equation 12-3 as

$$\mu_A = \nu_+\mu_+ + \nu_-\mu_- = \nu_+RT \ln (m_+\gamma_+\lambda_+^\theta) + \nu_-RT \ln (m_-\gamma_-\lambda_-^\theta) \tag{14-9}$$

or
$$\mu_A = RT \ln [(m_+\gamma_+\lambda_+^\theta)^{\nu_+}(m_-\gamma_-\lambda_-^\theta)^{\nu_-}]. \tag{14-10}$$

Since $A$ is neutral, equation 5 requires that $\gamma_+^{\nu_+}\gamma_-^{\nu_-} \longrightarrow 1$ as $m \longrightarrow 0$. Hence, this specification of the secondary reference state allows a certain combination of the $\lambda_i^\theta$'s to be determined

$$(\lambda_+^\theta)^{\nu_+}(\lambda_-^\theta)^{\nu_-} = \lim_{m \to 0} \frac{e^{\mu_A/RT}}{m_+^{\nu_+}m_-^{\nu_-}}. \tag{14-11}$$

This limiting process then allows the subsequent determination of the combination $\gamma_+^{\nu_+}\gamma_-^{\nu_-}$ at any nonzero value of $m$ by means of equation 10.

By a generalization of these thoughts we come to the following conclusions:

Combinations of the form

$$\prod_i (\lambda_i^\theta)^{\nu_i}\gamma_i^{\nu_i}$$

can be determined unambiguously for products whose exponents satisfy the Guggenheim condition

$$\sum_i z_i\nu_i = 0.$$

A choice of the secondary reference state, equation 5, thus allows the separate determination of the corresponding products of the forms

$$\prod_i (\lambda_i^\theta)^{\nu_i} \quad \text{and} \quad \prod_i \gamma_i^{\nu_i}.$$

These conclusions follow from the fact that the corresponding combinations of electrochemical potentials and absolute activities

$$\sum_i \nu_i\mu_i \quad \text{and} \quad \prod_i \lambda_i^{\nu_i}$$

are independent of the electrical state for neutral combinations of ions.

On the other hand, differences in $\mu_i$ and ratios of $\lambda_i$ between phases are well defined but depend upon the electrical states of the phases. The absolute values, in a single phase, are not defined because the primary reference state (say, the elements at $0°C$ and one atmosphere) involves no electrical reference state. Correspondingly, the secondary reference state also involves only neutral combinations of species. Consequently, the $\lambda_i^\theta$'s for ionic species are not uniquely determined, a situation which could be rectified by the arbitrary assignment of the value of $\lambda_i^\theta$ for *one* ionic species in each solvent at each temperature. However, in any application, the equations can be arranged so that only those products of $\lambda_i^\theta$'s (and also of $\gamma_i$'s) corresponding to neutral combinations of ions are ever needed.

Let us return to the solution of a single electrolyte. By convention, the mean activity coefficient $\gamma_{+-}$ (or $\gamma_\pm$) on the molal scale is defined by

$$\gamma_{+-}^\nu = \gamma_+^{\nu_+}\gamma_-^{\nu_-}, \tag{14-12}$$

where

$$\nu = \nu_+ + \nu_-. \tag{14-13}$$

The above discussion shows that this mean activity coefficient is unambiguously defined and independent of the electrical state of the solution. If we also define $\lambda_{+-}^\theta$ by

$$(\lambda_{+-}^\theta)^\nu = (\nu_+\lambda_+^\theta)^{\nu_+}(\nu_-\lambda_-^\theta)^{\nu_-}, \tag{14-14}$$

then we can write equation 10 in the form

$$\mu_A = \nu RT \ln (m\gamma_{+-}\lambda_{+-}^\theta). \tag{14-15}$$

It is this activity coefficient $\gamma_{+-}$ which is measured and tabulated for solutions of single electrolytes.

The thermodynamic properties of solutions of a single electrolyte, can, of course, be studied by nonelectrochemical means, and without detailed consideration of its state of dissociation into ions. For example, a study of the vapor pressure or the freezing point will yield the variation in the chemical potential $\mu_A$ with concentration. It is, in fact, one of the beauties of thermodynamics that it provides a framework to record the macroscopic properties of a system without a knowledge of its state of molecular aggregation, as long as the possible molecular species equilibrate rapidly with each other.

If we applied equation 2 to the electrolyte $A$, without regard for its dissociation, we would have

$$\mu_A = RT \ln (m\gamma_A \lambda_A^\theta). \tag{14-16}$$

This differs from equation 15 principally in the absence of the factor $\nu$. Consequently $\gamma_A$ and $\lambda_A^\theta$ must be different from $\gamma_{+-}$ and $\lambda_{+-}^\theta$, and $\gamma_A$ must have a concentration dependence considerably different from that of $\gamma_{+-}$. Specifically, we have

$$\gamma_A = \frac{m^{\nu-1}\gamma_{+-}^\nu (\lambda_{+-}^\theta)^\nu}{\lambda_A^\theta}. \tag{14-17}$$

Consequently $\gamma_A \longrightarrow 0$ as $m \longrightarrow 0$, and equation 5 cannot be applied to define a secondary reference state for $\lambda_A^\theta$ in equation 16. Thus, we conclude that a knowledge of the state of aggregation of the solute at infinite dilution is essential, and must be followed in choosing the components of a solution, if we are to be able to use equation 5 to define the secondary reference state. Except for this necessity to choose a different secondary reference state, it is legitimate from a strictly thermodynamic point of view to treat the electrolyte as undissociated, although this is seldom, if ever, done. By the activity coefficient, we shall thus mean the mean ionic activity coefficient of an electrolyte.

A parallel development can be carried through for the molar scale of concentration. On this scale, the mean activity coefficient of the electrolyte is defined by

$$f_{+-}^\nu = f_+^{\nu_+} f_-^{\nu_-}, \tag{14-18}$$

and if we also define $a_{+-}^\theta$ by

$$(a_{+-}^\theta)^\nu = (\nu_+ a_+^\theta)^{\nu_+} (\nu_- a_-^\theta)^{\nu_-}, \tag{14-19}$$

then we can express the chemical potential of the electrolyte as

$$\mu_A = \nu RT \ln (c f_{+-} a_{+-}^\theta). \tag{14-20}$$

The mean activity coefficients on the two scales are related by

$$f_{+-} = \gamma_{+-} \frac{\rho_0}{\rho - M_A c}, \tag{14-21}$$

where $M_A = \nu_+ M_+ + \nu_- M_-$ is the molecular weight of the electrolyte. This equation is restricted to solutions of a single electrolyte.

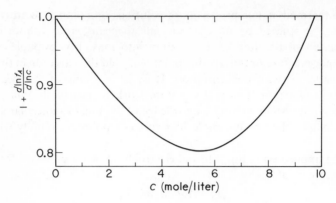

**Figure 14-1.** Variation of the molar activity coefficient of acetic acid with concentration. For $1 + d\ln f_{+-}/d\ln c$, divide the ordinate scale by 2. (Values taken from Vitagliano and Lyons, *J. Am. Chem. Soc.*, *78*, 4538 (1956).)

Various examples can be cited to illustrate the consequences of ionic dissociation. Lewis *et al.*[3] show the effect of considering hydrochloric acid to be dissociated or undissociated by plotting the partial pressure of HCl in equilibrium with the solution against both $m$ and $m^2$. They also plot the freezing point depressions of acetic acid, indicating that acetic acid behaves as a nonelectrolyte except at very low concentrations. The same thing is indicated in figure 1, which shows the variation with concentration of the activity coefficient $f_A$ on a molar scale (defined analogously with $\gamma_A$). The behavior is typical of a nonelectrolyte. However, if acetic acid truly dissociates into hydrogen ions and acetate ions at infinite dilution, $1 + d\ln f_A/d\ln c$ should approach 2 as $c$ approaches zero. The observed behavior of acetic acid is due to its small dissociation constant, as a consequence of which acetic acid is nearly undissociated except at very small concentrations.

The activity of the solvent could be expressed by equation 2. However, it is more common to relate deviations from ideal behavior to an osmotic coefficient, defined by:

$$\ln \frac{\lambda_0}{\lambda_0^0} = -\phi M_0 \sum_{i \neq 0} m_i, \qquad (14\text{-}22)$$

where $\lambda_0^0$ is the absolute activity of the pure solvent at the same temperature and pressure. For a solution of a single electrolyte, this becomes

$$\ln \frac{\lambda_0}{\lambda_0^0} = -vm\phi M_0. \qquad (14\text{-}23)$$

For a solution of a single electrolyte, the mean molal activity coefficient can be related to the osmotic coefficient by means of the Gibbs-Duhem

[3]Gilbert Newton Lewis, Merle Randall, Kenneth S. Pitzer, and Leo Brewer, *Thermodynamics* (New York: McGraw-Hill Book Company, Inc., 1961), pp. 313 and 300.

equation, which, for constant temperature and pressure, reads

$$c_0 \, d\mu_0 + c \, d\mu_A = 0. \tag{14-24}$$

Substitution of the relevant equations gives

$$d(m\phi) = m \, d \ln (m\gamma_{+-}) \tag{14-25}$$

or

$$dm(\phi - 1) = m \, d \ln \gamma_{+-}. \tag{14-26}$$

Integration from $m = 0$ to $m = m$ gives

$$- \ln \gamma_{+-} = \int_0^m \frac{\partial [m(1 - \phi)]}{\partial m} \frac{dm}{m}. \tag{14-27}$$

## 15. Cell with solution of uniform concentration

The cells discussed in this section are taken to contain a single electrolyte in a solution of uniform concentration throughout the cell. In most textbooks of physical chemistry, electrochemistry, or thermodynamics, a distinction is made between cells *without* transference and cells *with* transference, depending upon whether there are concentration gradients in the electrolytic solution. As we shall see, this division is somewhat subjective, except for the simplest cells. The division rather depends upon whether we are willing to ignore the concentration gradients that do exist.

An example of a cell with the same solution throughout is a cell with two electrodes of the same metal dipping into the same solution of a salt of the metal. To be specific, consider the cell:

$$\begin{array}{c|c|c|c|c} \alpha & \beta & \delta & \beta' & \alpha' \\ \mathrm{Pt}(s) & \mathrm{Cu}(s) & \mathrm{CuSO_4 \ soln} & \mathrm{Cu}(s) & \mathrm{Pt}(s). \end{array} \tag{15-1}$$

Phase equilibrium among the several phases is described by equation 12-5, for example,

$$\mu_{e^-}^\alpha = \mu_{e^-}^\beta, \tag{15-2}$$

$$\mu_{\mathrm{Cu^{++}}}^\beta = \mu_{\mathrm{Cu^{++}}}^\delta. \tag{15-3}$$

Similarly, cupric ions are equilibrated between the solution $\delta$ and the electrode $\beta'$, and electrons are equilibrated between the electrode $\beta'$ and the platinum lead $\alpha'$.

The cell potential $U$ is related to the electric potential difference between $\alpha$ and $\alpha'$ and, in turn, to the difference in the electrochemical potentials of electrons in these leads by

$$FU = z_e F(\Phi^\alpha - \Phi^{\alpha'}) = \mu_{e^-}^\alpha - \mu_{e^-}^{\alpha'}. \tag{15-4}$$

(In this chapter, $U$ will denote the potential of the right electrode relative to the left electrode.) In view of the conditions of phase equilibrium, this can be rewritten as

$$\begin{aligned} FU &= \mu_{e^-}^\beta - \mu_{e^-}^{\beta'} = \mu_{e^-}^\beta - \mu_{e^-}^{\beta'} + \tfrac{1}{2}\mu_{\mathrm{Cu^{++}}}^\delta - \tfrac{1}{2}\mu_{\mathrm{Cu^{++}}}^\delta \\ &= \mu_{e^-}^\beta + \tfrac{1}{2}\mu_{\mathrm{Cu^{++}}}^\delta - \mu_{e^-}^{\beta'} - \tfrac{1}{2}\mu_{\mathrm{Cu^{++}}}^{\beta'}. \end{aligned} \tag{15-5}$$

By means of equation 12-3, this can be written in terms of the chemical potential of copper in the two electrodes:

$$FU = \tfrac{1}{2}\mu_{Cu}^{\beta} - \tfrac{1}{2}\mu_{Cu}^{\beta'}. \tag{15-6}$$

If the electrodes are identical, the cell potential will be zero. If, however, one of the copper electrodes is alloyed with another metal, a nonzero cell potential will result. Here it is assumed that the only effect of introducing the foreign metal is to change the chemical potential of copper in that electrode; the foreign metal should have a negligible tendency to dissolve spontaneously in the solution, by replacing cupric ions.

Another cell of this type is

$$
\begin{array}{c|c|c|c|c}
\alpha & \beta & \delta & \epsilon & \alpha' \\
Pt(s) & Pb(s) & PbCl_2 \text{ in } H_2O & Pb(Hg) & Pt(s),
\end{array} \tag{15-7}
$$

in which Pb(Hg) is a lead amalgam. By a similar analysis, the cell potential is given by

$$FU = -F(\Phi^{\alpha} - \Phi^{\alpha'}) = \tfrac{1}{2}\mu_{Pb}^{\beta} - \tfrac{1}{2}\mu_{Pb}^{\epsilon}. \tag{15-8}$$

This cell provides the means for determining the thermodynamic properties of lead amalgam as a function of the amalgam composition, the most common application for this type of cell.

The above cells truly have a uniform concentration, and their electrical potentials and thermodynamic properties are connected by equation 6 or 8. The cell given in section 12 is usually considered to have the same electrolyte throughout. A particular system is

$$
\begin{array}{c|c|c|c|c|c}
\alpha & \beta & \delta & \epsilon & \phi & \alpha' \\
Pt(s) & Li(s) & LiCl \text{ in DMSO} & TlCl(s) & Tl(Hg) & Pt(s)
\end{array}, \tag{15-9}
$$

where the electrolytic solution is a solution of lithium chloride in the solvent dimethyl sulfoxide, TlCl is a salt which is sparingly soluble in DMSO, and Tl(Hg) is a thallium amalgam.

For the electrode on the right, the equilibria among phases $\delta$, $\epsilon$, and $\phi$ can be expressed as

$$\mu_{Cl^-}^{\delta} = \mu_{Cl^-}^{\epsilon} \quad \text{and} \quad \mu_{Tl^+}^{\epsilon} = \mu_{Tl^+}^{\phi}. \tag{15-10}$$

We proceed in the usual manner by expressing the cell potential in terms of the electrochemical potentials of electrons in the leads, using the conditions of phase equilibrium to relate these to conditions in the solutions adjacent to the electrodes, and finally collecting electrochemical potentials of charged species into chemical potentials of neutral species wherever possible. The result is

$$FU = -F(\Phi^{\alpha} - \Phi^{\alpha'}) = \mu_{Li}^{\beta} - \mu_{Tl}^{\phi} + \mu_{TlCl}^{\epsilon} - \mu_{LiCl}^{\delta}. \tag{15-11}$$

Thus, the cell potential is related to the thermodynamic properties of neutral species, even though the phase equilibria were expressed in terms of charged species. This will always be true for cells which can be treated by thermodynamics alone. The measurement of the potential of this cell can be used to study the thermodynamic properties of solutions of lithium chloride in dimethyl sulfoxide.

It would appear from the above treatment that the assumption of uniform composition is valid here. However, it is known that there will be thallous chloride in the solution, although the quantity will be small, and the presence of this added electrolyte will change the chemical potential of LiCl in the immediate vicinity of the thallium amalgam-thallous chloride electrode. One cannot allow the thallous chloride to saturate the entire solution because it will react spontaneously with the lithium metal. Thus, there must be a gradient of the concentration of TlCl and of the electrochemical potential of the chloride ion, and one can no longer assume that both electrodes are in contact with the same electrolytic solution. This picture can be represented as

$$
\begin{array}{c|c|c|c|c|c|c|c}
\alpha & \beta & \delta & \{\text{transition}\} & \delta' & \epsilon & \phi & \alpha' \\
\text{Pt}(s) & \text{Li}(s) & \begin{array}{c}\text{LiCl in}\\\text{DMSO}\end{array} & \{\text{region}\} & \begin{array}{c}\text{LiCl in}\\\text{DMSO}\end{array} & \text{TlCl}(s) & \text{Tl(Hg)} & \text{Pt}(s)
\end{array} \quad (15\text{-}12)
$$

Here phase $\delta'$ differs from phase $\delta$ due to the dissolved TlCl, and these are connected by a junction or transition region in which the concentration of thallous chloride varies with position.

The equilibria among phases $\delta'$, $\epsilon$, and $\phi$ can now be written:

$$\mu^{\delta'}_{\text{Cl}^-} = \mu^{\epsilon}_{\text{Cl}^-} \quad \text{and} \quad \mu^{\delta'}_{\text{Tl}^+} = \mu^{\epsilon}_{\text{Tl}^+} = \mu^{\phi}_{\text{Tl}^+}. \qquad (15\text{-}13)$$

The solution $\delta'$ will actually facilitate the equilibrium between the amalgam $\phi$ and the solid salt $\epsilon$. Instead of equation 11, we now obtain

$$FU = \mu^{\beta}_{\text{Li}} - \mu^{\phi}_{\text{Tl}} + \mu^{\epsilon}_{\text{TlCl}} - \mu^{\delta}_{\text{LiCl}} + (\mu^{\delta}_{\text{Cl}^-} - \mu^{\delta'}_{\text{Cl}^-}). \qquad (15\text{-}14)$$

We see that this equation will be identical with equation 11 if we are willing to ignore the difference in electrochemical potential of the chloride ion between the solutions adjacent to the two electrodes. Thermodynamics alone does not provide the means for evaluating this difference since the junction region basically involves the irreversible process of diffusion and must be treated by the laws of transport in electrolytic solutions. For this system, in the absence of current, the gradient of the electrochemical potential of the chloride ion can be expressed in terms of the neutral salts, as discussed in the next section,

$$\nabla \mu_{\text{Cl}^-} = t^0_{\text{Li}^+} \nabla \mu_{\text{LiCl}} + t^0_{\text{Tl}^+} \nabla \mu_{\text{TlCl}}, \qquad (15\text{-}15)$$

where $t^0_i$ is the transference number of species $i$ with respect to the solvent velocity. From this equation, we can perceive that the more insoluble the salt (here, TlCl), the smaller the value of $t^0_{\text{Tl}^+}$ will be and the more nearly the same the solution will be throughout. Then equation 14 can be adequately approximated by equation 11.

The purpose of this section has been to illustrate how to apply the phase-equilibrium conditions of section 12 to typical systems that involve an electrolytic solution of uniform composition throughout the cell. For the first two cases examined above, the thermodynamic properties of the solution do not influence the cell potential. The third example can be used to study the thermodynamic properties of the electrolytic solution, although, strictly speaking, this cell does not belong in this section. The assessment of the errors involved in ignoring the concentration gradients in such cells will be treated in section 18.

The results of the previous examples can be applied to cells with two (or more) electrolytes in solution at a uniform concentration under the following conditions:

1. The second electrolyte changes the thermodynamic properties of the first electrolyte in solution, but it does not react with it to form a precipitate or evolve a gas, and it does not react spontaneously with the electrodes.

2. The second electrolyte does not participate in the phase equilibria except to alter thermodynamic properties in the solution phase.

## 16. Transport processes in junction regions

The treatment of the open-circuit potentials of electrochemical cells involves first the description of phase equilibria between the electrodes and the solutions or solids adjacent to them (discussed briefly in section 12), followed by a consideration of the junction regions that are likely to exist between the solutions adjacent to the electrodes. We have found, in the previous section, a need to treat such regions.

We have indicated in section 13 that it is more appropriate to assess conditions in a junction region of nonuniform composition by means of electrochemical potentials rather than by our usual concept of an electric potential. For this purpose, we call upon an equation developed in part C, transport processes in electrolytic solutions (see section 84):

$$\frac{F}{\kappa}\mathbf{i} = - \sum_i \frac{t_i^0}{z_i} \nabla \mu_i, \tag{16-1}$$

where $\mathbf{i}$ is the current density, $\kappa$ is the conductivity, $t_i^0$ is the transference number of species $i$ relative to the velocity of species 0, and $\nabla \mu_i$ is the gradient of the electrochemical potential of species $i$.

The species 0 whose velocity is used as a reference can be any species in the solution, but it is usually taken to be the solvent, if a solvent is evident. The sum in equation 1 includes any neutral species which might be present, and the ratio $t_i^0/z_i$ is not generally zero for a neutral species. However, $t_i^0$ and $t_i^0/z_i$ are always zero for the reference species, and this is one reason why it is convenient to choose the solvent for this reference.

The current density $\mathbf{i}$ is supposed to be zero in the present application, the treatment of the open-circuit potentials of electrochemical cells; but its presence in equation 1 would aid in the assessment of the effects of those small currents which are unavoidable in the actual measurement of cell potentials. In this connection, and to give further insight into equation 1, we might note that it reduces to Ohm's law in a medium of uniform composition. Then the variations in the electrochemical potentials can be expressed by equation 13-2, and we have

$$\mathbf{i} = -\kappa \, \nabla \Phi \sum_i t_i^0 = -\kappa \, \nabla \Phi, \qquad (16\text{-}2)$$

the second equality following from the fact that the sum of the transference numbers of all species is equal to one.

To treat junction regions, it is convenient to rewrite equation 1 in the form

$$\frac{1}{z_n} \nabla \mu_n = -\frac{F}{\kappa} \mathbf{i} - \sum_i \frac{t_i^0}{z_i} \left( \nabla \mu_i - \frac{z_i}{z_n} \nabla \mu_n \right), \qquad (16\text{-}3)$$

where species $n$ can be any charged species in the solution. The combinations of electrochemical potentials in parentheses correspond to neutral combinations of species and hence are independent of the electrical state of the solution; these terms depend only on the spatial variation of the chemical composition of the medium (at uniform temperature and pressure). Hence, equation 3 permits the assessment of the variation of the electrochemical potential of one charged species in a region of nonuniform composition. In other words, the electrical states of different parts of a phase are related to each other because they are physically connected to each other. In a medium of uniform composition, this amounts to a determination of the ohmic potential drop; in a nonuniform medium, the variation of composition can also be accounted for.

For the solution of lithium chloride and thallous chloride in dimethyl sulfoxide, considered in section 15, equation 3 becomes

$$\nabla \mu_{\text{Cl}^-} = t_{\text{Li}^+}^0 \, \nabla \mu_{\text{LiCl}} + t_{\text{Tl}^+}^0 \, \nabla \mu_{\text{TlCl}}, \qquad (16\text{-}4)$$

if species $n$ is taken to be the chloride ion and if the current density is zero. In practice, this equation must now be integrated across the junction region for known concentration profiles of TlCl and LiCl.

In this section, only one equation has been presented—one roughly equivalent to Ohm's law. It is useful for assessing variations in electrical state across a junction region where the concentration profiles are known. It is not sufficient for the determination of these concentration profiles nor of the current density. These concentration profiles, which may change with time, are determined from the laws of diffusion and the method of forming the junction, topics which are beyond the scope of this section.

## 17. Cell with a single electrolyte of varying concentration

A cell in which the concentration of a single electrolyte varies with location in the cell is the simplest example of a so-called *cell with transference*. An example is

$$
\begin{array}{cccccccc}
\alpha & \beta & \delta & \{\text{transition}\{ & \epsilon & \beta' & \alpha' \\
\text{Pt}(s) & \text{Li}(s) & \text{LiCl in DMSO} \} \text{region} & \{ & \text{LiCl in DMSO} & \text{Li}(s) & \text{Pt}(s),
\end{array}
$$

(17-1)

in which the chemical composition of both platinum leads is identical, as is that of both lithium electrodes. The concentration of LiCl in the $\delta$-phase is different from that in the $\epsilon$-phase. The *transition region* is one in which concentration gradients exist, as the concentration varies from that in the $\delta$-phase to that in the $\epsilon$-phase. This region is sometimes called a *liquid junction*.

By means of the conditions of phase equilibrium (equation 12-5) and the definition of the potential difference between phases of identical composition (equation 13-2), the cell potential reduces to

$$FU = -F(\Phi^\alpha - \Phi^{\alpha'}) = \mu_{\text{Li}^+}^\epsilon - \mu_{\text{Li}^+}^\delta.$$

(17-2)

One can go no further in treating the cell by reversible thermodynamics, since diffusion is present. Equation 16-3 can be applied to evaluate the difference of electrochemical potentials appearing in equation 2. If we take the current density to be zero and lithium ions to be species $n$, equation 16-3 becomes

$$\nabla \mu_{\text{Li}^+} = t_{\text{Cl}^-}^0 \nabla \mu_{\text{LiCl}}.$$

(17-3)

This can be integrated across the junction region and the result substituted into equation 2 to yield

$$FU = \int_\delta^\epsilon t_{\text{Cl}^-}^0 \frac{\partial \mu_{\text{LiCl}}}{\partial x} \, dx.$$

(17-4)

In this particular case, both $t_{\text{Cl}^-}^0$ and $\mu_{\text{LiCl}}$ depend only on the concentration of LiCl, and the integral in equation 4 becomes independent of the detailed form of the concentration profile in the junction region:

$$FU = \int_\delta^\epsilon t_{\text{Cl}^-}^0 \, d\mu_{\text{LiCl}} = \int_\delta^\epsilon t_{\text{Cl}^-}^0 \frac{d\mu_{\text{LiCl}}}{dm} \, dm.$$

(17-5)

This equation is equivalent to the expressions which appear in most treatments of cells of the type considered here. However, the *virtual passage of current* used in most derivations has been avoided here, because such derivations give the impression that reversible thermodynamics is sufficient to yield the cell potential. The potential difference in equation 5 should not be called a *liquid junction potential*; rather it is *the potential of a cell with liquid junction*.

Equation 5 can be generalized to metals and electrolytes with different charge numbers, with the result

$$FU = \int_\delta^\epsilon \frac{t_-^0}{z_+\nu_+} \frac{d\mu_A}{dm} dm, \tag{17-6}$$

where $A$ denotes the single electrolyte in the solution. The equilibria between the electrodes and the solutions are assumed to involve only cations, and the electrodes are assumed to be identical. This would include the copper sulfate concentration cell in figure 6-1.

Upon introduction of activity coefficients by means of equation 14-15, equation 6 becomes

$$FU = \nu RT \int_\delta^\epsilon \frac{t_-^0}{z_+\nu_+} \left(1 + \frac{d \ln \gamma_{+-}}{d \ln m}\right) d \ln m. \tag{17-7}$$

If the concentration dependence of the transference number is ignored, we have the approximation

$$U = \frac{\nu}{z_+\nu_+} \frac{RT}{F} t_-^0 \ln \frac{(m\gamma_{+-})_\epsilon}{(m\gamma_{+-})_\delta} = \frac{\nu}{z_+\nu_+} \frac{RT}{F} t_-^0 \ln \frac{(cf_{+-})_\epsilon}{(cf_{+-})_\delta}. \tag{17-8}$$

The latter expression was the basis for the approximation in equation 6-1 for the potential of the copper concentration cell used as an example in chapter 1.

Finally, one might generalize these results for an arbitrary electrode reaction with a solution of a single electrolyte $A$. Let the cell be represented as

$$\begin{array}{c|c|c|c|c|c|c}
\alpha & \beta & \delta & \{\text{transition}\} & \epsilon & \beta' & \alpha' \\
\text{Pt}(s) & \text{electrode} & \text{solution} & \{\text{region}\} & \text{solution} & \text{electrode} & \text{Pt}(s).
\end{array} \tag{17-9}$$

If the "electrodes" involve an additional gas phase or sparingly soluble salt, the chemical potentials of these neutral species are taken to be the same on both sides of the cell, and the solubilities are taken to be small enough that the solution can still be regarded as that of the electrolyte alone in the solvent.

For a general electrode reaction, we can take

$$\sum_i s_i M_i^{z_i} \longrightarrow ne^-, \tag{17-10}$$

where $s_i$ is the stoichiometric coefficient of species $i$ and $M_i$ is a symbol for the chemical formula of species $i$. A generalization of the formulas for phase equilibrium for this reaction would be

$$\sum_i s_i \mu_i = n\mu_{e^-}, \tag{17-11}$$

where superscripts for the appropriate phases in which the species exist should be added. Since the extraneous phases are neutral and absorb no current, it follows that the cations and anions of the electrolyte must be responsible for the electrons produced or consumed at the electrodes:

$$s_+ z_+ + s_- z_- = -n. \tag{17-12}$$

In this case, we choose to treat the cell by means of a reference electrode, perhaps imaginary, of the same kind as the main electrodes and which can be moved, together with its extraneous neutral phases, through the solution in the region between the electrodes. The variation of the potential of this electrode with position is given by the gradient of equation 11:

$$s_+ \nabla \mu_+ + s_- \nabla \mu_- + s_0 \nabla \mu_0 = n \nabla \mu_{e^-} = -nF \nabla \Phi, \qquad (17\text{-}13)$$

it being presumed that only the properties of the solution vary with position. Equation 16-3 allows us to relate variations of electrochemical potentials of ions to variations of the chemical potential of the electrolyte:

$$\frac{1}{z_+} \nabla \mu_+ = \frac{t_-^0}{z_+ \nu_+} \nabla \mu_A. \qquad (17\text{-}14)$$

The Gibbs-Duhem equation 14-24 allows us to relate the variation of the chemical potential of the solvent to that of the electrolyte:

$$\nabla \mu_0 + M_0 m \nabla \mu_A = 0. \qquad (17\text{-}15)$$

Combination of equations 12, 13, and 14 yields

$$F \nabla \Phi = \left( \frac{t_-^0}{z_+ \nu_+} - \frac{s_-}{n \nu_-} + \frac{s_0 M_0}{n} m \right) \nabla \mu_A. \qquad (17\text{-}16)$$

Integration across the junction region gives

$$FU = -F(\Phi^\alpha - \Phi^{\alpha'}) = \int_\delta^\epsilon \left( \frac{t_-^0}{z_+ \nu_+} - \frac{s_-}{n \nu_-} + \frac{s_0 M_0}{n} m \right) \frac{d\mu_A}{dm} dm$$

$$= \nu RT \int_\delta^\epsilon \left( \frac{t_-^0}{z_+ \nu_+} - \frac{s_-}{n \nu_-} + \frac{s_0 M_0}{n} m \right) \left( 1 + \frac{d \ln \gamma_{+-}}{d \ln m} \right) d \ln m. \quad (17\text{-}17)$$

If $t_-^0$ is independent of concentration, this becomes

$$FU = \nu RT \left( \frac{t_-^0}{z_+ \nu_+} - \frac{s_-}{n \nu_-} \right) \ln \frac{(m \gamma_{+-})_\epsilon}{(m \gamma_{+-})_\delta} + \nu RT \frac{s_0 M_0}{n} \int_\delta^\epsilon \left( 1 + \frac{d \ln \gamma_{+-}}{d \ln m} \right) dm.$$
$$(17\text{-}18)$$

In the case where the anion and the solvent do not participate in the electrode reaction, equation 17 is seen to coincide with equation 7.

We have given here, in various degrees of generality, a treatment of concentration cells, where the electrodes are identical but are placed in different solutions of a single electrolyte which are joined by a junction region where the concentration varies. The measured cell potential is found to depend not only on the thermodynamic properties but also on the transport properties of the solutions in the cell. Such cells are useful for determining the activity coefficient if the transference number is known and the transference number if the activity coefficient is known. Both types of determination are common practice.

Let us consider some additional examples. For the cell

$$\begin{array}{c|c|c|c|c}
\alpha & \epsilon & \text{transition} & \lambda & \alpha' \\
\text{Pt}(s), \text{H}_2(g) & \text{HCl in H}_2\text{O} & \text{region} & \text{HCl in H}_2\text{O} & \text{Pt}(s), \text{H}_2(g)
\end{array},$$
$$(17\text{-}19)$$

the cell potential is

$$FU = -F(\Phi^\alpha - \Phi^{\alpha'}) = \frac{1}{2}RT \ln \left( p_{H_2}^\alpha / p_{H_2}^{\alpha'} \right) - \mu_{H^+}^\epsilon + \mu_{H^+}^\lambda$$

$$= \frac{1}{2}RT \ln \frac{p_{H_2}^\alpha}{p_{H_2}^{\alpha'}} + \int_\epsilon^\lambda t_{Cl^-}^0 \, d\mu_{HCl}. \tag{17-20}$$

This agrees with equation 17 if the partial pressure of hydrogen is the same near the two platinum electrodes.

If the hydrogen electrodes are replaced by silver-silver chloride electrodes, we have the cell

| $\alpha$ | $\beta$ | $\delta$ | $\epsilon$ | | | $\lambda$ | $\delta'$ | $\beta'$ | $\alpha'$ |
|---|---|---|---|---|---|---|---|---|---|
| Pt(s) | Ag(s) | AgCl(s) | HCl in H$_2$O | $\}$ transition $\{$ | region | $\}$ HCl in H$_2$O | AgCl(s) | Ag(s) | Pt(s) |

$$\tag{17-21}$$

for which the cell potential is

$$FU = -F(\Phi^\alpha - \Phi^{\alpha'}) = \mu_{Cl^-}^\epsilon - \mu_{Cl^-}^\lambda = \int_\lambda^\epsilon t_{H^+}^0 \, d\mu_{HCl}, \tag{17-22}$$

if the silver electrodes are identical and the solid silver chloride is the same on both sides of the cell. The potential of this cell is opposite in sign to that of the preceding cell, and the magnitudes are quite different since $t_{H^+}^0 \approx 0.82$, whereas $t_{Cl^-}^0 \approx 0.18$. Equation 22 agrees with equation 17 if we use the values $s_- = 1$, $s_0 = 0$, and $n = 1$.

Finally, for the cell

| $\alpha$ | $\beta$ | $\delta$ | $\epsilon$ | | | $\lambda$ | $\delta'$ | $\beta'$ | $\alpha'$ |
|---|---|---|---|---|---|---|---|---|---|
| Pt(s) | Hg(l) | HgO(s) | KOH in H$_2$O | $\}$ transition $\{$ | region | $\}$ KOH in H$_2$O | HgO(s) | Hg(l) | Pt(s) |

$$\tag{17-23}$$

the reaction at the mercury-mercuric oxide electrodes is

$$Hg + 2\,OH^- \longrightarrow HgO + H_2O + 2e^-, \tag{17-24}$$

for which we have

$$n = 2, \qquad s_- = 2, \qquad s_0 = -1. \tag{17-25}$$

Consequently, the cell potential is given by

$$FU = -F(\Phi^\alpha - \Phi^{\alpha'}) = \int_\lambda^\epsilon (t_{K^+}^0 + \tfrac{1}{2}M_0 m) \, d\mu_{KOH}. \tag{17-26}$$

## 18. Cell with two electrolytes, one of nearly uniform concentration

The cells of this section have two electrolytes in solution; one of the electrolytes is of nearly uniform concentration throughout, while the concentration of the other varies with position in the cell. An example of this type of cell has been discussed in section 15:

$$
\begin{array}{c|c|c|c|c|c|c|c}
\alpha & \beta & \delta & \{ & \{ & \delta' & \epsilon & \phi & \alpha' \\
\text{Pt}(s) & \text{Li}(s) & \text{LiCl in} & \text{transition} & \text{LiCl in} & \text{TlCl}(s) & \text{Tl}(\text{Hg}) & \text{Pt}(s) \\
& & \text{DMSO} & \text{region} & \text{DMSO} & & & 
\end{array} \quad (18\text{-}1)
$$

in which the transition region denotes the region of variable concentration of TlCl. The lithium chloride is of nearly uniform concentration, although its value may be changed slightly near the thallous chloride salt.

The potential of this cell was expressed as

$$
FU = \mu_{\text{Li}}^{\beta} - \mu_{\text{Tl}}^{\phi} + \mu_{\text{TlCl}}^{\epsilon} - \mu_{\text{LiCl}}^{\delta} + (\mu_{\text{Cl}^-}^{\delta} - \mu_{\text{Cl}^-}^{\delta'}). \quad (18\text{-}2)
$$

Just as for the cells of section 17, this expression contains a difference in the electrochemical potential of an ionic species between two points in the solution. However, in this case, this difference is regarded as a small error term in an expression which otherwise relates the cell potential to thermodynamic quantities, whereas in section 17, such differences represent a significant contribution to the cell potential. The estimation of these errors has been treated by Smyrl and Tobias,[4] whose development we follow here.

Equation 15-15, as repeated below,

$$
\nabla \mu_{\text{Cl}^-} = t_{\text{Li}^+}^0 \, \nabla \mu_{\text{LiCl}} + t_{\text{Tl}^+}^0 \, \nabla \mu_{\text{TlCl}}, \quad (18\text{-}3)
$$

provides the means for assessing the magnitude of this difference. Thus,

$$
\mu_{\text{Cl}^-}^{\delta} - \mu_{\text{Cl}^-}^{\delta'} = \int_{\delta'}^{\delta} \left( t_{\text{Li}^+}^0 \frac{d\mu_{\text{LiCl}}}{dx} + t_{\text{Tl}^+}^0 \frac{d\mu_{\text{TlCl}}}{dx} \right) dx. \quad (18\text{-}4)
$$

In contrast to the situation in section 17, this integral cannot be evaluated exactly without a knowledge of the detailed concentration profiles in the junction region, since the transference numbers and chemical potentials depend on the concentrations of both LiCl and TlCl. But here we only want to assess the magnitude of a term which is usually neglected, and for this purpose we make approximations appropriate to dilute solutions and take the mobilities of the ions to be equal.

For the transference numbers, these approximations become

$$
t_{\text{Li}^+}^0 = \frac{m_{\text{Li}^+}}{2m_{\text{Cl}^-}} \quad \text{and} \quad t_{\text{Tl}^+}^0 = \frac{m_{\text{Tl}^+}}{2m_{\text{Cl}^-}}. \quad (18\text{-}5)
$$

From equation 14-10, the variations of the chemical potential of thallous chloride can be expressed as

$$
\nabla \mu_{\text{TlCl}} = RT \nabla \ln (m_{\text{Tl}^+} m_{\text{Cl}^-} \gamma_{\text{TlCl}}^2), \quad (18\text{-}6)
$$

and a similar expression applies to the lithium chloride. Substitution of these equations into equation 3 gives

$$
\nabla \mu_{\text{Cl}^-} = RT \nabla \ln m_{\text{Cl}^-} + RT \left( \frac{m_{\text{Li}^+}}{m_{\text{Cl}^-}} \nabla \ln \gamma_{\text{LiCl}} + \frac{m_{\text{Tl}^+}}{m_{\text{Cl}^-}} \nabla \ln \gamma_{\text{TlCl}} \right). \quad (18\text{-}7)
$$

[4]W. H. Smyrl and C. W. Tobias, "The Effect of Diffusion of a Sparingly Soluble Salt on the EMF of a Cell without Transference," *Electrochimica Acta*, 13 (1968), 1581–1589.

For dilute solutions, the Debye-Hückel expression can be used for the activity coefficients (see section 28):

$$\ln \gamma_{\text{LiCl}} = \ln \gamma_{\text{TlCl}} = -\alpha I^{1/2} = -\alpha m_{\text{Cl}^-}^{1/2}, \qquad (18\text{-}8)$$

where $\alpha$ is the Debye-Hückel constant, equal to $1.176 \text{ kg}^{1/2}/\text{mole}^{1/2}$ for water and 2.57 for dimethyl sulfoxide[4] at 25°C. In the Debye-Hückel approximation, the activity coefficients are the same for electrolytes of the same charge type and depend only on the ionic strength $I$, which here is equal to $m_{\text{Cl}^-}$.

Equation 7 now becomes

$$\nabla \mu_{\text{Cl}^-} = RT \, \nabla \ln m_{\text{Cl}^-} - \alpha \, RT \, \nabla m_{\text{Cl}^-}^{1/2}, \qquad (18\text{-}9)$$

and integration allows us to write equation 4 as

$$\mu_{\text{Cl}^-}^{\delta} - \mu_{\text{Cl}^-}^{\delta'} = -RT \ln \frac{m_{\text{Cl}^-}^{\delta'}}{m} + \alpha \, RT[(m_{\text{Cl}^-}^{\delta'})^{1/2} - m^{1/2}], \qquad (18\text{-}10)$$

where $m$ is the molality of lithium chloride in phase $\delta$.

In order to measure activity coefficients and standard cell potentials by using cells of this type, one makes measurements at low concentrations of LiCl and extrapolates to infinite dilution of this electrolyte, thereby establishing the reference state given by equation 14-5. These measurements at low concentrations of LiCl are, therefore, the most important, thermodynamically, and are also the measurements which are most subject to errors from liquid junctions such as treated here. We can approximate the concentration change from solution $\delta$ to $\delta'$ as

$$m_{\text{Cl}^-}^{\delta'} - m \approx m_{\text{Tl}^+}^{\delta'} = \frac{K_{sp}}{m_{\text{Cl}^-}^{\delta'}}, \qquad (18\text{-}11)$$

where $K_{sp}$ is the solubility product of thallous chloride. For the corresponding uncertainty in the measured cell potential to be less than 10 $\mu$V for $m = 10^{-3}$ mole/kg, the solubility product of the sparingly soluble salt must be less than about $4 \times 10^{-10}$ (mole/kg)$^2$.

Another cell of this type, in aqueous solution, is

$$
\begin{array}{c|c|c|c|c|c|c}
\alpha & \beta & & \delta & \epsilon & \lambda & \alpha' \\
\text{Pt}(s), & \text{HCl in} & \text{transition} & \text{HCl in} & \text{AgCl}(s) & \text{Ag}(s) & \text{Pt}(s) \\
\text{H}_2(g) & \text{H}_2\text{O} & \text{region} & \text{H}_2\text{O} & & &
\end{array} \qquad (18\text{-}12)
$$

where solutions $\beta$ and $\delta$ are different primarily because solution $\delta$ is saturated with silver chloride. The expression for the cell potential is

$$FU = -F(\Phi^\alpha - \Phi^{\alpha'}) = \tfrac{1}{2}\mu_{\text{H}_2}^\alpha - \mu_{\text{HCl}}^\beta - \mu_{\text{Ag}}^\lambda + \mu_{\text{AgCl}}^\epsilon + (\mu_{\text{Cl}^-}^\beta - \mu_{\text{Cl}^-}^\delta). \qquad (18\text{-}13)$$

Table 18-1 gives values of $(\mu_{\text{Cl}^-}^\beta - \mu_{\text{Cl}^-}^\delta)/F$ for $K_{sp} = 10^{-10}$ (mole/kg)$^2$, calculated by a method[5] to be outlined in a later chapter. Values obtained from equations 10 and 11 are given for comparison. The effect is much larger at practical concentrations for the nonaqueous systems treated by Smyrl and

[5]William H. Smyrl and John Newman, " Potentials of Cells with Liquid Junctions," *Journal of Physical Chemistry*, 72 (1968), 4660–4671.

Tobias, since the effect becomes important for bulk concentrations on the order of the square root of the solubility product.

The effect of the sparingly soluble salt on the cell potential, treated above, was ignored for some of the cells in section 17, specifically cells 17-21 and 17-23.

TABLE 18-1. EFFECT OF SOLUBILITY OF SILVER CHLORIDE FOR DECREASING VALUES OF BULK HCl CONCENTRATION.

| $m$, mole/kg | $m_{Cl^-}^{\delta}/m_{Cl^-}^{\beta}$ | $(\mu_{Cl^-}^{\beta} - \mu_{Cl^-}^{\delta})/F$, mV | |
| | | calculated | formula[a] |
| --- | --- | --- | --- |
| $10^{-4}$ | 1.00961 | −0.226 | −0.252 |
| $5 \times 10^{-5}$ | 1.0392 | −0.914 | −0.967 |
| $2 \times 10^{-5}$ | 1.200 | −4.32 | −4.82 |
| $10^{-5}$ | 1.604 | −11.22 | −12.34 |
| $5 \times 10^{-6}$ | 2.539 | −22.16 | −24.13 |
| $2 \times 10^{-6}$ | 5.499 | −40.58 | −43.86 |

[a]From equations 10 and 11 for $K_{sp} = 10^{-10}$ (mole/kg)$^2$.

Under the classification of this section, cells with two electrolytes, one of nearly uniform concentration, we could logically include cells with an inert electrolyte of nearly uniform concentration, where the species which react at the electrodes are present at much smaller concentrations. An example is

$$\begin{array}{c|c|c|c|c|c|c} \alpha & \beta & \delta & \{\text{transition}\} & \epsilon & \chi & \alpha' \\ \text{Pt}(s) & \text{Na(Hg)} & \begin{array}{c}\text{NaNO}_3, \\ \text{KNO}_3 \\ \text{in H}_2\text{O}\end{array} & \begin{array}{c}\{\text{region}\} \\ \{\text{KNO}_3 \text{ in}\} \\ \{\text{H}_2\text{O}\}\end{array} & \begin{array}{c}\text{AgNO}_3, \\ \text{KNO}_3 \text{ in} \\ \text{H}_2\text{O}\end{array} & \text{Ag}(s) & \text{Pt}(s), \end{array} \quad (18\text{-}14)$$

in which $KNO_3$ is present throughout the cell at the same, or nearly the same, concentration. The transition region contains concentration gradients of both $NaNO_3$ and $AgNO_3$. The cell potential can be expressed as

$$FU = -F(\Phi^{\alpha} - \Phi^{\alpha'}) = \mu_{Na}^{\beta} - \mu_{NaNO_3}^{\delta} - \mu_{Ag}^{\chi} + \mu_{AgNO_3}^{\epsilon} + (\mu_{NO_3^-}^{\delta} - \mu_{NO_3^-}^{\epsilon}),$$
$$(18\text{-}15)$$

on the assumption that $KNO_3$ is not involved in the phase equilibria at the electrodes, except to alter the chemical potentials of the other electrolytes in the solution.

With the same approximations used above, we can write

$$FU = \mu_{Na}^{\beta} - \mu_{Ag}^{\chi} + RT \ln \frac{\lambda_{Ag^+}^{\theta}}{\lambda_{Na^+}^{\theta}} + RT \ln \frac{m_{Ag^+}^{\epsilon}}{m_{Na^+}^{\delta}}$$
$$- \alpha RT [(m_{NO_3^-}^{\epsilon})^{1/2} - (m_{NO_3^-}^{\delta})^{1/2}]. \quad (18\text{-}16)$$

Variations in the cell potential are thus due primarily to changes in the concentrations of silver and sodium ions. If the ionic strength is reasonably uniform, the last term, related to activity coefficient corrections, can be ignored. We notice that the ratio $\lambda_{Ag^+}^\theta/\lambda_{Na^+}^\theta$ is uniquely determined according to the considerations of section 14.

When activity coefficients are ignored, equation 16 is a form of the so-called Nernst equation, relating cell potentials to the logarithms of ionic concentrations. It is frequently written in terms of molar concentrations; for example, equation 10-1 is based on similar considerations. The Nernst equation can be used when there is an excess of inert electrolyte of nearly uniform concentration and the reactant species are present at much smaller concentrations. The assessment of the errors involved will be considered again in chapter 6.

We have seen that cell potentials frequently depend upon the transport properties of the electrolytic solutions as well as the thermodynamic properties, and they also depend upon the detailed form of the concentration profiles in the junction region. Under certain conditions, such as those considered in this section, approximations can be introduced so that the cell potential is expressed in terms of thermodynamic properties alone. Whether these approximations are sufficiently accurate depends on the intended application.

## 19.  Cell with two electrolytes, both of varying concentration

Cells of this type may still be divided into two groups according to whether the two electrolytes have an ion in common. A cell with a junction between solutions of $CuSO_4$ and $ZnSO_4$ is an example where there is a common ion; a junction between NaCl and $HClO_4$ is an example where there is not. The former class will be discussed first.

Consider the cell

$$
\begin{array}{c|c|c|c|c|c|c|c}
\alpha & \beta & \delta & \epsilon & \lambda & \delta' & \beta' & \alpha' \\
Pt(s) & Ag(s) & AgCl(s) & \begin{array}{c} HCl\ in \\ H_2O \end{array} & \left\{\begin{array}{c} transition \\ region \end{array}\right\} & \left\{\begin{array}{c} KCl\ in \\ H_2O \end{array}\right. & AgCl(s) & Ag(s) & Pt(s)
\end{array}
$$

$$(19\text{-}1)$$

for which the cell potential is

$$FU = -F(\Phi^\alpha - \Phi^{\alpha'}) = \mu_{Cl^-}^\epsilon - \mu_{Cl^-}^\lambda. \tag{19-2}$$

The effect of the nonzero solubility of AgCl, discussed in section 18, can be ignored here. Integration of equation 16-3 for this case gives

$$\mu_{Cl^-}^\epsilon - \mu_{Cl^-}^\lambda = \int_\lambda^\epsilon \left( t_{H^+}^0 \frac{\partial \mu_{HCl}}{\partial x} + t_{K^+}^0 \frac{\partial \mu_{KCl}}{\partial x} \right) dx. \tag{19-3}$$

Here, as with equation 18-4, and in contrast to equation 17-4, the integral depends on the detailed form of the concentration profiles in the junction region. We are not yet in a position to consider approximations possible in the evaluation of this integral.

We should also like to know how to treat the potentials of cells containing two electrolytes of varying concentration but with no common ion. Such a cell is

$$
\begin{array}{c|c|c|c|c|c|c}
\alpha & \beta & \text{transition} & \delta & \epsilon & \lambda & \alpha' \\
\text{Pt}(s), & \text{HClO}_4 & \text{transition} & \text{NaCl in} & \text{AgCl}(s) & \text{Ag}(s) & \text{Pt}(s) \\
\text{H}_2(g) & \text{in H}_2\text{O} & \text{region} & \text{H}_2\text{O} & & &
\end{array}
\qquad (19\text{-}4)
$$

The transition region contains the solutions which vary in composition from $\beta$ to $\delta$. From the conditions of phase equilibria at the electrodes and the definitions of the chemical potentials of neutral species, the cell potential can be written

$$
FU = -F(\Phi^\alpha - \Phi^{\alpha'}) = \tfrac{1}{2}\mu_{\text{H}_2}^\alpha - \mu_{\text{Ag}}^\lambda + \mu_{\text{AgCl}}^\epsilon - (\mu_{\text{H}^+}^\beta + \mu_{\text{Cl}^-}^\delta). \quad (19\text{-}5)
$$

The cell potential is again related to the thermodynamic properties of electrically neutral components, but a new term has appeared. Instead of the difference of electrochemical potential of a single ion between the two solutions, there is now a combination of electrochemical potentials of two ions. One way to analyze this more complicated situation is to select an intermediate point $I$ in the junction where both ions are present. If the concentration profiles are known, equation 16-3 can be used to evaluate the differences $\mu_{\text{H}^+}^\beta - \mu_{\text{H}^+}^I$ and $\mu_{\text{Cl}^-}^\delta - \mu_{\text{Cl}^-}^I$. The cell potential can then be written,

$$
FU = \tfrac{1}{2}\mu_{\text{H}_2}^\alpha - \mu_{\text{Ag}}^\lambda + \mu_{\text{AgCl}}^\epsilon - (\mu_{\text{H}^+}^\beta - \mu_{\text{H}^+}^I) - (\mu_{\text{Cl}^-}^\delta - \mu_{\text{Cl}^-}^I) - \mu_{\text{HCl}}^I.
$$
$$(19\text{-}6)$$

The last term, $\mu_{\text{HCl}}^I$, can be evaluated if the activity coefficient of HCl is known for the solution of $\text{HClO}_4$ and NaCl at the point $I$. The cell potential is, of course, independent of the choice of the intermediate solution $I$.

Although none of the examples has been carried through here, it should be apparent that the potential of such cells can, in principle, be treated. The concentration profiles are determined from the laws of diffusion and the method of forming the junction. The expression for the cell potential will involve the difference in electrochemical potential of an ion or of two different ions in different solutions, and this difference can be evaluated from equation 16-3, perhaps with the selection of an intermediate point, because the solutions are connected to each other through the junction. The development is brought to fruition in chapter 6.

## 20.  Standard cell potential and activity coefficients

The prediction of cell potentials will generally require a knowledge of the thermodynamic and transport properties of the system. The necessary thermodynamic properties usually are tabulated in terms of standard cell potentials and activity coefficients. The standard cell potential is the composition-independent part of the cell potential and contains information about the free energy change related to the overall cell reaction under certain very special conditions. The activity coefficients describe deviations from a pre-scribed composition dependence for the chemical potentials in the various phases. The standard cell potential and the activity coefficients depend upon the choice of the secondary reference state in a manner such that the cell potential itself should be independent of this choice.

For example, for the cell 15-9 the potential was expressed as

$$FU = \mu_{Li}^{\beta} - \mu_{Tl}^{\phi} + \mu_{TlCl}^{\epsilon} - \mu_{LiCl}^{\delta}. \tag{20-1}$$

By means of equation 14-15, this can be written as

$$FU = FU^{\theta} - 2RT \ln (m_{LiCl}^{\delta} \gamma_{LiCl}^{\delta}) - RT \ln a_{Tl}^{\phi}, \tag{20-2}$$

where the standard cell potential is given by

$$FU^{\theta} = \mu_{Li}^{0} - \mu_{Tl}^{0} + \mu_{TlCl}^{0} - 2RT \ln \lambda_{LiCl}^{\theta}, \tag{20-3}$$

and where the relative activity of thallium in the thallium amalgam is

$$a_{Tl}^{\phi} = \frac{\lambda_{Tl}^{\phi}}{\lambda_{Tl}^{0}}. \tag{20-4}$$

The superscripts 0 denote pure phases at the same temperature and pressure. It has been assumed that the lithium electrode and the thallous chloride salt are pure phases. Equation 4 defines a secondary reference state for the thallium in the thallium amalgam, namely, pure thallium.

The standard cell potential $U^{\theta}$ is independent of the concentration of electrolyte in the solution but does depend on the components, including the solvent, as well as on the temperature and pressure. It also depends on the choice of the secondary reference states 4 and 14-5, the latter being necessary for the separate determination of $\lambda_{LiCl}^{\theta}$ and $\gamma_{LiCl}^{\delta}$. We notice that $\lambda_{LiCl}^{\theta}$ is unique-ly determined according to the considerations of section 14. The standard cell potential also has an unfortunate dependence on the concentration scale and the units chosen. In this application, $\lambda_{LiCl}^{\theta}$ is invariably expressed in kg/mole. Correspondingly, $m_{LiCl}^{\delta}$ in equation 2 must be expressed in mole/kg to be consistent.

If the values of $U^{\theta}$ and $\gamma_{LiCl}$ are chosen for a particular thermodynamic state of the system, the value of $\gamma_{LiCl}$ may be determined in other states by potential measurements and the use of equation 2. Correspondingly, values of cell potentials can be reproduced from tabulated values of $\gamma_{LiCl}$. The reference state used is expressed by equation 14-5, even though this reference state is difficult to apply experimentally. These difficulties are asso-ciated with the larger relative effect of impurities in dilute solutions, inaccu-racies in the determination of the concentration of lithium chloride, the rapid variation of the activity coefficient in dilute solutions, and the effect of the solubility of the thallous chloride, which can become significant in dilute solutions. Nevertheless, this reference state is used because no other logical possibility presents itself. It has the advantage that the reference state is essentially the same for all the solute species, and activity coefficient expres-sions for multicomponent solutions are thereby simplified.

For the cells of section 17, with a single electrolyte of varying concentration, identical electrodes were used on the two sides of the cell. Consequently no standard cell potential is needed. One sees, for example, from equation 17-7, that a tabulation of transference numbers and activity coefficients would be sufficient to reproduce the experiemental cell potentials. If the activity coefficients are known from measurements on cells without transference or from vapor pressure measurements, the transference number can be calculated from the experimental data on cells of the type described in section 17. On the other hand, if the transference number is known from moving boundary or Hittorf measurements, information about the activity coefficients can be obtained from these cells. The latter procedure is particularly useful in dilute solutions where the transference numbers may be well known and are relatively independent of concentration.

The cells discussed in section 18 still will be useful for the determination of thermodynamic properties if the error in the estimation of the nonthermodynamic terms can be made sufficiently small. For example, for cell 18-14 the approximate expression 18-16 for the cell potential can be written

$$FU = FU^\theta + RT \ln a_{Na}^\beta + RT \ln \frac{m_{Ag^+}^\epsilon}{m_{Na^+}^\delta} - \alpha RT[(m_{NO_3^-}^\epsilon)^{1/2} - (m_{NO_3^-}^\delta)^{1/2}],$$

(20-5)

where the standard cell potential is given by

$$FU^\theta = \mu_{Na}^0 - \mu_{Ag}^0 + RT \ln \frac{\lambda_{Ag^+}^\theta}{\lambda_{Na^+}^\theta}$$

(20-6)

and the relative activity of sodium in the sodium amalgam is

$$a_{Na}^\beta = \frac{\lambda_{Na}^\beta}{\lambda_{Na}^0}.$$

(20-7)

This cell can be used to obtain the standard cell potential, even though it does not yield any useful information about the activity coefficients.

The method of applying the secondary reference state 14-5 to the above cells will be considered again in chapter 4 after we have discussed the behavior of activity coefficients in dilute solutions. One wants to extrapolate to infinite dilution in a way that provides the greatest accuracy.

For the cell 19-4, the cell potential can be written

$$FU = FU^\theta + \tfrac{1}{2}RT \ln p_{H_2}^\alpha - (\mu_{H^+}^\beta - \mu_{H^+}^I) - (\mu_{Cl^-}^\delta - \mu_{Cl^-}^I) \\ - RT \ln [m_{H^+}^I \cdot m_{Cl^-}^I (\gamma_{HCl}^I)^2],$$

(20-8)

where the standard cell potential is

$$FU^\theta = \tfrac{1}{2}\mu_{H_2}^* - \mu_{Ag}^0 + \mu_{AgCl}^0 - 2RT \ln \lambda_{HCl}^\theta$$

(20-9)

and the fugacity of hydrogen and $\mu_{H_2}^*$ are defined by (see problem 14)

$$\mu_{H_2} = \mu_{H_2}^* + RT \ln p_{H_2}.$$

(20-10)

Here, the secondary reference state is specified by the requirement that, as the pressure approaches zero, $p_{H_2}$ approaches the partial pressure of hydrogen expressed in atmospheres. The standard cell potential is a collection of thermodynamic quantities, independent of the concentrations in the cell, and is also equal to the standard cell potential of cell 18-12. However, in this case, the nonthermodynamic terms in equation 8 are not negligible; they are difficult to evaluate accurately because they require a knowledge of transference numbers and activity coefficients in multicomponent solutions of moderate concentration, as well as a knowledge of how the junction was formed.

As the above example demonstrates, the cells of section 19 are not particularly useful for the precise determination of thermodynamic or transport properties of solutions, or of standard cell potentials. Cell 18-12 will yield the same standard cell potential as cell 19-4 but with less uncertainty. Cells such as those treated in section 19 are encountered in practice, and the prediction of their potentials is more a test of our ability to treat junction regions. Tabulated values of standard cell potentials, as well as activity coefficients and transference numbers, do find application in this endeavor.

In compiling the standard cell potentials of many cells, it is desirable to tabulate as few details as possible without being ambiguous. Of $n$ possible electrodes, one can make measurements on $\frac{1}{2}n(n-1)$ different combinations of these electrodes taken two at a time. Only $n-1$ of these combinations are independent, and the others can be obtained by appropriate addition and subtraction of the $n-1$ independent combinations. Hence, one can report the standard cell potentials of $n-1$ possible electrodes against the other possible electrode, and the standard cell potential of other combinations can be obtained from these.

By convention, the hydrogen electrode is used for this reference point. Table 20-1 gives values for selected standard electrode potentials in water relative to the hydrogen electrode. To emphasize the thermodynamic nature of these quantitites, the explicit expressions in terms of chemical potentials in the secondary reference states are also given. A number of remarks can be made about the entries in this table.

The sign convention gives the standard potential of the electrode *relative to* the hydrogen electrode. For example, a silver-silver chloride electrode in 1 $M$ HCl would be expected to be positive relative to a hydrogen electrode. The expression for the standard electrode potential involves only those species which are involved in the overall cell reaction. For entry 15 the overall cell reaction is the electrolysis of water and involves no ions at all. The chemical potentials with a superscript 0 denote elements or compounds in the pure state. The values of $\lambda_i^\theta$ depend upon the extrapolation to infinite dilution and therefore depend on the particular solvent involved. Consequently, the table of electrode potentials would be different in a different solvent. Values of $\lambda_i^\theta$ for ions appear only in those combinations which can be unambiguously determined according to the considerations of section 14.

As noted earlier, $\lambda_i^\theta$ should be expressed in kg/mole, although this is

TABLE 20-1.  SELECTED STANDARD ELECTRODE POTENTIALS REFERRED TO THE HYDROGEN ELECTRODE. AQUEOUS SOLUTIONS AT 25°C.

| reaction | $FU^\theta$ | $U^\theta$, volt |
|---|---|---|
| 1  $K \longrightarrow K^+ + e^-$ | $\tfrac{1}{2}\mu^*_{H_2} - \mu^0_K + RT \ln (\lambda^\theta_{K^+}/\lambda^\theta_{H^+})$ | $-2.925$ |
| 2  $Pb + SO_4^= \longrightarrow PbSO_4 + 2e^-$ | $\tfrac{1}{2}(\mu^*_{H_2} + \mu^0_{PbSO_4} - \mu^0_{Pb}) - \tfrac{1}{2}RT \ln [(\lambda^\theta_{H^+})^2 \lambda^\theta_{SO_4^=}]$ | $-0.356$ |
| 3  $Pb \longrightarrow Pb^{++} + 2e^-$ | $\tfrac{1}{2}\mu^*_{H_2} - \tfrac{1}{2}\mu^0_{Pb} + \tfrac{1}{2}RT \ln [\lambda^\theta_{Pb^{++}}/(\lambda^\theta_{H^+})^2]$ | $-0.126$ |
| 4  $H_2 \longrightarrow 2H^+ + 2e^-$ | — | 0 |
| 5  $Hg + 2OH^- \longrightarrow HgO + H_2O + 2e^-$ | $\tfrac{1}{2}(\mu^*_{H_2} + \mu^0_{HgO} + \mu^0_{H_2O} - \mu^0_{Hg}) - RT \ln (\lambda^\theta_{H^+}\cdot\lambda^\theta_{OH^-})$ | 0.098 |
| 6  $Cu^+ \longrightarrow Cu^{++} + e^-$ | $\tfrac{1}{2}\mu^*_{H_2} + RT \ln (\lambda^\theta_{Cu^{++}}/\lambda^\theta_{H^+}\cdot\lambda^\theta_{Cu^+})$ | 0.153 |
| 7  $Ag + Cl^- \longrightarrow AgCl + e^-$ | $\tfrac{1}{2}\mu^*_{H_2} + \mu^0_{AgCl} - \mu^0_{Ag} - RT \ln (\lambda^\theta_{H^+}\cdot\lambda^\theta_{Cl^-})$ | 0.222 |
| 8  $Cu \longrightarrow Cu^{++} + 2e^-$ | $\tfrac{1}{2}\mu^*_{H_2} - \tfrac{1}{2}\mu^0_{Cu} + \tfrac{1}{2}RT \ln [\lambda^\theta_{Cu^{++}}/(\lambda^\theta_{H^+})^2]$ | 0.337 |
| 9  $4OH^- \longrightarrow O_2 + 2H_2O + 4e^-$ | $\tfrac{1}{2}\mu^*_{H_2} + \tfrac{1}{4}\mu^0_{O_2} + \tfrac{1}{2}\mu^0_{H_2O} - RT \ln (\lambda^\theta_{H^+}\cdot\lambda^\theta_{OH^-})$ | 0.401 |
| 10  $Cu \longrightarrow Cu^+ + e^-$ | $\tfrac{1}{2}\mu^*_{H_2} - \mu^0_{Cu} + RT \ln (\lambda^\theta_{Cu^+}/\lambda^\theta_{H^+})$ | 0.521 |
| 11  $2I^- \longrightarrow I_2(s) + 2e^-$ | $\tfrac{1}{2}\mu^*_{H_2} + \tfrac{1}{2}\mu^0_{I_2} - RT \ln (\lambda^\theta_{H^+}\cdot\lambda^\theta_{Cl^-})$ | 0.5355 |
| 12  $3I^- \longrightarrow I_3^- + 2e^-$ | $\tfrac{1}{2}\mu^*_{H_2} + \tfrac{1}{2}RT \ln \lambda^\theta_{I_3^-} - \tfrac{3}{2}RT \ln \lambda^\theta_{I^-} - RT \ln \lambda^\theta_{H^+}$ | 0.536 |
| 13  $Fe^{++} \longrightarrow Fe^{3+} + e^-$ | $\tfrac{1}{2}\mu^*_{H_2} + RT \ln (\lambda^\theta_{Fe^{3+}}/\lambda^\theta_{Fe^{++}}\cdot\lambda^\theta_{H^+})$ | 0.771 |
| 14  $Au + 4Cl^- \longrightarrow AuCl_4^- + 3e^-$ | $\tfrac{1}{2}\mu^*_{H_2} - \tfrac{1}{3}\mu^0_{Au} + \tfrac{1}{3}RT \ln \lambda^\theta_{AuCl_4^-} - \tfrac{4}{3}RT \ln \lambda^\theta_{Cl^-} - RT \ln \lambda^\theta_{H^+}$ | 1.00 |
| 15  $2H_2O \longrightarrow O_2 + 4H^+ + 4e^-$ | $\tfrac{1}{2}\mu^*_{H_2} + \tfrac{1}{4}\mu^0_{O_2} - \tfrac{1}{2}\mu^0_{H_2O}$ | 1.229 |
| 16  $2Cl^- \longrightarrow Cl_2(g) + 2e^-$ | $\tfrac{1}{2}\mu^*_{H_2} + \tfrac{1}{2}\mu^*_{Cl_2} - RT \ln (\lambda^\theta_{H^+}\cdot\lambda^\theta_{Cl^-})$ | 1.3595 |
| 17  $PbSO_4 + 2H_2O \longrightarrow PbO_2 + SO_4^= + 4H^+ + 2e^-$ | $\tfrac{1}{2}(\mu^*_{H_2} + \mu^0_{PbO_2} - \mu^0_{PbSO_4}) - \mu^0_{H_2O} + \tfrac{1}{2}RT \ln [(\lambda^\theta_{H^+})^2 \lambda^\theta_{SO_4^=}]$ | 1.685 |

$\lambda^\theta_i$ must be expressed in kg/mole. The superscript 0 denotes the pure element or compound at 25°C and one atmosphere. The superscript * denotes the chemical potential of gases in an ideal standard state. See problem 14. Values are from Latimer, *Oxidation Potentials*, 2nd ed., © 1952, with permission of Prentice-Hall, Inc.

not critical for entries 1, 10, and 15. For entries 2 and 17, one should observe that $(\lambda_{H^+}^\theta)^2 \lambda_{SO_4^=}^\theta = \frac{1}{4}(\lambda_{H_2SO_4}^\theta)^3$ according to equation 14-14. Thus, for the cell

$$\begin{array}{c|c|c|c|c}
\alpha & \beta & \delta & \epsilon & \alpha' \\
Pt(s), H_2(g) & H_2SO_4 \text{ in } H_2O & PbSO_4(s) & PbO_2(s) & Pt(s),
\end{array} \quad (20\text{-}11)$$

the expression for the cell potential becomes

$$FU = FU^\theta + \tfrac{1}{2}RT \ln p_{H_2}^\alpha + \tfrac{3}{2}RT \ln (m_{H_2SO_4}^\beta \gamma_{H_2SO_4}^\beta)$$
$$- RT \ln a_{H_2O}^\beta + RT \ln 2. \quad (20\text{-}12)$$

Some authors avoid the explicit appearance of the last term by defining the mean ionic molality of a single electrolyte as

$$m_{+-}^\nu = m_+^{\nu_+} m_-^{\nu_-}. \quad (20\text{-}13)$$

Entries 9 and 15 give

$$\frac{\lambda_{H^+}^\theta \lambda_{OH^-}^\theta}{\lambda_{H_2O}^0} = \exp\left(\frac{1.229 - 0.401}{RT/F}\right) = 10^{14} \frac{kg^2}{mole^2}. \quad (20\text{-}14)$$

Since

$$\lambda_{H^+} \lambda_{OH^-} = \lambda_{H_2O}, \quad (20\text{-}15)$$

this corresponds to a dissociation constant for water of

$$K_w = \frac{m_{H^+} m_{OH^-} \gamma_{H^+} \gamma_{OH^-}}{a_{H_2O}} = 10^{-14} \frac{mole^2}{kg^2}. \quad (20\text{-}16)$$

Entries 6, 8, and 10 are not independent. Specifically, two times entry 8 is equal to the sum of entries 6 and 10, either for the numerical values of $U^\theta$ or for the expressions for $FU^\theta$. These entries also describe the equilibrium concentrations of cuprous ions present in the solution of a cupric salt. The cuprous ions can disproportionate according to the reaction

$$2\,Cu^+ \longrightarrow Cu + Cu^{++}, \quad (20\text{-}17)$$

for which the equilibrium condition is

$$\lambda_{Cu^+}^2 = \lambda_{Cu} \lambda_{Cu^{++}}. \quad (20\text{-}18)$$

From entries 6 and 10,

$$\frac{(\lambda_{Cu^+}^\theta)^2}{\lambda_{Cu}^0 \lambda_{Cu^{++}}^0} = \exp\left(\frac{0.521 - 0.153}{RT/F}\right) = 1.67 \times 10^6 \frac{kg}{mole}. \quad (20\text{-}19)$$

Hence the equilibrium constant can be expressed as

$$\frac{m_{Cu^+}^2 \gamma_{Cu^+}^2}{m_{Cu^{++}} \gamma_{Cu^{++}}} = 0.6 \times 10^{-6} \frac{mole}{kg}. \quad (20\text{-}20)$$

By ignoring the activity coefficients, which, incidentally, appear in a combination which is determined unambiguously according to the considerations of section 14, we estimate the equilibrium concentration of cuprous ions to be $2.4 \times 10^{-4}$ mole/kg in a 0.1 $M$ $CuSO_4$ solution. We see from this example that two or more electrode processes can be simultaneously in equilibrium at the same electrode at the same potential.

Entries 2 and 3 of table 20-1 yield

$$\frac{\lambda_{Pb^{++}}^\theta \lambda_{SO_4^=}^\theta}{\lambda_{PbSO_4}^0} = \exp\left(\frac{0.356 - 0.126}{RT/2F}\right) = 6 \times 10^7 \frac{kg^2}{mole^2}. \quad (20\text{-}21)$$

Hence, the solubility product of $PbSO_4$ is

$$K_{sp} = m_{Pb^{..}} m_{SO_4^{..}} \gamma^2_{PbSO_4} = 1.7 \times 10^{-8} \frac{mole^2}{kg^2}. \tag{20-22}$$

Table 2 gives additional values of standard electrode potentials in aqueous solutions at 25°C. Detailed expressions for $FU^\theta$, as given in table 20-1,

TABLE 20-2 ADDITIONAL STANDARD ELECTRODE POTENTIALS
IN AQUEOUS SOLUTIONS AT 25°C.

|  | reaction | $U^\theta$, volt |
|---|---|---|
| 1 | $Li \longrightarrow Li^+ + e^-$ | $-3.045$ |
| 2 | $Na \longrightarrow Na^+ + e^-$ | $-2.714$ |
| 3 | $Al \longrightarrow Al^{3+} + 3e^-$ | $-1.66$ |
| 4 | $Mn \longrightarrow Mn^{++} + 2e^-$ | $-1.18$ |
| $5^a$ | $\frac{1}{2} H_2 + OH^- \longrightarrow H_2O + e^-$ | $-0.828$ |
| 6 | $Zn \longrightarrow Zn^{++} + 2e^-$ | $-0.763$ |
| 7 | $Cr \longrightarrow Cr^{3+} + 3e^-$ | $-0.74$ |
| 8 | $Cr \longrightarrow Cr^{++} + 2e^-$ | $-0.91$ |
| 9 | $Fe \longrightarrow Fe^{++} + 2e^-$ | $-0.440$ |
| 10 | $Cr^{++} \longrightarrow Cr^{3+} + e^-$ | $-0.41$ |
| 11 | $H_2 \longrightarrow 2 H^+ + 2e^-$ | $0$ |
| 12 | $2 Hg + 2 Cl^- \longrightarrow Hg_2Cl_2 + 2e^-$ | $0.2676$ |
| 13 | $Fe(CN)_6^{4-} \longrightarrow Fe(CN)_6^{3-} + e^-$ | $0.36$ |
| 14 | $2 Hg \longrightarrow Hg_2^{++} + 2e^-$ | $0.789$ |
| 15 | $Ag \longrightarrow Ag^+ + e^-$ | $0.7991$ |
| 16 | $Hg_2^{++} \longrightarrow 2 Hg^{++} + 2e^-$ | $0.920$ |
| 17 | $2 Br^- \longrightarrow Br_2(l) + 2e^-$ | $1.0652$ |
| 18 | $Ag^+ \longrightarrow Ag^{++} + e^-$ | $1.98$ |
| 19 | $2 F^- \longrightarrow F_2(g) + 2e^-$ | $2.87$ |

[a]Represents a hydrogen electrode in a basic medium relative to a hydrogen electrode in an acidic medium. Here

$$FU^\theta = \mu^0_{H_2O} - RT \ln (\lambda^\theta_H \cdot \lambda^\theta_{OH^-}).$$

are not given here. The reader can reproduce them by reference to the examples in table 20-1. Latimer[6] prepared the classic tabulation of standard electrode potentials in aqueous solutions. He also discusses how to obtain these by thermochemical calculations as well as by direct measurements on galvanic cells. Extensive tabulations are also given by Milazzo,[7] in Gmelins Handbuch,[8] and by de Bethune and Loud.[9]

[6]Wendell M. Latimer, *The Oxidation States of the Elements and Their Potentials in Aqueous Solutions* (Englewood Cliffs, N.J.: Prentice-Hall, Inc., 1952).

[7]Giulio Milazzo, *Electrochemistry* (Amsterdam: Elsevier Publishing Company, 1963).

[8]*Gmelins Handbuch der anorganischen Chemie*, eighth ed. (Weinheim: Verlag Chemie, 1950—).

[9]Andre J. de Bethune and Nancy Swendeman Loud, "Table of Standard Aqueous Electrode Potentials and Temperature Coefficients at 25°C," Clifford A. Hampel, ed., *The Encyclopedia of Electrochemistry* (New York: Reinhold Publishing Corporation, 1964), pp. 414–426.

Standard cell potentials and activity coefficients are tabulated separately. Activity coefficients are characteristic of the solutions and can be measured by nonelectrochemical methods, such as vapor pressure measurements, or by the cells of section 17, which do not involve standard cell potentials. On the other hand, standard cell potentials are associated with the overall cell reaction, for which all the ionic species in the solution need not be specified. Because of the diverse sources and applications of standard cell potentials and activity coefficients, their separate tabulation is dictated and amounts to the briefest way to collect the results of many experiments on many different systems. Consequently, in an attempt to reproduce the potentials of a particular cell with the use of these separate tabulations, the errors associated with the extrapolation to infinite dilution may not cancel exactly.

## 21. Pressure dependence of activity coefficients

The definition of activity coefficients involves the use of a secondary reference state for the solute, for example, that given by equation 14-5. The value of $\lambda_i^\theta$ depends upon the choice of the secondary reference state in a complementary manner, so that the product $\gamma_i \lambda_i^\theta$ is independent of this choice. If equation 14-5 is to be applied at each temperature and pressure, then $\lambda_i^\theta$ will also depend on temperature and pressure. If, on the other hand, equation 14-5 is applied at each temperature but only at one atmosphere pressure, then $\lambda_i^\theta$ depends only on temperature. The mean ionic activity coefficient of neutral combinations of ions will then approach unity at infinite dilution only at this pressure of one atmosphere.

The variation of activity coefficients with pressure is determined by the fact that the derivative of the chemical potential with respect to pressure is equal to the partial molar volume:

$$\left(\frac{\partial \mu_A}{\partial p}\right)_{T,\,n_i} = \left(\frac{\partial V}{\partial n_A}\right)_{T,\,p,\,n_B \atop B \neq A} = \bar{V}_A. \qquad (21\text{-}1)$$

We shall apply this equation only to neutral electrolytes, for which the method of obtaining partial molar volumes from density determinations is given in appendix A. For an electrolyte,

$$\mu_A = RT \ln\left[(m_+ \lambda_+^\theta)^{\nu_+} \cdot (m_- \lambda_-^\theta)^{\nu_-} \gamma_{+-}^\nu\right]. \qquad (21\text{-}2)$$

If $\lambda_i^\theta$ is taken to be independent of pressure, then equation 1 yields

$$\left(\frac{\partial \ln \gamma_{+-}}{\partial p}\right)_{T,\,m_i} = \frac{\bar{V}_A}{\nu RT} \quad (\lambda_i^\theta \text{ independent of } p). \qquad (21\text{-}3)$$

For $\bar{V}_A = 27 \text{ cm}^3/\text{mole}$ and a 1–1 electrolyte, the pressure variation of $\gamma_{+-}$ at 25°C is

$$\frac{\partial \ln \gamma_{+-}}{\partial p} = 5.5 \times 10^{-4} \text{ atm}^{-1}. \qquad (21\text{-}4)$$

Thus, a pressure change of 18 atmospheres is necessary to change $\gamma_{+-}$ by 1 percent. This will apply to the value of $\gamma_{+-}$ at infinite dilution as well.

If, on the other hand, equation 14-5 is applied at each pressure and $\lambda_i^\theta$ depends on pressure, then application of equation 1 to equation 2 gives

$$\frac{\partial \ln \lambda_+^{\theta v_+} \cdot \lambda_-^{\theta v_-}}{\partial p} + \nu \frac{\partial \ln \gamma_{+-}}{\partial p} = \frac{\bar{V}_A}{RT}. \tag{21-5}$$

Under these conditions, $\gamma_{+-}$ is independent of pressure at infinite dilution, and consequently

$$\frac{\partial \ln \lambda_+^{\theta v_+} \cdot \lambda_-^{\theta v_-}}{\partial p} = \frac{\bar{V}_A^0}{RT}, \tag{21-6}$$

where the superscript 0 denotes the value at infinite dilution. At other concentrations, equation 5 then becomes

$$\left(\frac{\partial \ln \gamma_{+-}}{\partial p}\right)_{T, m_i} = \frac{\bar{V}_A - \bar{V}_A^0}{\nu RT} \quad (\gamma_{+-} \longrightarrow 1 \text{ at infinite dilution}). \tag{21-7}$$

Since the partial molar volume does not change very much with concentration, the activity coefficient so defined changes even less with pressure than the variation indicated by equation 3. For example, if $\bar{V}_A = 32$ at 3 $M$ and 27 cm$^3$/mole at infinite dilution, then for a 1–1 electrolyte at 25°C

$$\frac{\partial \ln \gamma_{+-}}{\partial p} = 10^{-4} \text{ atm}^{-1}, \tag{21-8}$$

and a pressure change of 97 atmospheres is required for a 1 percent change in $\gamma_{+-}$.

From these examples, one can perhaps perceive why the pressure dependence of activity coefficients is of little concern.

## PROBLEMS

**1.** For theoretical treatments of diffusion, it is sometimes desirable to use a diffusion coefficient $\mathfrak{D}$ for which the driving force is based on a gradient of chemical potential. For a solution of a single electrolyte, this is related to the measured value $D$ (based on a concentration driving force) by

$$D = \mathfrak{D}(1 + \nu M_0 m)\left(1 + \frac{d \ln \gamma_{+-}}{d \ln m}\right).$$

(a) Show that the factor $1 + d \ln \gamma_{+-}/d \ln m$ is independent of the choice of the secondary reference state and is therefore characteristic of the solution at the concentration $m$. In other words, this factor can be used even for solutions below the freezing point of the pure solvent, where equation 14-5 can be used only by consideration of supercooled liquids.

(b) Show that osmotic coefficients can be used directly for this conversion by deriving the relationship

$$1 + \frac{d \ln \gamma_{+-}}{d \ln m} = \phi + \frac{d\phi}{d \ln m}.$$

(c) Obtain the following relation, where activity coefficients based on molar concentrations are used:

$$D = \mathfrak{D}\, \frac{\dfrac{1 + \nu M_0 m}{1 + M_A m}\left(1 + \dfrac{d \ln f_{+-}}{d \ln c}\right)}{\left(1 - \dfrac{d \ln \rho}{d \ln c}\right)}.$$

(d) Describe the behavior of $1 + d \ln \gamma_{+-}/d \ln m$ in a concentration range which includes the pure molten salt. What is the behavior of the expression in part (c)?

(e) Express the relationship between $\mathfrak{D}$ and $D$ in terms of the activity coefficient $\gamma_A$, for which the dissociation of the electrolyte is ignored. On the other hand, for a nonelectrolyte one expects to use the relation

$$D = \mathfrak{D}(1 + M_0 m)\left(1 + \frac{d \ln \gamma_A}{d \ln m}\right).$$

Since the two expressions are different and since the measured value $D$ is independent of the state of molecular aggregation, one concludes that $\mathfrak{D}$, like $\gamma$, is an idealized quantity whose definition depends on whether the solute is to be regarded as dissociated or undissociated.

**2.** Obtain an expression for the potential of the cell

| $\alpha$ | $\beta$ | $\delta$ | $\epsilon$ | $\phi$ | $\alpha'$ |
|---|---|---|---|---|---|
| $Pt(s)$ | $Ni(s), O_2(g)$ | $KOH$ in $H_2O$ | $HgO(s)$ | $Hg(l)$ | $Pt(s)$. |

The nickel electrode can be regarded as inert, and the solubility of oxygen and mercuric oxide in the solution can be ignored. Thus, the solution can be regarded to be of uniform concentration. The electrode reactions are

$$O_2 + 4e^- + 2\,H_2O \longrightarrow 4\,OH^-$$

and
$$Hg + 2\,OH^- \longrightarrow HgO + H_2O + 2e^-.$$

Actually it is difficult to achieve a reversible potential for the oxygen electrode. What is the expression for the standard cell potential, and what is its value?

**3.** Obtain an expression for the potential of the cell

| $\alpha$ | $\beta$ | $\delta$ | $\epsilon$ | $\phi$ | $\alpha'$ |
|---|---|---|---|---|---|
| $Pt(s)$ | $K(Hg)$ | $KOH$ in $H_2O$ | $HgO(s)$ | $Hg(l)$ | $Pt(s)$. |

The solution can be treated as one of uniform concentration. What is the expression for the standard cell potential, and what is its value? Here the cell potential depends on the chemical potentials of both the potassium hydroxide and the water in the solution.

**4.** Obtain an expression for the potential of the copper concentration cell in figure 6-1. What is the standard cell potential? Introduce approximations sufficient to obtain equation 6-1.

**5.** Treat the potential of a copper concentration cell with an excess of sulfuric acid of nearly uniform concentration. What is the standard cell potential? Introduce approximations sufficient to justify equation 10-1 as an expression for the concentration overpotential for the rotating cylinder cell with an excess of sulfuric acid as a supporting electrolyte.

6. Obtain an expression for the potential of the cell

$$
\begin{array}{c|c|c|c|c|c|c|c}
\alpha & \beta & \delta & \epsilon & \phi & \epsilon' & \beta' & \alpha' \\
Pt(s) & Pb(s) & PbO_2(s) & PbSO_4(s) & H_2SO_4 \text{ in } H_2O & PbSO_4(s) & Pb(s) & Pt(s),
\end{array}
$$

for which the electrode reactions are

$$PbO_2 + SO_4^= + 4\,H^+ + 2e^- \longrightarrow PbSO_4 + 2\,H_2O$$

and     $$Pb + SO_4^= \longrightarrow PbSO_4 + 2e^-.$$

This is the common lead-acid battery. For the electrode at the left, electrons are equilibrated among phases $\alpha$, $\beta$, and $\delta$, while the lead dioxide protects the lead from contact with the solution. What is the expression for the standard cell potential, and what is its value?

7. Treat the potential of the cell

$$
\begin{array}{c|c|c|c|c|c|c}
\alpha & \beta & \delta & & \epsilon & \phi & \alpha' \\
Pt(s) & Cu(s) & Cu(ClO_4)_2 \text{ in } H_2O & \text{transition region} & AgNO_3 \text{ in } H_2O & Ag(s) & Pt(s).
\end{array}
$$

What is the expression for the standard cell potential, and what is its value?

8. Treat the potential of the cell

$$
\begin{array}{c|c|c|c|c|c}
\alpha & \beta & & \delta & \epsilon & \alpha' \\
Pt(s), H_2(g) & HCl \text{ in } H_2O & \text{transition region} & KNO_3 \text{ in } H_2O & K(Hg) & Pt(s).
\end{array}
$$

What is the expression for the standard cell potential, and what is its value?

9. Treat the potential of a cell in which the solution is saturated throughout with a component. Pick one of the cells 17-21 or 17-19 or that of problem 6.

10. Derive the appropriate form of the Nernst equation for the following cells:

|         |         |
|---------|---------|
| (a) 15-9 | (h) 19-1 |
| (b) 15-12 | (i) 19-4 |
| (c) 17-1 | (j) problem 2 |
| (d) 17-19 | (k) problem 3 |
| (e) 17-21 | (l) problem 6 |
| (f) 17-23 | (m) problem 7 |
| (g) 18-12 | (n) problem 8 |

11. Treat the potential of the cell

$$
\begin{array}{c|c|c|c|c|c|c}
\alpha & \beta & & \delta & \epsilon & \phi & \alpha' \\
Pt(s), H_2(g) & HCl \text{ in } H_2O & \text{transition region} & KOH \text{ in } H_2O & HgO(s) & Hg(l) & Pt(s)
\end{array}
$$

Here, the electrolytes can react in the junction region to form KCl and $H_2O$. Assume that the junction region is maintained at a uniform temperature. What is the expression for the standard cell potential, and what is its value?

**12.** Treat the potential of the cell

| $\alpha$ | $\delta$ | $\epsilon$ | $\phi$ | $\alpha'$ |
|---|---|---|---|---|
| Pt(s), H$_2$(g) | KOH in H$_2$O | HgO(s) | Hg(l) | Pt(s), |

where the reaction for the hydrogen electrode in alkaline media is regarded to be

$$2\,H_2O + 2e^- \longrightarrow H_2 + 2\,OH^-.$$

What is the expression for the standard cell potential, and what is its value?

**13.** From the entries in tables 20-1 and 20-2, determine the solubility product of silver chloride in water at 25°C.

**14.** In setting up tables of standard electrode potentials, the chemical potentials of gases are referred to the *ideal gas state* for the secondary reference state. Thus, for hydrogen gas,

$$\mu_{H_2} = \mu_{H_2}^* + RT \ln p_{H_2} = RT \ln p_{H_2} \lambda_{H_2}^*,$$

where $p_{H_2}$ is the partial pressure or *fugacity* of hydrogen, expressed in atmospheres. The secondary reference state is defined such that

$$p_{H_2} \longrightarrow x_{H_2} p \quad \text{as} \quad p \longrightarrow 0,$$

where $x_{H_2}$ is the mole fraction of hydrogen in the gas. Consequently, $\lambda_{H_2}^*$ is expressed in atm$^{-1}$, and the value of $\mu_{H_2}^*$ depends upon this choice of units:

$$\mu_{H_2}^* = RT \ln \lambda_{H_2}^*.$$

Show that the fugacity of pure hydrogen gas can be expressed as

$$p_{H_2} = p \exp \frac{Bp}{RT},$$

where $B$ is the second virial coefficient appearing in the equation of state

$$\frac{pV}{nRT} = 1 + \frac{Bp}{RT}.$$

The second virial coefficient for hydrogen can be expressed as

$$B = 17.42 - 314.7\,T^{-1} - 211100\,T^{-2} \text{ (cm}^3/\text{mole)},$$

where $T$ is the temperature (°K). Show that the difference between using $\mu_{H_2}^0$ and $\mu_{H_2}^*$ in the tables of standard electrode potentials amounts to 7.3 $\mu$V in the tabulated values (at 25°C).

## NOTATION

| | |
|---|---|
| $a_i$ | relative activity of species $i$ |
| $a_i^\theta$ | property expressing secondary reference state, liter/mole |
| $A$ | Helmholtz free energy, J |
| $c$ | molar concentration of a single electrolyte, mole/liter |
| $c_i$ | molar concentration of species $i$, mole/liter |
| $f_i$ | molar activity coefficient of species $i$ |
| $f_{+-}$ | mean molar activity coefficient of an electrolyte |

| | |
|---|---|
| $F$ | Faraday's constant, 96,487 C/equiv |
| $G$ | Gibbs free energy, J |
| $H$ | enthalpy, J |
| $\mathbf{i}$ | current density, A/cm² |
| $I$ | ionic strength, mole/kg |
| $K_{sp}$ | solubility product of sparingly soluble salt, mole²/kg² |
| $K_w$ | dissociation constant for water, mole²/kg² |
| $m$ | molality of a single electrolyte, mole/kg |
| $m_i$ | molality of species $i$, mole/kg |
| $M_i$ | symbol for the chemical formula of species $i$ |
| $M_i$ | molecular weight of species $i$, g/mole |
| $n$ | number of electrons involved in electrode reaction |
| $n_i$ | number of moles of species $i$, mole |
| $p$ | pressure, dyne/cm² |
| $p_i$ | partial pressure of species $i$, atm |
| $R$ | universal gas constant, 8.3143 J/mole-deg |
| $s_i$ | stoichiometric coefficient of species $i$ in an electrode reaction |
| $S$ | entropy, J/deg |
| $t_i^0$ | transference number of species $i$ with respect to the velocity of species 0 |
| $T$ | absolute temperature, deg K |
| $U$ | open-circuit cell potential, V |
| $U$ | internal energy, J |
| $V$ | volume, cm³ |
| $\bar{V}_i$ | partial molar volume of species $i$, cm³/mole |
| $w$ | work of transfer |
| $z_i$ | charge number of species $i$ |
| $\alpha$ | Debye-Hückel constant, (kg/mole)$^{1/2}$ |
| $\gamma_i$ | molal activity coefficient of species $i$ |
| $\gamma_{+-}$ | mean molal activity coefficient of an electrolyte |
| $\kappa$ | conductivity, mho/cm |
| $\lambda_i$ | absolute activity of species $i$ |
| $\lambda_i^\theta$ | property expressing secondary reference state, kg/mole |
| $\mu_i$ | electrochemical potential of species $i$, J/mole |
| $\nu$ | number of moles of ions into which a mole of electrolyte dissociates |
| $\nu_+, \nu_-$ | numbers of cations and anions into which a mole of electrolyte dissociates |
| $\rho$ | density, g/cm³ |
| $\rho_0$ | density of pure solvent, g/cm³ |
| $\phi$ | osmotic coefficient |
| $\Phi$ | electric potential, V |

superscripts

| | |
|---|---|
| $^0$ | pure state |
| $^0$ | relative to velocity of species 0 |
| $^\theta$ | secondary reference state at infinite dilution |
| $^*$ | ideal-gas secondary reference state |

subscripts

| | |
|---|---|
| $_0$ | solvent |
| $_+$ | cation |
| $_-$ | anion |

# The Electric Potential 3

In the preceding chapter, we have discussed the thermodynamics of electrochemical cells without the introduction of electric potentials except the potential difference between two phases of identical composition, namely, the terminals of the cell. Much of the electrochemical literature is written in terms of electrical potentials of various kinds, and it is necessary to set our minds straight on these matters and to investigate how potentials might be used in electrochemistry. Much of the confusion in electrochemistry arises from uncertainty in the use of these concepts.

## 22. The electrostatic potential

Electrostatic theory deals with purely electrical forces between bodies and not with any specific chemical forces such as exist between molecules. The systems treated are usually macroscopic bodies separated by a vacuum, and the specific forces are not important. For this reason, the concepts developed in electrostatic theory are not directly applicable to energetic relationships within condensed phases.

The electric force $f$ between two bodies of charges $q_1$ and $q_2$, separated by a distance $r$, is given by Coulomb's law

$$f = \frac{q_1 q_2}{4\pi \epsilon r^2}, \tag{22-1}$$

where $\epsilon$ is the permittivity of the medium surrounding the bodies. The force

66

is directed along the line joining the bodies and is repulsive if the two charges are of the same sign and attractive if the charges are of opposite sign.

The permittivity $\epsilon_0$ of a vacuum has the value $\epsilon_0 = 8.8542 \times 10^{-14}$ farad/cm $= 8.8542 \times 10^{-14}$ C/V-cm. If the bodies are immersed in a dielectric medium composed of polarizable matter, the force between them will be different from that in a vacuum. The ratio $\epsilon/\epsilon_0$ is called the relative dielectric constant of the medium. For water, the value of this ratio is 78.303 at 25°C.

The force on a body is the sum of the forces exerted on it by all the other bodies in the system. For the development of the theory, it is convenient to define the electric field $\mathbf{E}$ so that the electric force acting on a charge $q$ is given by

$$\mathbf{f} = q\mathbf{E}. \tag{22-2}$$

The electric field is defined at all points in the medium by supposing that a *test* charge $q$ can be introduced to test the force, and hence the electric field $\mathbf{E}$ by means of equation 2, without disturbing the other charges comprising the system.

In this manner, it is possible to obtain differential equations describing the variation of the electric field. For example, the curl of the electric field is zero:

$$\nabla \times \mathbf{E} = 0. \tag{22-3}$$

This allows us to introduce the electrostatic potential $\Phi$, so that the electric field can be expressed as the negative gradient of this scalar quantity:

$$\mathbf{E} = -\nabla\Phi. \tag{22-4}$$

This is permissible since the curl of the gradient of any scalar field is zero:

$$\nabla \times \nabla\Phi = 0. \tag{22-5}$$

The variation in the electric field is also related to the charge distribution in the system by Poisson's equation

$$\nabla\cdot(\epsilon\mathbf{E}) = -\nabla\cdot(\epsilon\,\nabla\Phi) = \rho_e, \tag{22-6}$$

where $\rho_e$ is the electric charge density per unit volume. For a medium of uniform dielectric constant, this is equivalent to the expression of the electrostatic potential in terms of the charges:

$$\Phi(r) = \sum_i \frac{q_i}{4\pi\,\epsilon\,|\mathbf{r} - \mathbf{r}_i|}, \tag{22-7}$$

where the sum includes all the charges in the system.

Equation 6 provides a differential equation for the determination of the electrostatic potential in terms of the charge distribution. For a medium of uniform dielectric constant, this becomes

$$\nabla^2\Phi = -\frac{\rho_e}{\epsilon}, \tag{22-8}$$

and in a medium with no free charges, it reduces to Laplace's equation

$$\nabla^2 \Phi = 0. \tag{22-9}$$

At the interface between two phases, the tangential component of the electric field is continuous. The relationship between the normal components of the electric field in the two phases can be obtained by applying equation 6 to a *pill box* enclosing a portion of the interface (see figure 22-1). We

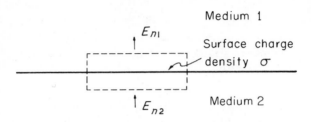

**Figure 22-1.** Normal components of the electric field at an interface. The interface may have a charge $\sigma$ per unit area.

include the possibility that the surface charge density at the interface is not zero. By means of the divergence theorem, equation 6 can be written in terms of integrals over the surface and the volume of an arbitrary region:

$$\oint \epsilon \mathbf{E} \cdot d\mathbf{S} = \int \rho_e \, dV. \tag{22-10}$$

This is an expression of Gauss's law, which says that the integral of the outward normal component of $\epsilon \mathbf{E}$ over the surface of a closed region is equal to the charge enclosed. Application of this result to the interface in figure 22-1 gives the relationship between the normal components of the electric field:

$$\epsilon_1 E_{n1} - \epsilon_2 E_{n2} = \sigma, \tag{22-11}$$

where $\sigma$ is the charge per unit area at the interface.

A considerable body of electrostatic theory has been developed,[1,2,3] concerned with the solution of equations 8 and 9 for a variety of geometries and boundary conditions. We have not discussed the magnetic effects which arise when the electric field varies in time and electric currents are present.

We conclude this section with an example. Consider two metal spheres, each 1 cm in radius and with a distance of 10 cm between their centers (see figure 22-2). We want to charge each sphere to an average of 10 $\mu$C/cm² by

[1] James Clerk Maxwell, *A Treatise on Electricity and Magnetism*, Vol. I (Oxford: Clarendon Press, 1892).

[2] K. J. Binns and P. J. Lawrenson, *Analysis and Computation of Electric and Magnetic Field Problems* (New York: The Macmillan Company, 1963).

[3] John David Jackson, *Classical Electrodynamics* (New York: John Wiley & Sons, Inc., 1962).

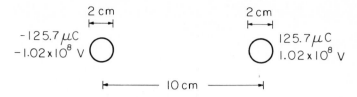

**Figure 22-2.** Potential difference between two metal spheres for an average surface charge of 10 $\mu C/cm^2$.

transporting 125.7 $\mu C$ or 1.3 $\times$ 10$^{-9}$ moles of electrons from one sphere to the other. The capacity of this system should be about 0.618 $\times$ 10$^{-12}$ farad. Hence, we can estimate that the final potential difference between the spheres will be 2.04 $\times$ 10$^8$ V.

This example shows that large potentials are required to effect a modest separation of electrical charge.

## 23. Intermolecular forces

The intermolecular potential energy relation for two ions is given an idealized representation in figure 23-1. The corresponding intermolecular force is shown in figure 23-2. The noteworthy feature of these interactions is how slowly they go to zero at large distances of separation.

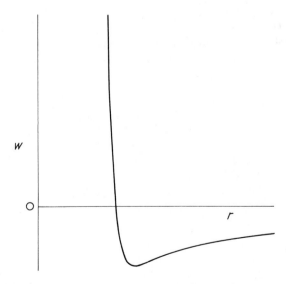

**Figure 23-1.** Intermolecular potential energy for two ions a distance $r$ apart.

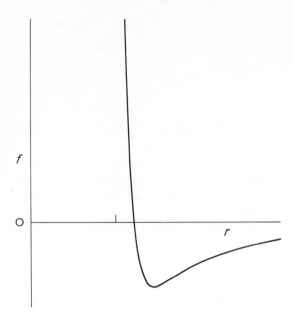

**Figure 23-2.** Intermolecular force between two ions.

$$\left. \begin{array}{l} f = \dfrac{q_1 q_2}{4\pi\,\epsilon r^2} \\[2mm] w = \displaystyle\int_r^\infty f\,dr = \dfrac{q_1 q_2}{4\pi\,\epsilon r} \end{array} \right\} \quad \text{as} \quad r \longrightarrow \infty. \qquad (23\text{-}1)$$

Because they obey the Coulomb force law at large distances, they are called *ions*, to distinguish them from neutral nonpolar molecules or neutral polar molecules, for which the interaction energy is

$$w = \frac{-\text{constant}}{r^6} \quad \text{as} \quad r \longrightarrow \infty. \qquad (23\text{-}2)$$

At close distances of approach, the interionic interaction depicted in figures 23-1 and 23-2 departs from equations 1. Since the nature of this deviation at short distances depends on the specific nature of the ions, this deviation is sometimes referred to as the *short-range specific interaction force*. Of course, the decomposition of the single curves of figures 23-1 and 23-2 into coulombic parts and specific-interaction parts is artificial. It is the intermolecular interaction itself which determines the behavior of ions.

The asymptotic behavior expressed by equations 1 is sufficiently common and causes such difficulties as to warrant a special treatment. A large body of knowledge included in electrostatic and electromagnetic theory has been developed to deal with coulombic interactions. This involves, for example, the electrostatic potential given by equation 22-7.

The results and methods of electrostatic theory can be applied most directly to dilute, ionized gases. The restriction to a dilute, ionized gas is useful for two reasons:

1. There is a large fraction of free space, and we can imagine inserting a probe among the ions without actually disturbing them. In a condensed phase, on the other hand, all space is occupied.

2. All the ions are widely separated, so that an ion interacts at close range with no more than one other ion at a time. This close range interaction can be called a *collision*. An un-ionized gas can be treated by consideration of binary interactions or binary collisions.[4] In an ionized gas where the inverse square law applies, an ion interacts at all times with all other ions in the system. Some progress can be made by introducing an electric potential which accounts for interactions between distant ions. A binary collision is then an interaction with another ion at small distances where the coulombic force law no longer holds, and this can be treated separately.

Even in this case, however, it remains a fact that the separation of the intermolecular force law into different parts, say, an electric part and a specific chemical part, is without any physical basis. The same difficulty arises in attempts to separate electrochemical potentials, used in chapter 2, into chemical potentials and electric potentials.

The electrostatic theory can be modified somewhat to handle condensed phases which involve a dilute dispersion of charged particles in an otherwise uniform, dielectric medium. In this case, the electric field due to the charged particles induces dipoles in the dielectric medium or causes a preferential orientation of permanent dipoles. The presence of all these charges induced in the medium can be accounted for by introducing an averaged electric field and an averaged electrostatic potential. The sum in equation 22-7 then extends over the permanently charged particles but not over the charges induced in the dielectric medium. The average of $E$ and $\Phi$ is taken over spatial regions of at least molecular dimensions and accounts for the dipole charges not included in the sum by the use of the permittivity $\epsilon$ of the dielectric medium rather than the permittivity $\epsilon_0$ of free space.

This treatment of a condensed phase can be justified only when the charged particles are so far apart that the coulombic interaction applies, just as for a dilute, ionized gas. Also, there should be enough dielectric between charged particles to permit the averaging. Thus, both $\Phi$ and $\epsilon$ are macroscopic concepts, and it is not meaningful to discuss variations of $\Phi$ and $\epsilon$ over distances of molecular dimensions. The evaluation of the energy required to move a charged particle from one phase to another requires, in addition, consideration of the energy of interaction with the dielectric solvent and the nature of the charge distribution near the interface.

[4]Sydney Chapman and T. G. Cowling, *The Mathematical Theory of Non-Uniform Gases* (Cambridge: University Press, 1939).

Although concentrated ionic solutions defy exact treatment on an electrostatic basis, the coulombic inverse square law has many important consequences. For example, the coulombic attraction between charges is so strong that large departures from electrical neutrality are precluded in electrolytic solutions. This conclusion remains valid even though the interionic forces depart from Coulomb's law at small distances.

The electrostatic model of a dilute dispersion of charged particles in a dielectric medium is applied fruitfully in the theory of Debye and Hückel for the activity coefficients of extremely dilute electrolytic solutions. Although the result and the model are applicable only in the limit of infinite dilution, the strong departures from ideal-solution behavior can be clearly attributed to the inverse square law governing the force between ions at large distances. The same model can be applied to determine the consequences of these long-range forces on other properties of the solution, notably the conductivity, at high dilutions.

Long-range forces are also important in electrode reactions. In a nonequilibrium double layer, there can be extremely large forces acting on an ion; and one is led to say that there is a very large gradient of potential in the double layer at an electrode surface even though the rigorous definition of this potential may be difficult.

## 24.  Outer and inner potentials

The *outer* potential, also called the *cavity* or *Volta* potential, is frequently used in electrochemistry. It is defined in terms of the energy required to move a charged particle from a point just outside one phase to a point just outside another phase:

$$w_i = z_i F(\Phi^\beta - \Phi^\alpha). \tag{24-1}$$

In order to avoid the influence of external fields, it is imagined that the charged particle is moved from a macroscopic cavity in one phase to a macroscopic cavity in the other phase (see figure 24-1), and this is the origin of the term *cavity potential*.

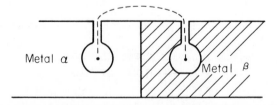

**Figure 24-1.** Movement of a charged particle from a cavity in one phase to a cavity in another phase. This thought experiment is used to define the Volta potential and the contact potential difference between two metals.

By moving the particle only to a point *outside* each phase, only the long-range forces of charges in that phase are able to act on the test particle; the short-range, specific forces are not encountered. For this reason, the outer potential is independent of the ion type used for the test particle. Differences in outer potentials are measurable since they are differences in the potential of two points in a phase of uniform composition, namely, the external medium. However, the accurate measurement of differences in outer potentials is a difficult experimental undertaking. Outer potentials can be used to characterize the electrical state of a phase, but they do not have any direct thermodynamic relevance and require difficult experimental measurements which are not necessary for any thermodynamic discussion.

The *inner* potential, also called the *Galvani* potential, relates to the energy required to move an idealized charged particle to a point *inside* a phase. It is generally conceded that inner potentials are not measurable, and they do not need to occupy our attention further. The difference between the inner potential and the outer potential is called the *surface* potential, another unmeasurable quantity.

## 25. Potentials of reference electrodes

From a practical point of view, the potential that is measured in an electrochemical system in order to assess the electrical state of part of the system is that of a reference electrode. For our present discussion, we mean an electrode that can be inserted directly into the solution at any point. In particular, we exclude reference electrodes which are in a separate vessel and connected to the point in question by a capillary probe, unless the solution in the capillary probe and the auxiliary vessel is of the same composition as the solution at the point where we intend to insert the probe. This restriction is imposed in order to avoid the uncertainties associated with liquid junctions. In practical measurements, this restriction cannot always be observed. Reference electrodes are discussed in more detail in chapter 5.

A reference electrode should behave reversibly with respect to one of the ions in the solution. The material presented in chapter 2 shows that the reference electrode will provide, in essence, a measure of the electrochemical potential of that ion. Figure 25-1 shows how Ag–AgCl electrodes might be used to measure the potential difference between two solutions. Equation 19-2 indicates that the measured potential difference is related to the difference in electrochemical potential of chloride ions between the two points $\alpha$ and $\beta$:

$$-F(\Phi^\alpha - \Phi^\beta) = \mu_{Cl^-}^\alpha - \mu_{Cl^-}^\beta. \tag{25-1}$$

In figure 25-2, the two reference electrodes are placed in the same vessel. By moving electrode $\alpha$ in the solution, one can investigate the spatial variation of the electrochemical potential of chloride ions even though the con-

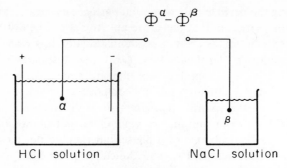

Figure 25-1. Use of reference electrodes to investigate the potential in a solution. Silver-silver chloride reference electrodes are represented by $\alpha$ and $\beta$. The vessel on the left also contains two working electrodes.

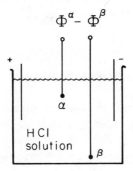

Figure 25-2. Use of reference electrodes to investigate potential variations within a solution.

centration of electrolyte is not uniform and a current passes between the two working electrodes.

Reference electrodes provide the most convenient means for assessing the electrical state of an electrolytic solution.

## 26. The electric potential in thermodynamics

The potential sought in thermodynamics is one related to the energy required for the *reversible* transfer of ions from one phase to another. This, of course, is the electrochemical potential of an ionic species. The electrostatic potential, aside from the problems involved in its definition in condensed phases, is not directly related to reversible work. Although the electrostatic potential can be avoided in thermodynamics, electrochemical potentials being used in its stead, there does remain a need to characterize the electrical state of a phase.

Frequently, the electrochemical potential of an ionic species is split into an electrical term and a "chemical" term,

$$\mu_i = \mu_i^{\text{chem}} + z_i F \Phi = RT \ln \lambda_i^\theta m_i \Gamma_i + z_i F \Phi, \qquad (26\text{-}1)$$

where $\Phi$ is the "electrostatic" potential and $\Gamma_i$ is an activity coefficient which now is supposed to be independent of the electrical state of the phase. We should first note that this decomposition is not necessary, since the relevant formulae of thermodynamic significance have already been derived in chapter 2.

The electrostatic potential $\Phi$ could be defined so that it is measurable or unmeasurable. Depending upon how well defined $\Phi$ is, so $\Gamma_i$ is just as well or poorly defined. It is possible to proceed with only a vague concept of the electrostatic potential, as supplied by electrostatic theory, and never bother to define carefully just what is meant. If the analysis is consistent, physically meaningful results can be obtained by recombining poorly defined terms at the end of the analysis.

Any definition of $\Phi$ which is chosen should satisfy one condition. It should reduce to the definition 13-2 used for the difference in electric potential between phases of identical composition. Thus, if $\alpha$ and $\beta$ are phases of identical composition, then

$$\mu_i^\alpha - \mu_i^\beta = z_i F(\Phi^\alpha - \Phi^\beta). \qquad (26\text{-}2)$$

The potential $\Phi$ thus provides a quantitative measure of the electrical state of one phase relative to a second phase of identical composition. A number of possible definitions of $\Phi$ satisfy this condition.

The outer potential, which is measurable in principle, can be used for $\Phi$. It has the disadvantages of difficulty of measurement and lack of relevance in thermodynamic calculations. It has the advantage of giving a definite meaning to $\Phi$, and, at the end of the analysis, its definition cancels out so that its value need never be actually measured.

A second possibility is to use the potential of a suitable reference electrode. Since the reference electrode is reversible to an ion in the solution, this is equivalent to using the electrochemical potential of an ion, or $\mu_i/z_i F$. The arbitrariness of this definition is apparent from the need to select a particular reference electrode or ionic species for the definition. This choice has the added disadvantage that $\mu_i$ is equal to minus infinity for a solution in which this ionic species is absent. Thus, it does not conform to our usual concept of an electrostatic potential; this is because $\mu_i$ relates to reversible work. This choice of the potential does have the advantage of being related to a measurement, with reference electrodes, commonly made in electrochemistry.

A third possibility should occupy our attention here. Select an ionic species $n$ and define the potential $\Phi$ as follows:

$$\mu_n = RT \ln c_n + z_n F \Phi. \tag{26-3}$$

Then, the electrochemical potential of any other species can be expressed as

$$\mu_i = RT \ln c_i + z_i F \Phi + RT \left( \ln f_i - \frac{z_i}{z_n} \ln f_n \right)$$
$$+ RT \left( \ln a_i^\theta - \frac{z_i}{z_n} \ln a_n^\theta \right). \tag{26-4}$$

One should recognize that the combination of $f_i$'s and $a_i^\theta$'s in parentheses are well defined and independent of the electrical state, according to the rules laid down in section 14. At constant temperature, the gradient of the electrochemical potential is then

$$\nabla \mu_i = RT \, \nabla \ln c_i + z_i F \, \nabla \Phi + RT \, \nabla \left( \ln f_i - \frac{z_i}{z_n} \ln f_n \right). \tag{26-5}$$

The arbitrariness of this definition of $\Phi$ is again apparent from the need to select a particular ionic species $n$. This definition of $\Phi$ has the advantage of being related unambiguously to the electrochemical potentials, and it conforms to our usual concept of an electrostatic potential. It can be used in a solution of vanishing concentration of species $n$ because of the presence of the term $RT \ln c_n$ in equation 3.

In the limit of infinitely dilute solutions, the activity coefficient terms in equations 4 and 5 disappear due to the choice of the secondary reference state 14-6. In this limit, the definition of $\Phi$ becomes independent of the choice of the reference ion $n$. This forms the basis of what should be termed the *dilute-solution theory* of electrolytic solutions. At the same time, equations 4 and 5 show how to apply activity-coefficient corrections to dilute-solution theory without the utilization of the activity coefficients of individual ions. In infinitely dilute solutions, the lack of dependence on the ion type $n$ is related to the ability to measure differences in electric potential between phases of identical composition. Such solutions are of essentially the same composition in the sense that an ion in solution is subject only to interactions with the solvent, and even the long-range electrical interactions with other ions are not felt.

The introduction of such an electric potential is useful in the calculation of transport processes in electrolytic solutions.[5] Smyrl and Newman use the term *quasi-electrostatic potential* for the potential so defined.

We have discussed here possible ways of using the electric potential in electrochemical thermodynamics. The use of the potential in transport theory is basically the same as its use in thermodynamics. By using electrochemical potentials it is possible to avoid the electric potential, although its introduction may be useful or convenient. In the kinetics of electrode proc-

[5] William H. Smyrl and John Newman. "Potentials of Cells with Liquid Junctions," *Journal of Physical Chemistry*, 72 (1968), 4660–4671.

esses, it is possible to use a free energy change as a driving force for the reaction. This is equivalent to the use of the surface overpotential defined in section 8.

An electric potential also finds use in microscopic models, such as the Debye-Hückel theory alluded to earlier and developed in the next chapter. Such a potential cannot always be defined with rigor. A clear distinction should always be made between macroscopic theories—such as thermodynamics, transport phenomena, and fluid mechanics—and microscopic theories—such as statistical mechanics and the kinetic theory of gases and liquids. Microscopic theories explain the behavior of, predict the values of, and provide means to correlate macroscopic properties, such as activity coefficients and diffusion coefficients, on the basis of molecular or ionic properties. Quantitative success is seldom achieved without some additional empiricism. The macroscopic theories, on the other hand, provide both the framework for the economical measurement and tabulation of macroscopic properties and the means for using these results to predict the behavior of macroscopic systems.

# NOTATION

| | |
|---|---|
| $a_i^\theta$ | property expressing secondary reference state, liter/mole |
| $c_i$ | concentration of species $i$, mole/liter |
| $\mathbf{E}$ | electric field, V/cm |
| $f$ | force, J/cm |
| $f_i$ | molar activity coefficient of species $i$ |
| $F$ | Faraday's constant, 96,487 C/equiv |
| $m_i$ | molality of species $i$, mole/kg |
| $q$ | electric charge, C |
| $r$ | distance or position, cm |
| $R$ | universal gas constant, 8.3143 J/mole-deg |
| $S$ | surface area, cm² |
| $T$ | absolute temperature, deg K |
| $V$ | volume, cm³ |
| $w$ | work or interaction energy |
| $z_i$ | charge number of species $i$ |
| $\Gamma_i$ | activity coefficient |
| $\epsilon$ | permittivity, f/cm |
| $\epsilon_0$ | permittivity of free space, $8.8542 \times 10^{-14}$ f/cm |
| $\lambda_i^\theta$ | property expressing secondary reference state, kg/mole |
| $\mu_i$ | electrochemical potential of species $i$, J/mole |
| $\rho_e$ | electric charge density, C/cm³ |
| $\sigma$ | surface charge density, C/cm² |
| $\Phi$ | electric potential, V |

# *Activity Coefficients*  **4**

Due to long-range coulombic interactions between ions, the activity coefficients of electrolytes in dilute solutions behave differently from those of nonelectrolytes, as explained by the electrostatic theory of Debye and Hückel. This theory forms the basis for the empirical correlation of activity coefficients over a range of concentration, including the activity coefficients of solutions of several electrolytes.

## 27. Ionic distributions in dilute solutions

When an electrolyte dissolves in a solvent, it dissociates into cations and anions; this is the source of the electrical conductivity of the solution. This dissociation also manifests itself in the thermodynamic properties of the solution, as discussed in section 14. For example, it leads to the presence of the factor $v$ in equation 14-15, and it is responsible for the large freezing point depression and vapor pressure lowering of solutions of electrolytes.

The distribution of ions is not completely random even in dilute solutions, because of the attractive and repulsive electrical forces between ions. Consequently, the thermodynamic properties show further departures from those of nonelectrolytic solutions. Debye and Hückel[1] used an electrostatic model to describe these ionic distributions quantitatively.

---

[1] P. Debye and E. Hückel, "Zur Theorie der Elektrolyte," *Physikalische Zeitschrift*, **24** (1923), 185–206.

Suppose that an ion of valence $z_c$ is at the origin of coordinates. Ions of opposite charge to this *central ion* will be attracted toward the origin, and ions of like sign will be repelled. Random thermal motion of the ions and the solvent molecules tends to counteract this electric effect and promote a random distribution of ions. The balance of these competing effects can be expressed by a Boltzmann distribution of ionic concentrations:

$$c_i = c_{i\infty} \exp\left(-\frac{z_i F \Phi}{RT}\right), \tag{27-1}$$

where $c_{i\infty}$ is the average concentration of species $i$ and $\Phi$ is the electrostatic potential established around the central ion. The electrical interaction energy per mole is expressed as $z_i F \Phi$, and other contributions to the interaction energy are ignored. Far from the central ion, the potential $\Phi$ approaches zero, and, consequently, $c_i$ approaches $c_{i\infty}$.

The potential $\Phi$ results not only from the central ion but also from other ions that are attracted toward or repelled from the origin. Its distribution is governed by Poisson's equation 22-8

$$\nabla^2 \Phi = -\frac{\rho_e}{\epsilon} = -\frac{F}{\epsilon} \sum_i z_i c_i. \tag{27-2}$$

In a solution at equilibrium, radial symmetry prevails. In spherical coordinates, with the central ion at the origin, equation 2 becomes

$$\frac{1}{r^2} \frac{d}{dr}\left(r^2 \frac{d\Phi}{dr}\right) = -\frac{\rho_e}{\epsilon} = -\frac{F}{\epsilon} \sum_i z_i c_{i\infty} \exp\left(-\frac{z_i F \Phi}{RT}\right). \tag{27-3}$$

The centers of other ions are supposed to be precluded from approaching within a distance $a$ of the central ion because of the short-range repulsive forces. Thus equation 1, and hence equation 3, applies only for $r > a$. The boundary condition at $r = a$ can be found by applying Gauss's law 22-10 to the region within $r = a$:

$$\frac{d\Phi}{dr} = -\frac{z_c e}{4\pi \epsilon a^2} \quad \text{at} \quad r = a. \tag{27-4}$$

This is equivalent to the statement that the charge distribution around the central ion exactly counterbalances the charge on that ion. Thus, integration of equation 3 from $a$ to $\infty$ gives

$$r^2 \frac{d\Phi}{dr}\bigg|_a^\infty = -\int_a^\infty \frac{\rho_e r^2}{\epsilon} dr \tag{27-5}$$

or

$$\int_a^\infty \rho_e 4\pi r^2 dr = 4\pi \epsilon a^2 \frac{d\Phi}{dr}\bigg|_{r=a} = -z_c e. \tag{27-6}$$

On the left is the integral of the charge density $\rho_e$ over the volume outside $r = a$, and this is equated to the negative of the charge $z_c e$ on the central ion.

Equation 3 now describes the potential distribution near the central ion in terms of the known average concentrations $c_{i\infty}$ and other known parameters. To effect a solution, Debye and Hückel approximated the exponential terms in equation 3 as though the exponents were small

$$\exp\left(-\frac{z_i F\Phi}{RT}\right) \approx 1 - \frac{z_i F\Phi}{RT}. \tag{27-7}$$

Equation 3 becomes

$$\frac{1}{r^2}\frac{d}{dr}\left(r^2\frac{d\Phi}{dr}\right) = \frac{\Phi}{\lambda^2}, \tag{27-8}$$

where $\lambda$ is the Debye length given by

$$\lambda = \left(\frac{\epsilon RT}{F^2 \sum_i z_i^2 c_{i\infty}}\right)^{1/2} \tag{27-9}$$

The term involving $\sum_i z_i c_{i\infty}$ is zero because the solution is electrically neutral on the average.

The solution of equation 8 satisfying the boundary condition 4 and the condition that $\Phi$ approaches zero as $r$ approaches infinity is

$$\Phi = \frac{z_c e}{4\pi \epsilon r}\frac{e^{(a-r)/\lambda}}{1 + a/\lambda}. \tag{27-10}$$

The potential due to the central ion alone would be $z_c e/4\pi \epsilon r$. Equation 10 therefore reveals that the other ions attracted toward the origin, being of opposite sign to the central ion, lower the magnitude of the potential and cause it to vanish rapidly at large distances from the central ion. Thus, ions at some distance are shielded from the charge of the central ion by these other ions.

The Debye length $\lambda$ is an important parameter describing the potential distribution. It has the value $\lambda = 9.6 \times 10^{-8}$ cm $= 9.6$ Å in a 0.1 $M$ aqueous solution of a uni-univalent electrolyte at 25°C. It is inversely proportional to ionic charges and the square root of concentration and directly proportional to the square roots of the permittivity and the absolute temperature. Figure 27-1 shows the distribution of anions and cations near a central cation as calculated from equations 3 and 10 for a 0.1 $M$ aqueous solution of a uni-univalent electrolyte at 25°C. The shielding ions form an *ion cloud* around the central ion, with a thickness on the order of the Debye length $\lambda$.

The parameter $a$ is generally regarded as an average value of the sum of the radii of pairs of hydrated ions. For a solution containing only one kind of anion and one kind of cation, the sum of the radii of a cation and an anion dominate this average, since the long-range repulsive forces tend to prevent ions of like sign from interacting at short distances.

## 28.  Electrical contribution to the free energy

For an ideal solution of dissociated ions, the activity coefficient of neutral combinations of ions would be equal to unity. Important departures from ideality result from the coulombic electrical forces between ions. This effect can be assessed by means of the following thought experiment. Start with a given volume of solvent and also a reservoir, which is a large volume of solution containing ions at high dilution. Now imagine that the charge can be removed from the ions, and let this involve an amount of work $w_1$. The *discharged* ions are now to be transferred reversibly at constant volume from the reservoir to the given volume of solvent. The work involved in this process is the ideal or nonelectrical contribution to the change in the Helmholtz free energy $A$. The average concentration of a solute species in the new solution is now $c_i$.

Finally, the ions are to be recharged simultaneously to their appropriate charge levels, an amount of work $w_2$ being expended in this process. This work $w_2$ is different from $w_1$ because the ions are now at a high enough concentration that they shield each other, and the potential distribution around a central ion is given by equation 27-10. The electrical contribution

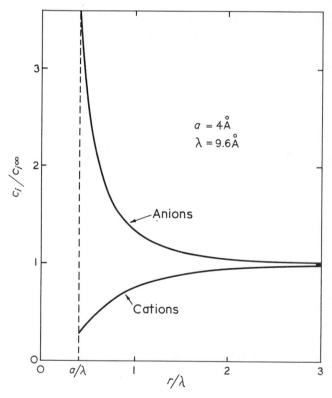

**Figure 27-1.** Ionic distributions near a central cation, according to the theory of Debye and Hückel, for a 0.1 $M$ aqueous solution of a uni-univalent electrolyte at 25°C.

to the Helmholtz free energy is taken to be the difference between these two work terms:

$$A_{el} = w_2 - w_1. \tag{28-1}$$

It now remains to express this quantitatively.

The work required to bring an element of charge $dq$ from infinity to a distance $r = a$ from a central ion is $\Phi(r = a)dq$. Since all the ions are to be charged simultaneously, it is expeditious to let $\xi$ denote the fraction of the final charge carried by any ion at any time during the charging process. Then, the charge on an ion is $z_i e\xi$, and $dq = z_i e\, d\xi$, and $\xi$ varies from 0 to 1 during the charging process. The quantity $a/\lambda$ in equation 27-10 should be expressed as $\xi a/\lambda$, since $\lambda$ is inversely proportional to the charge level.

The contribution to $w_2 - w_1$ from charging a central ion of valence $z_j$ is therefore

$$\int_0^1 \left[ \frac{z_j e\xi}{4\pi\,\epsilon a} \frac{1}{1 + \xi a/\lambda} - \frac{z_j e\xi}{4\pi\,\epsilon a} \right] z_j e\, d\xi = -\frac{z_j^2 e^2}{4\pi\,\epsilon\lambda} \int_0^1 \frac{\xi^2\, d\xi}{1 + \xi a/\lambda}. \tag{28-2}$$

The term in brackets is the difference between the potential at $r = a$ for the charging processes corresponding to $w_2$ and $w_1$. The electrical contribution to the Helmholtz free energy is now the sum of the contributions from charging the individual ions.

$$A_{el} = w_2 - w_1 = -\sum_j Ln_j \frac{z_j^2 e^2}{4\pi\,\epsilon\lambda} \int_0^1 \frac{\xi^2\, d\xi}{1 + \xi a/\lambda}, \tag{28-3}$$

where $n_j$ is the number of moles of species $j$ and $L$ is Avogadro's number. Integration gives

$$A_{el} = -\frac{Fe}{12\pi\,\epsilon\lambda} \tau\left(\frac{a}{\lambda}\right) \sum_j z_j^2 n_j, \tag{28-4}$$

where

$$\tau(x) = \frac{3}{x^3}\left[ \ln(1 + x) - x + \frac{1}{2}x^2 \right]. \tag{28-5}$$

The chemical potential of a species can be obtained by suitable differentiation of the Helmholtz free energy (see equation 13-1):

$$\mu_i = \left(\frac{\partial A}{\partial n_i}\right)_{T,V,n_j \atop j\neq i} = \left(\frac{\partial A}{\partial c_i}\right)_{T,V,c_j \atop j\neq i} \left(\frac{\partial c_i}{\partial n_i}\right)_{T,V,n_j \atop j\neq i} = \frac{1}{V}\left(\frac{\partial A}{\partial c_i}\right)_{T,V,c_j \atop j\neq i}. \tag{28-6}$$

From equation 4,

$$\frac{A_{el}}{V} = -\frac{Fe}{12\pi\,\epsilon\lambda} \tau\left(\frac{a}{\lambda}\right) \sum_j z_j^2 c_j. \tag{28-7}$$

The electrical contribution to the chemical potential, therefore, is

$$\mu_{i,el} = -\frac{Fe}{12\pi\,\epsilon} \frac{\tau(a/\lambda)}{\lambda} z_i^2 - \frac{Fe}{12\pi\,\epsilon} \sum_j z_j^2 c_j \frac{dx\,\tau(x)}{dx}\bigg|_{x=a/\lambda} \left(\frac{\partial 1/\lambda}{\partial c_i}\right)_{T,V,c_j \atop j\neq i}. \tag{28-8}$$

The second term arises from the fact that $\lambda$ depends on $c_i$. It should also be borne in mind that only neutral combinations of these chemical potentials are independent of the electrical state of the phase. Carrying out the differentiation in equation 8, we obtain

$$\mu_{i,\,el} = -\frac{z_i^2 Fe}{8\pi\,\epsilon\lambda}\frac{1}{1 + a/\lambda}. \qquad (28\text{-}9)$$

The above method of obtaining the electrical effect on the thermodynamic properties is known as the *Debye charging process*. The electrical contribution to the chemical potential can be arrived at more directly by means of the so-called *Güntelberg charging process*. Equation 6 shows that the chemical potential of a species is equal to the reversible work of transferring, at constant temperature and volume, one mole of the species to a large volume of the solution. The electrical contribution to $\mu_i$ then comes from charging one ion or a mole of ions in a solution in which all the other ions are *already* charged. The Debye length $\lambda$ is consequently treated as a constant in this process, and we have

$$\mu_{i,\,el} = L\int_0^1\left[\frac{z_i e\xi}{4\pi\,\epsilon a}\frac{1}{1 + a/\lambda} - \frac{z_i e\xi}{4\pi\,\epsilon a}\right]z_i e\,d\xi = -\frac{z_i^2 Fe}{8\pi\,\epsilon\lambda}\frac{1}{1 + a/\lambda}. \qquad (28\text{-}10)$$

This result agrees with equation 9.

For dilute solutions, the activity coefficient can now be expressed as

$$\ln f_i = \ln\gamma_i = -\frac{z_i^2 Fe}{8\pi\,\epsilon RT\lambda}\frac{1}{1 + a/\lambda}. \qquad (28\text{-}11)$$

The activity coefficient depends on the composition of the solution through the Debye length $\lambda$. Consequently, it is convenient to introduce the molar ionic strength $I'$ of the solution:

$$I' = \tfrac{1}{2}\sum_i z_i^2 c_i. \qquad (28\text{-}12)$$

Then equation 11 becomes

$$\ln f_i = \ln\gamma_i = -\frac{z_i^2\alpha'\sqrt{I'}}{1 + B'a\sqrt{I'}}, \qquad (28\text{-}13)$$

where

$$B' = \frac{F}{\sqrt{\epsilon RT/2}}, \qquad (28\text{-}14)$$

$$\alpha' = \frac{Fe}{8\pi\,\epsilon RT}B' = \frac{F^2 e\sqrt{2}}{8\pi(\epsilon RT)^{3/2}}. \qquad (28\text{-}15)$$

In applications, molal concentrations are frequently employed, and the molal ionic strength is defined as

$$I = \frac{1}{2}\sum_i z_i^2 m_i = \frac{I'}{c_0 M_0}. \qquad (28\text{-}16)$$

In dilute solutions, $c_0 M_0$ is approximately equal to $\rho_0$, the density of the pure solvent, and one writes $I' = \rho_0 I$. Consequently, in terms of the molal ionic strength, the activity coefficients are given by

$$\ln f_i = \ln \gamma_i = -\frac{z_i^2 \alpha \sqrt{I}}{1 + Ba\sqrt{I}}, \tag{28-17}$$

where

$$\alpha = \alpha' \sqrt{\rho_0} \quad \text{and} \quad B = B' \sqrt{\rho_0}. \tag{28-18}$$

Values of these parameters for aqueous solutions are given in table 28-1.

TABLE 28-1. DEBYE-HÜCKEL PARAMETERS FOR AQUEOUS SOLUTIONS

| T, deg C | 0 | 25 | 50 | 75 |
|---|---|---|---|---|
| $\alpha'$, (liter/mole)$^{1/2}$ | 1.1325 | 1.1779 | 1.2374 | 1.3115 |
| $\alpha$, (kg/mole)$^{1/2}$ | 1.1324 | 1.1762 | 1.2300 | 1.2949 |
| $B'$, (liter/mole)$^{1/2}$/Å | 0.3249 | 0.3291 | 0.3346 | 0.3411 |
| $B$, (kg/mole)$^{1/2}$/Å | 0.3248 | 0.3287 | 0.3326 | 0.3368 |

The Debye-Hückel limiting law is the limiting form of equation 13 or 17 as the ionic strength goes to zero.

$$\ln f_i = \ln \gamma_i \longrightarrow -z_i^2 \alpha' \sqrt{I'} = -z_i^2 \alpha \sqrt{I}. \tag{28-19}$$

This form, which has been verified experimentally, shows that the logarithm of the activity coefficient is proportional to the square root of the ionic strength. This is a stronger concentration dependence than one encounters in solutions of nonelectrolytes. The limiting law is independent of the parameter $a$. All the quantities entering into $\alpha$ and $\alpha'$ can be measured independently. The nonelectrical contributions to the logarithm of the activity coefficient should be proportional to the concentration or the ionic strength to the first power in dilute solutions. Thus, the limiting law of Debye and Hückel is valid because the effect of long-range electrical forces is so much larger than the effects usually encountered. Other effects are not, however, included in the theory of Debye and Hückel, and its validity is therefore restricted to dilute solutions.

Equation 6 gives no direct information on the electrical contribution to the chemical potential of the solvent. However, the Gibbs-Duhem equation can be used for this purpose. For variations at constant temperature and pressure, it reads

$$\sum_i c_i \, d\mu_i = 0 \tag{28-20}$$

or

$$-d\mu_0 = M_0 \sum_{i \neq 0} m_i \, d\mu_i = RTM_0 \sum_{i \neq 0} m_i \, d(\ln m_i \gamma_i). \tag{28-21}$$

Substitution of equation 17 gives

$$-d\mu_0 = RTM_0 \left[ \sum_{i \neq 0} dm_i - 2\alpha I \, d\left(\frac{\sqrt{I}}{1 + Ba\sqrt{I}}\right) \right]$$

$$= RTM_0 \left[ \sum_{i \neq 0} dm_i - \frac{2\alpha I}{(1 + Ba\sqrt{I})^2} \, d\sqrt{I} \right]. \tag{28-22}$$

Integration gives

$$\frac{\mu_0^0 - \mu_0}{RT} = -\ln\left(\frac{\lambda_0}{\lambda_0^0}\right) = M_0 \sum_{i \neq 0} m_i - \frac{2}{3}\alpha M_0 I^{3/2}\sigma(Ba\sqrt{I}), \quad (28\text{-}23)$$

where

$$\sigma(x) = \frac{3}{x^3}\left[x - 2\ln(1 + x) - \frac{1}{1 + x} + 1\right]. \quad (28\text{-}24)$$

Comparison with equation 14-22 shows that the osmotic coefficient $\phi$ is

$$\phi = 1 - \frac{\frac{2}{3}\alpha I^{3/2}\sigma(Ba\sqrt{I})}{\sum_{i \neq 0} m_i}. \quad (28\text{-}25)$$

Since $\sigma$ approaches one as the ionic strength approaches zero, $1 - \phi$ is proportional to the square root of the ionic strength in dilute solutions of electrolytes.

## 29. Shortcomings of the Debye-Hückel model

The expression of Debye and Hückel for the activity coefficients of ionic solutions is valid only in dilute solutions. This restricted range of validity can be discussed in terms of neglected factors which would be important even in solutions of nonelectrolytes, in terms of the mathematical approximation 27-7, and from the point of view of sound application of the principles of statistical mechanics.

The theory of Debye and Hückel gives specific consideration of only the long-range electrical interactions between ions. Even here, physical properties, such as the dielectric constant, are given values appropriate to the pure solvent. At higher concentrations, ion-solvent interactions and short-range interactions between ions become important. Solvation and association should not be ignored. These effects give contributions to the logarithm of the activity coefficient which are proportional to the solute concentration even in solutions of nonelectrolytes. Consequently, at concentrations where such terms are comparable to the square-root term, the Debye-Hückel theory can no longer adequately describe the thermodynamic properties. Refinement of the electrical contributions is not very useful unless these noncoulombic interactions are also accounted for.

The only significant mathematical approximation introduced by Debye and Hückel is that of equation 27-7. Its validity depends on the magnitude of $z_i F\Phi/RT$ being small compared to unity, and this means that $z_i z_c eF/4\pi \epsilon RTa$ should be small. However, this ratio is larger than unity for uni-univalent electrolytes in water, and the situation becomes worse for higher valence types and for solvents other than water. Furthermore, this ratio is independent of concentration, and it is not immediately clear that even the Debye-Hückel limiting law is free from error introduced by this approximation.

Fortunately, the Debye-Hückel limiting law can be substantiated by a singular-perturbation treatment of the problem. What enters into equation 28-2 or 28-10 is the potential at $r = a$ due to the *ion cloud*. For extremely dilute solutions, the Debye length becomes very large. This means that most of the ions comprising the ion cloud are at a considerable distance from the central ion, where the potential due to the central ion is greatly reduced and the approximation 27-7 is valid. Consequently, a valid approximation to the concentration distributions is obtained in the region where most of the counterbalancing charge is found; and this, in turn, yields a correct value for the potential at $r = a$ due to the ion cloud. This result is obtained even though there is always a region near the central ion where the approximation 27-7 is not valid.

It is of interest to note that even though the parameter $a$ does not appear in the Debye-Hückel limiting law, no solution of the nonlinearized problem will be found, except with the ion cloud concentrated at the origin, unless a nonzero value for $a$ is assumed.

Equation 27-3 is known as the Poisson-Boltzmann equation, and considerable effort has been devoted to its solution without the approximation 27-7 of Debye and Hückel. Gronwall *et al.*[2] and La Mer *et al.*[3] obtained series expansions for small values of parameter $z_i z_c eF/4\pi \epsilon RTa$. This is, of course, not the same as a series expansion for small values of the concentration. More recently, Guggenheim[4, 5, 6, 7] has reported the results of computer solutions of the Poisson-Boltzmann equation. He concludes that Gronwall's expansions do not give a significant improvement over Debye and Hückel's solution for aqueous solutions of uni-univalent electrolytes and that the terms reported by Gronwall are not sufficient for higher valence types. Guggenheim gives some hope that accurate solutions of the Poisson-Boltzmann equation would give substantial improvement, for higher valence types, over the result of Debye and Hückel at low concentrations where noncoulombic effects are not yet considerable.

[2]T. H. Gronwall, Victor K. La Mer, and Karl Sandved, "Über den Einfluss der sogenannten höheren Glieder in der Debye-Hückelschen Theorie der Lösungen starker Elektrolyte," *Physikalische Zeitschrift*, 29 (1928), 358–393.

[3]Victor K. La Mer, T. H. Gronwall, and Lotti J. Greiff, "The Influence of Higher Terms of the Debye-Hückel Theory in the Case of Unsymmetric Valence Type Electrolytes," *The Journal of Physical Chemistry*, 35 (1931), 2245–2288.

[4]E. A. Guggenheim, "The Accurate Numerical Solution of the Poisson-Boltzmann Equation," *Transactions of the Faraday Society*, 55 (1959), 1714–1724.

[5]E. A. Guggenheim, "Activity Coefficients and Osmotic Coefficients of 2:2 Electrolytes," *Transactions of the Faraday Society*, 56 (1960), 1152–1158.

[6]E. A. Guggenheim, "Activity Coefficients of 2:1 Electrolytes," *Transactions of the Faraday Society*, 58 (1962), 86–87.

[7]E. A. Guggenheim, *Applications of Statistical Mechanics* (Oxford: Clarendon Press, 1966).

Finally, one should note that the theory of Debye and Hückel is not a straightforward application of the principles of statistical mechanics. One may even marvel that the charging process gives a correct electrical contribution to the Helmholtz free energy. The inconsistencies in the model of Debye and Hückel first showed up when refined calculations gave different results for $\mu_{i,\text{el}}$ according to the Debye and Güntelberg charging processes. These problems have been discussed clearly by Onsager.[8] The interaction energy which should enter into the Boltzmann factor in equation 27-1 is the *potential of mean force*, the intergal of the average force associated with virtual displacements of an ion when all interactions with the solvent and other ions are considered. This is not necessarily equal to $z_i F\Phi$.

An example due to Onsager illustrates this contradiction. Let the potential around a central ion of type $j$ be denoted by $\Phi_j$. The probability of finding an ion of type $j$ within a volume element at the origin is proportional to $c_{j\infty}$. The conditional probability of finding an ion of type $i$ within a volume element at a distance $r$, when it is known that an ion of type $j$ is at the origin, is, according to the model of Debye and Hückel, proportional to $c_{i\infty} \exp(-z_i F\Phi_j/RT)$. Hence, the probability of finding an ion of type $j$ at a point and an ion of type $i$ at a point at a distance $r$ is proportional to $c_{j\infty} c_{i\infty} \exp(-z_i F\Phi_j/RT)$. However, this probability must be the same, independent of which ion is regarded as the central ion. Thus, we must have

$$c_{j\infty} c_{i\infty} \exp\left(-\frac{z_i F\Phi_j}{RT}\right) = c_{i\infty} c_{j\infty} \exp\left(-\frac{z_j F\Phi_i}{RT}\right), \tag{29-1}$$

where the term on the right is Debye and Hückel's expression for this probability when the ion of type $i$ is regarded as the central ion. Equality of these expressions requires that

$$z_i \Phi_j(r) = z_j \Phi_i(r). \tag{29-2}$$

In order to complete the demonstration, one needs to show that equation 2 is violated for some case where the potentials $\Phi_j(r)$ and $\Phi_i(r)$ are determined by the solution of the Poisson-Boltzmann equation. The simplest case is that of an unsymmetric electrolyte (see problem 1). Equation 2 does not happen to be violated for symmetric electrolytes. However, this does not mean that the basic model is free from objection in this case. The Debye and Güntelberg charging processes still give different results for $\mu_{i,\text{el}}$.

It is frequently stated erroneously that the model of Debye and Hückel is inconsistent because it violates the principle of superposition of electrostatics. This principle is embodied in Poisson's equation 27-2; the potential due to a given charge distribution will be everywhere doubled if all the charges

[8] Lars Onsager, "Theories of Concentrated Electrolytes," *Chemical Reviews*, *13* (1933), 73–89.

are doubled but remain in the same positions. Instead, the inconsistency arises from an improper statistical treatment of the problem.

The difficulty pinpointed by Onsager has been overcome by Mayer.[9] He has applied the principles of statistical mechanics to the physical model of Debye and Hückel, namely, hard-sphere ions of diameter $a$ moving in a continuous dielectric fluid. The basic statistical methods, involving cluster integrals, found in chapter 13 of Mayer and Mayer's book[10] and in the article by McMillan and Mayer,[11] form the starting point for Mayer's work on ionic solutions. Because of the long-range nature of coulombic forces, this extension is not easy, and the cluster method itself is far from simple.

Mayer is able to obtain the Debye-Hückel limiting law without invoking anything comparable to the approximation 27-7. He also collects the expressions for the evaluation of the logarithm of the activity coefficient, accurate through terms of order $c^{3/2}$. He recommends evaluation of these terms without expansion for small values of $c$, although there is no rigorous justification for expecting better accuracy.

For the reasons stated at the beginning of this section, higher order corrections to the activity coefficient cannot be obtained without consideration of noncoulombic effects, and the statistical methods will not be pursued further here. Résibois[12] states, "Disappointingly enough, it must be admitted that the rigorous justification of the D-H theory is the most important progress that has been made by the recent developments in the field of electrolyte theory."

## 30. Binary solutions

In this section, we consider solutions of a single electrolyte which dissociates into $v_+$ cations of charge number $z_+$ and $v_-$ anions of charge number $z_-$. The thermodynamic properties of such solutions have been studied extensively; activity and osmotic coefficients are summarized by Lewis and Randall[13] and by Robinson and Stokes.[14]

[9]Joseph E. Mayer, "The Theory of Ionic Solutions," *The Journal of Chemical Physics*, *18* (1950), 1426–1436.

[10]Joseph Edward Mayer and Maria Goeppert Mayer, *Statistical Mechanics* (New York: John Wiley & Sons, Inc., 1940).

[11]William G. McMillan, Jr., and Joseph E. Mayer, "The Statistical Thermodynamics of Multicomponent Systems," *The Journal of Chemical Physics*, *13* (1945), 276–305.

[12]Pierre M. V. Résibois, *Electrolyte Theory* (New York: Harper & Row, Publishers, 1968).

[13]Gilbert Newton Lewis and Merle Randall, revised by Kenneth S. Pitzer and Leo Brewer, *Thermodynamics* (New York: McGraw-Hill Book Company, Inc., 1961).

[14]R. A. Robinson and R. H. Stokes, *Electrolyte Solutions* (New York: Academic Press Inc., 1959).

Since individual ionic activity coefficients can depend on the electrical state of the phase, activity coefficients from equation 28-17 should be combined into the mean molal activity coefficient $\gamma_{+-}$:

$$\ln \gamma_{+-} = \frac{z_+ z_- \alpha \sqrt{I}}{1 + Ba\sqrt{I}}. \tag{30-1}$$

From equation 28-25, the corresponding form of the osmotic coefficient is

$$\phi = 1 + \tfrac{1}{3} z_+ z_- \alpha \sqrt{I}\, \sigma(Ba\sqrt{I}). \tag{30-2}$$

In order to account for effects neglected in the theory of Debye and Hückel, additional terms can be added to these expressions

$$\ln \gamma_{+-} = \frac{z_+ z_- \alpha \sqrt{I}}{1 + Ba\sqrt{I}} + A_2 m + A_3 m^{3/2} + A_4 m^2 + \cdots + A_n m^{n/2} + \cdots, \tag{30-3}$$

$$\phi = 1 + \frac{1}{3} z_+ z_- \alpha \sqrt{I}\, \sigma(Ba\sqrt{I})$$
$$+ \frac{1}{2} A_2 m + \frac{3}{5} A_3 m^{3/2} + \frac{2}{3} A_4 m^2 + \cdots + \frac{n}{n+2} A_n m^{n/2} + \cdots, \tag{30-4}$$

where equation 4 has been derived from equation 3 by means of the Gibbs-Duhem equation 14-26. The values of the $A$'s can be determined empirically by fitting equation 3 to activity coefficient data or equation 4 to osmotic coefficient data. Enough terms are carried to ensure a good fit, but no special significance should be attached to the values of the $A$'s so obtained.

For many purposes, it is sufficient to drop the terms beyond that in $A_2 m$. Guggenheim[15] then writes $A_2$ in the form

$$A_2 = \frac{4 v_+ v_-}{v_+ + v_-} \beta. \tag{30-5}$$

It is further recommended that $Ba$ be given the same value for all electrolytes [say, $Ba = 1$ (kg/mole)$^{1/2}$, which corresponds to $a = 3.04$ Å for aqueous solutions at 25°C]. One advantage of this procedure is that for each electrolyte there is now only one adjustable parameter, which can be fit by linear regression.

Equation 1 shows that, with the Debye-Hückel expression, the activity coefficient would have the same value for all electrolytes of the same charge type. Departures from this rule must be associated with specific interactions of the ions with the solvent and with each other. The parameter $\beta$ can thus be considered to account for these specific interactions. According to Brønsted's principle of specific interaction of ions,[16] ions of like charge

[15] E. A. Guggenheim, *Thermodynamics* (Amsterdam: North-Holland Publishing Company, 1959).

[16] J. N. Brönsted, "Studies on Solubility. IV. The Principle of the Specific Interaction of Ions," *The Journal of the American Chemical Society*, 44 (1922), 877–898.

will repel each other to such an extent that their interaction will be non-specific. This is the basis of the treatment of multicomponent solutions in the next section, and $A_2$ is expressed by equation 5 with a view toward this goal. That treatment also requires that $Ba$ be given the same value for all electrolytes.

Table 30-1 gives values of $\beta$ for uni-univalent electrolytes at 25°C. Table 30-2 gives values for 2–1 and 1–2 electrolytes.

TABLE 30-1. VALUES OF $\beta$ FOR 1–1 ELECTROLYTES AT 25°C AND FOR $Ba = 1$ (kg/mole)$^{1/2}$ (taken from reference 17, unless otherwise indicated, with permission of the Faraday Society).

$\beta$, kg/mole

| | | | | | | | |
|---|---|---|---|---|---|---|---|
| HCl | 0.27 | NaOH | 0.06 | KOH | 0.13 | RbCl | 0.06 |
| HBr | 0.33 | NaF | 0.07 | KF | 0.13 | RbBr | 0.05 |
| HI | 0.36 | NaCl | 0.15 | KCl | 0.10 | RbI | 0.04 |
| HClO$_4$ | 0.30 | NaBr | 0.17 | KBr | 0.11 | RbNO$_3$ | −0.14 |
| HNO$_3$ | 0.20[a] | NaI | 0.21 | KI | 0.15 | RbC$_2$H$_3$O$_2$ | 0.26 |
| LiOH | −0.21[b] | NaClO$_3$ | 0.10 | KClO$_3$ | −0.04 | CsOH | 0.35 |
| LiCl | 0.22 | NaClO$_4$ | 0.13 | KBrO$_3$ | −0.07 | CsCl | 0.00 |
| LiBr | 0.26 | NaBrO$_3$ | 0.01 | KIO$_3$ | −0.07 | CsBr | 0.00 |
| LiI | 0.35 | NaNO$_3$ | 0.04 | KNO$_3$ | −0.11 | CsI | −0.01 |
| LiClO$_4$ | 0.34 | NaC$_2$H$_3$O$_2$ | 0.23 | KC$_2$H$_3$O$_2$ | 0.26 | CsNO$_3$ | −0.15 |
| LiNO$_3$ | 0.21 | NaCNS | 0.20 | KCNS | 0.09 | CsC$_2$H$_3$O$_2$ | 0.28 |
| LiC$_2$H$_3$O$_2$ | 0.18 | NaH$_2$PO$_4$ | −0.06 | KH$_2$PO$_4$ | −0.16 | TlClO$_4$ | −0.17 |
| NH$_4$Cl | 0.10[b] | NH$_4$NO$_3$ | −0.10[b] | AgNO$_3$ | −0.14 | TlNO$_3$ | −0.36 |
| | | | | | | TlC$_2$H$_3$O$_2$ | −0.04 |

[a]Derived from reference 7.
[b]Derived from reference 13.

Pitzer and Brewer[13] use one electrolyte as a reference and treat $\ln (\gamma_{+-}/\gamma_{KCl})$ or $\ln (\gamma_{+-}/\gamma_{CaCl_2})$. In these ratios, the electrical or coulombic effects should largely cancel; and the logarithm of the ratio should be proportional to $m$ in dilute solutions, if the reference electrolyte is of the same valence type as the electrolyte in question. Thus, Pitzer and Brewer use KCl as a reference for 1–1 electrolytes and CaCl$_2$ as a reference for 2–1 and 1–2 electrolytes. Earlier, Lewis and Randall[18] had used a similar comparison to aid in extrapolations to infinite dilution for electrolytes for which data at low concentrations were absent or unreliable. Pitzer and Brewer did not find suitable reference electrolytes of higher valence types for which concordant data were available.

[17]E. A. Guggenheim and J. C. Turgeon, "Specific Interaction of Ions," *Transactions of the Faraday Society*, *51* (1955), 747–761.

[18]Gilbert N. Lewis and Merle Randall, "The Activity Coefficient of Strong Electrolytes," *The Journal of the American Chemical Society*, *43* (1921), 1112–1154.

TABLE 30-2. VALUES OF $\beta$ AND $Ba$ FOR 2–1 and 1–2 ELECTROLYTES AT 25°C
(taken from reference 19, with permission of the Faraday Society).

|  | $\sqrt{3}\,Ba$, (kg/mole)$^{1/2}$ | $8\beta/3 \ln 10$, kg/mole |  | $\sqrt{3}\,Ba$, (kg/mole)$^{1/2}$ | $8\beta/3 \ln 10$, kg/mole |
|---|---|---|---|---|---|
| $MgCl_2$ | 2.75 | 0.206 | $Mg(NO_3)_2$ | 2.65 | 0.208 |
| $CaCl_2$ | 2.66 | 0.169 | $Ca(NO_3)_2$ | 2.40 | 0.052 |
| $SrCl_2$ | 2.70 | 0.125 | $Sr(NO_3)_2$ | 2.40 | $-0.043$ |
| $BaCl_2$ | 2.70 | 0.066 | $Co(NO_3)_2$ | 2.75 | 0.148 |
| $MnCl_2$ | 2.70 | 0.148 | $Cu(NO_3)_2$ | 2.65 | 0.122 |
| $FeCl_2$ | 2.70 | 0.162 | $Zn(NO_3)_2$ | 2.85 | 0.172 |
| $CoCl_2$ | 2.70 | 0.188 | $Cd(NO_3)_2$ | 2.75 | 0.104 |
| $NiCl_2$ | 2.70 | 0.188 | $Mg(ClO_4)_2$ | 3.25 | 0.356 |
| $CuCl_2$ | 2.70 | 0.084 | $Li_2SO_4$ | 2.45 | $-0.065$ |
| $MgBr_2$ | 2.80 | 0.291 | $Na_2SO_4$ | 2.20 | $-0.165$ |
| $CaBr_2$ | 2.80 | 0.212 | $K_2SO_4$ | 1.85 | $-0.087$ |
| $SrBr_2$ | 2.80 | 0.176 | $Rb_2SO_4$ | 2.30 | $-0.148$ |
| $BaBr_2$ | 2.70 | 0.140 | $Cs_2SO_4$ | 2.30 | $-0.113$ |

To illustrate the behavior of activity coefficients of electrolytes, figure 30-1 shows $\gamma_{+-}$ for HCl and $HNO_3$. The logarithm of $\gamma_{+-}$ shows a linear dependence on $\sqrt{m}$ at low concentrations, as predicted by equation 1 or 3, and the values of $\gamma_{+-}$ for HCl and $HNO_3$ approach each other at low concentrations. Hence, the electrical effects tend to cancel in the ratio $\gamma_{HCl}/\gamma_{HNO_3}$, and this ratio shows a linear dependence on $m$ at low concentrations, as indicated by figure 30-2. The slope of this curve should be proportional to $\beta_{HCl} - \beta_{HNO_3}$.

Robinson and Stokes[14] give a chemical model for hydration, etc., and arrive at a two-parameter equation which can be fit to activity coefficient data.

## 31.  Multicomponent solutions

Equations 30-3 and 30-4 express the composition dependence of the activity and osmotic coefficients of a binary solution. The parameters $\alpha$, $Ba$, and the $A$'s are taken to be constant at a given temperature and pressure; that is, they are independent of the solute concentration. The activity and osmotic coefficients are not independent; equation 30-4 could be derived from equation 30-3 by applying the Gibbs-Duhem relation at constant temperature and pressure. Since the chemical potentials of the components of

[19]E. A. Guggenheim and R. H. Stokes, "Activity Coefficients of 2:1 and 1:2 Electrolytes in Aqueous Solution from Isopiestic Data," *Transactions of the Faraday Society*, 54 (1958), 1646–1649.

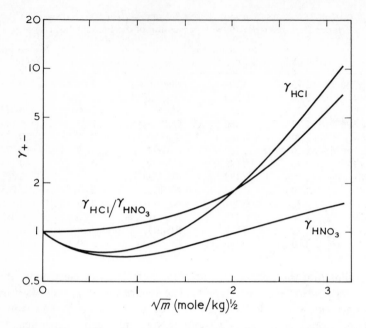

**Figure 30-1.** Mean molal activity coefficients of HCl (from reference 13) and $HNO_3$ (from reference 20) and the ratio of the activity coefficients of the two acids.

a mixture are not independent, inconsistencies are most easily avoided by deriving the chemical potentials from an expression for the total free energy. The Gibbs free energy is used instead of the Helmholtz free energy so that chemical potentials can be obtained by differentiation at constant pressure instead of constant volume (see equation 13-1), and the Gibbs-Duhem equation can be more easily applied if the parameters are constant at a given pressure.

For moderately dilute solutions containing several electrolytes, let us express the Gibbs free energy as[21,22]

$$\frac{G}{RT} = \frac{n_0 \mu_0^0}{RT} + \sum_{j \neq 0} n_j \left[ \ln \left( m_j \lambda_j^\theta \right) - 1 \right]$$

$$- \frac{2}{3} \alpha \sqrt{I} \, \tau (Ba\sqrt{I}) \sum_j z_j^2 n_j + \sum_{i \neq 0} \sum_{j \neq 0} \beta_{i,j} n_i m_j, \qquad (31\text{-}1)$$

[20] W. A. Roth and K. Scheel, eds., *Landolt-Börnstein Physikalisch-chemische Tabellen*, fifth edition, supplement 3, part 3 (Berlin: Springer Verlag, 1936), p. 2145.

[21] E. A. Guggenheim, "The Specific Thermodynamic Properties of Aqueous Solutions of Strong Electrolytes," *Philosophical Magazine*, Ser. 7, *19* (1935), 588–643.

[22] George Scatchard, "Concentrated Solutions of Strong Electrolytes," *Chemical Reviews, 19* (1936), 309–327.

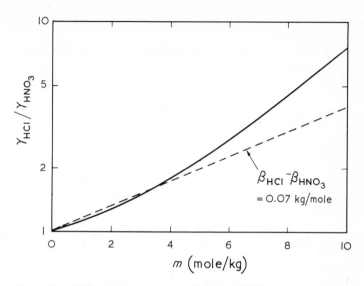

**Figure 30-2.** Ratio of the activity coefficients of HCl and $HNO_3$ plotted against the molality.

where $\mu_0^0$ is the chemical potential of the pure solvent at the same temperature and pressure.

The first two terms on the right in equation 1 can be regarded as an *ideal* contribution to the free energy, but it should be realized that this is only a manner of speaking. An ideal contribution should be expressed in terms of mole fractions, particle fractions, volume fractions, or concentrations, since the molality approaches infinity as the concentration of the solvent approaches zero. Notice that the combination of $\lambda_j^\theta$'s in the second term is uniquely defined according to the principles of section 14, since the solution as a whole is electrically neutral.

The third term on the right in equation 1 represents the electrical contribution expressed by equation 28-4, $Ba\sqrt{I}$ approaching $a/\lambda$ as the ionic strength approaches zero. Recall that $a$ represents an average value of the sum of the radii of pairs of hydrated ions and that only a single value of $a$ can apply to a given solution.

The last term in equation 1 is a first approximation to the effect of specific interactions between pairs of ions, and we take $\beta_{i,j} = \beta_{j,i}$. (This is no loss in generality since $n_i m_j = n_j m_i$.) In accordance with Brønsted's principle of specific interaction of ions, $\beta_{i,j}$ is set equal to zero if $z_i z_j > 0$; that is, for a pair of ions $i$ and $j$ with charges of the same sign.

Molalities and the molal ionic strength $I$ are employed in equation 1 because this simplifies the differentiation and facilitates the use of the Gibbs-Duhem equation. Furthermore, Scatchard[22] argues on a theoretical basis that the molal ionic strength is more appropriate for the electrical contribution to the free energy.

Differentiation of equation 1 according to equation 13-1 gives, for the chemical potential of the solute species,

$$\frac{\mu_k}{RT} = \ln (m_k \lambda_k^\theta) - \frac{z_k^2 \alpha \sqrt{I}}{1 + Ba \sqrt{I}} + 2 \sum_{j \neq 0} \beta_{k,j} m_j \qquad (31\text{-}2)$$

and, for the chemical potential of the solvent,

$$\frac{\mu_0}{RT} = \frac{\mu_0^0}{RT} - M_0 \sum_{j \neq 0} m_j$$

$$+ \frac{2}{3} \alpha M_0 I^{3/2} \sigma(Ba \sqrt{I}) - M_0 \sum_{i \neq 0} \sum_{j \neq 0} \beta_{i,j} m_i m_j. \qquad (31\text{-}3)$$

Hence, the solute activity coefficients are

$$\ln \gamma_k = -\frac{z_k^2 \alpha \sqrt{I}}{1 + Ba \sqrt{I}} + 2 \sum_{j \neq 0} \beta_{k,j} m_j, \qquad (31\text{-}4)$$

and the osmotic coefficient is

$$\phi = 1 - \frac{\frac{2}{3} \alpha I^{3/2} \sigma(Ba \sqrt{I}) - \sum_{i \neq 0} \sum_{j \neq 0} \beta_{i,j} m_i m_j}{\sum_{i \neq 0} m_i}. \qquad (31\text{-}5)$$

Although equations 2 and 4 apply to individual ions, the dependence on the electrical state of the phase has not been included. Consequently, these expressions should be used only for electrically neutral combinations of ions, as discussed in section 14.

As a consequence of Brønsted's principle of specific interaction of ions, $\beta_{i,j} = 0$ for a pair of ions of like charge. This means that each $\beta$ which appears in these equations for multicomponent solutions corresponds to one cation and one anion and can be determined from activity-coefficient or osmotic-coefficient measurements in a binary solution of this electrolyte. Such values are given in tables 30-1 and 30-2. However, the same value of $Ba$ must be used in fitting the data for the binary solutions. For the systems in table 30-2, an average value of $Ba$ should be selected, and values of $\beta$ recomputed from the original data. The same value of $Ba$ should then be used to compute $\beta$ values for the systems of table 30-1. The tables of Pitzer and Brewer[13] would be helpful in these calculations, since they give activity coefficients relative to a reference electrolyte, KCl or $CaCl_2$.

The theory presented above is valuable because it allows the thermodynamic properties of multicomponent solutions to be predicted from measurements on binary solutions. For a solution of two electrolytes, the logarithms of the activity coefficients of the electrolytes are predicted to vary linearly with the molality of one of the electrolytes when the ionic strength $I$ is maintained constant. Such behavior is commonly observed.

When the above treatment proves to be inadequate, more terms can be added to the expression 1 for the free energy. For uni-univalent electro-lytes, Guggenheim[7,23] relaxes Brønsted's principle and allows $\beta_{i,j}$ to be different from zero for pairs of ions of like charge. ($\beta_{i,j}$ is still zero for $i = j$.) The new values of $\beta$'s (see table 31-1) are not applicable to binary solutions and are somewhat smaller than those for pairs of ions of unlike charge, in partial accord with Brønsted's principle. The modification does not greatly affect the activity coefficients in mixed electrolytes and is perhaps more important in the correlation of enthalpies of mixing.

In addition to the terms in equation 1, Scatchard[22,24] generally carries terms of cubic and higher order in the molalities.

TABLE 31-1. VALUES OF $\beta$ FOR IONS OF LIKE
CHARGE (from reference 7, with
permission of Oxford University Press).

| ions | $\beta$, kg/mole |
|------|------------------|
| $H^+$—$Li^+$ | 0.021 |
| $H^+$—$Na^+$ | 0.015 |
| $H^+$—$K^+$ | −0.010 |
| $Li^+$—$Na^+$ | 0.006 |
| $Li^+$—$K^+$ | −0.022 |
| $Cl^-$—$NO_3^-$ | 0.01 |

Harned and Robinson[25] have reviewed the equilibrium properties of solutions of several electrolytes. They discuss the behavior of activity coeffi-cients corresponding to several representations of the free energy, and they present literature references for those systems where measurements have been made.

## 32. Measurement of activity coefficients

The activity coefficients of electrolytic solutions can be determined by methods applicable to solutions of nonelectrolytes. We mention, in particular, freezing-point measurements, vapor-pressure measurements, and isopiestic methods. Other methods involve the measurement of the potentials of gal-vanic cells and are peculiar to electrolytic solutions. Robinson and Stokes[14] and Lewis and Randall[13] review these and other methods.

Measurement of freezing point depressions is useful if the solution can

[23] E. A. Guggenheim, "Mixtures of 1 : 1 Electrolytes," *Transactions of the Faraday Society*, *62* (1966), 3446–3450.

[24] George Scatchard, "Osmotic Coefficients and Activity Coefficients in Mixed Electro-lyte Solutions," *Journal of the American Chemical Society*, *83* (1961), 2636–2642.

[25] H. S. Harned and R. A. Robinson, *Multicomponent Electrolyte Solutions* (Oxford: Pergamon Press, 1968).

exist in equilibrium with the pure solid solvent. Then, at the freezing point
of the solution,

$$\mu_0(T) = \mu_{0,s}(T),$$    (32-1)

where $\mu_{0,s}$ denotes the chemical potential of the solid solvent. Next, it is
necessary to correct the chemical potentials to a constant temperature (since
the data are at the freezing points of the solutions), usually to the temperature
$T_f$ of the freezing point of the pure solvent. This is done on the basis of the
relation

$$\frac{\partial \mu_i / T}{\partial T} = -\frac{\bar{H}_i}{T^2},$$    (32-2)

where $\bar{H}_i$ is the partial molar enthalpy of component $i$ in the phase in question. Then

$$\frac{\mu_0(T_f)}{T_f} = \frac{\mu_0(T)}{T} - \int_T^{T_f} \frac{\bar{H}_0}{T^2} dT,$$    (32-3)

and    $$\frac{\mu_{0,s}(T_f)}{T_f} = \frac{\mu_0^0(T_f)}{T_f} = \frac{\mu_{0,s}(T)}{T} - \int_T^{T_f} \frac{\bar{H}_{0,s}}{T^2} dT,$$    (32-4)

where we have noted that at $T_f$ the pure solvent is in equilibrium with the
solid solvent. Hence, we have

$$\frac{\mu_0(T_f) - \mu_0^0(T_f)}{T_f} = R \ln \frac{\lambda_0}{\lambda_0^0} = -\phi R M_0 \nu m$$

$$= -\int_T^{T_f} \frac{\bar{H}_0 - \bar{H}_{0,s}}{T^2} dT.$$    (32-5)

The measurements give the freezing points of solutions of molality $m$.
The above calculation then yields the osmotic coefficient for the solution of
molality $m$ at the freezing point $T_f$ of the pure solvent. Finally, osmotic coefficients can be converted to activity coefficients by means of the Gibbs-Duhem
equation 14-24. Measurements of the freezing-point depressions are an important source of accurate values of activity coefficients of electrolytic solutions, particularly dilute solutions. The calculation procedure was developed
by Lewis and Randall[19] and is described in detail by Pitzer and Brewer,[13]
including the treatment of the thermal properties of the solution.

Many electrolytes are nonvolatile, and measurement of the vapor pressure yields the chemical potential of the solvent in the solution. This gives
directly the osmotic coefficient. Measurement of the absolute vapor pressure
is necessary for only one system. Subsequently, only isopiestic measurements
for other systems are needed. This amounts to the determination of the solution concentration which has the same vapor pressure as a given solution of
the standard system for which the solvent activity is already known. The
osmotic coefficients of solutions of many electrolytes have been determined
by the isopiestic method, although the accuracy is not adequate below about
0.1 molal.

The open-circuit potentials of electrochemical cells were treated extensively in chapter 2. These provide a means for determining the chemical potential of the solute. Consider the cell 18-12:

$$\begin{array}{c|c|c|c|c|c|c}
\alpha & \beta & & \delta & \epsilon & \lambda & \alpha' \\
\text{Pt}(s),\ \text{H}_2(g) & \text{HCl in} & \text{transition} & \text{HCl in} & \text{AgCl}(s) & \text{Ag}(s) & \text{Pt}(s). \\
& \text{H}_2\text{O} & \text{region} & \text{H}_2\text{O} & & &
\end{array}$$

$$(32\text{-}6)$$

If the difference in electrochemical potential of chloride ions between solutions $\beta$ and $\delta$ is ignored, the cell potential can be expressed as (see equation 18-13)

$$FU = -F(\Phi^\alpha - \Phi^{\alpha'}) = FU^\theta + \tfrac{1}{2}RT \ln p_{\text{H}_2}^\alpha - 2RT \ln (m_{\text{HCl}}^\beta \gamma_{\text{HCl}}^\beta),$$

$$(32\text{-}7)$$

where
$$FU^\theta = \tfrac{1}{2}\mu_{\text{H}_2}^* - \mu_{\text{Ag}}^0 + \mu_{\text{AgCl}}^0 - 2RT \ln \lambda_{\text{HCl}}^\theta.$$

$$(32\text{-}8)$$

A system should be chosen such that the neglected term $\mu_{\text{Cl}^-}^\beta - \mu_{\text{Cl}^-}^\delta$ is as small as possible. For the present system, where the solubility of silver chloride is low, table 18-1 shows that the error amounts to less than 0.23 mV for HCl concentrations greater than $10^{-4}$ molal.

Equation 7 can be used to calculate $\gamma_{\text{HCl}}$ directly from the measured cell potentials. However, this requires that the standard cell potential $U^\theta$ be known, and this is determined by the definition 14-5 of the secondary reference state. In order to extrapolate most accurately to infinite dilution, one makes use of the fact that in dilute solutions the activity coefficient behaves like (see equations 30-3 and 30-5)

$$\ln \gamma_{\text{HCl}} = -\frac{\alpha \sqrt{m}}{1 + \sqrt{m}} + 2\beta_{\text{HCl}} m,$$

$$(32\text{-}9)$$

where $\alpha$ is the Debye-Hückel constant and is known (see equation 28-18). Consequently, one defines a secondary quantity $U'$ by

$$FU' = FU - \frac{1}{2} RT \ln p_{\text{H}_2}^\alpha + 2RT \ln m - \frac{2RT\alpha\sqrt{m}}{1 + \sqrt{m}},$$

$$(32\text{-}10)$$

where $m$ signifies $m_{\text{HCl}}^\beta$. Then, to the extent that equation 9 is applicable, $U'$ behaves in dilute solutions like

$$U' = U^\theta - \frac{4RT\beta_{\text{HCl}} m}{F}.$$

$$(32\text{-}11)$$

A plot of $U'$ versus $m$ should be linear near $m = 0$. The intercept of this straight line is then $U^\theta$, and the slope is $-4RT\beta_{\text{HCl}}/F$. This procedure, which allows the most accurate extrapolation of cell data to infinite dilution, was apparently first used by Schumb et al.[26] Equation 7 can now be used to cal-

[26]Walter C. Schumb, Miles S. Sherrill, and Sumner B. Sweetser, "The Measurement of the Molal Ferric-Ferrous Electrode Potential," *Journal of the American Chemical Society*, 59 (1937), 2360–2365.

culate values of $\gamma_{HCl}$ which correspond to the secondary reference state 14-5.

For some electrolytes, electrodes reversible to both ions may be impossible to find or difficult to work with. A cell with liquid junction, such as cell 17-23:

$$
\begin{array}{c|c|c|c|c|c|c|c|c}
\alpha & \beta & \delta & \epsilon & & \lambda & \delta' & \beta' & \alpha' \\
\mathrm{Pt}(s) & \mathrm{Hg}(l) & \mathrm{HgO}(s) & \begin{array}{c}\mathrm{KOH\ in}\\ \mathrm{H_2O}\end{array} & \begin{array}{c}\mathrm{transition}\\ \mathrm{region}\end{array} & \begin{array}{c}\mathrm{KOH\ in}\\ \mathrm{H_2O}\end{array} & \mathrm{HgO}(s) & \mathrm{Hg}(l) & \mathrm{Pt}(s),
\end{array}
$$

$$(32\text{-}12)$$

may then be set up with only one type of electrode. The potential of this cell is given by

$$FU = -F(\Phi^\alpha - \Phi^{\alpha'}) = \int_\lambda^\epsilon (t^0_{K^+} + \tfrac{1}{2}M_0 m)\, d\mu_{KOH}. \qquad (32\text{-}13)$$

If the transference number $t^0_{K^+}$ is known, measurements of the potential of this cell can be used to study the activity and osmotic coefficients of this electrolytic solution.

For thermodynamic measurements, cells with liquid junction are generally to be avoided. However, when the junction involves solutions of only one electrolyte, the cell potential is independent of the method of forming the junction and can be expressed unambiguously in terms of the thermodynamic and transport properties according to equation 13.

## 33. Weak electrolytes

Aqueous solutions of sulfuric acid involve the equilibrium between sulfate and bisulfate ions:

$$\mathrm{H^+ + SO_4^= \rightleftharpoons HSO_4^-}. \qquad (33\text{-}1)$$

Because the equilibrium is rapid, there are only two independent components, which can be taken to be $H_2SO_4$ and $H_2O$. In treating the activity coefficient of the solute, one can consider the solution to be composed of one of the following sets of species:

(a) $H_2SO_4$ and $H_2O$
(b) $H^+$, $HSO_4^-$, and $H_2O$
(c) $H^+$, $SO_4^=$, and $H_2O$.

In the last two cases, one deals with the mean activity coefficient of hydrogen and bisulfate ions or the mean activity coefficient of hydrogen and sulfate ions.

In section 14 we dealt with the difference between the treatments of the electrolyte as dissociated or without regard for its dissociation. Any one of the above three formulations provides a consistent basis for describing all the relevant thermodynamic properties of the system, although formulations (a) and (b) have the disadvantage that they do not take into account the state of aggregation of the solute at infinite dilution and thus preclude the application of equations 14-5 and 14-6 to define the secondary reference state.

One can attempt to take into explicit account the equilibrium 1 by writing

$$\mu_{H^+} + \mu_{SO_4^=} = \mu_{HSO_4^-} \tag{33-2}$$

and defining an equilibrium constant, for example,

$$K = \frac{c_{H^+}^* c_{SO_4^=}^* f_{H^+}^* f_{SO_4^=}^*}{c_{HSO_4^-}^* f_{HSO_4^-}^*} = \frac{a_{HSO_4^-}^*}{a_{H^+}^* a_{SO_4^=}^*}. \tag{33-3}$$

We have put asterisks on these quantities to denote the fact that they refer to a view of the solution as composed of water molecules and hydrogen, bisulfate, and sulfate ions. For example, the concentrations $c_i^*$ are not the stoichiometric concentrations of independent components which can be assessed by ordinary analytical means.

The thermodynamic constant $K$ is independent of concentration. In addition to this constant, we have introduced one extra concentration and one extra activity coefficient. In problem 6, we show how to calculate $K$, $f_{H^+}^* f_{HSO_4^-}^*$ and $(f_{H^+}^*)^2 f_{SO_4^=}^*$, if a microscopic concentration, say $c_{HSO_4^-}^*$, is known as a function of stoichiometric concentration. It should be emphasized, however, that such a procedure is outside the strict bounds of thermodynamics.

In going beyond thermodynamic means to investigate the equilibrium of a weak electrolyte, one could adopt the microscopic model, assume that the activity coefficients of the microscopic species are given by some suitably simple theory (see section 31), and fit the macroscopic thermodynamic quantities to determine the best values of $K$ and any other parameters of the microscopic model. Transport porperties also can be used to help determine the values of these parameters.

This approach can be valuable to predict the behavior of macroscopic quantities in the absence of complete data; for example, for the activities of phenolic compounds in water. For sulfuric acid, the data are quite complete, but this approach can give a guide to the correlation of the concentration dependence of the activity coefficient (compare reference 14, p. 213).

A second approach is to use, for example, Raman spectra to give independent evidence of the microscopic species concentrations. Then one can obtain a reliable value for $K$, determine values for the microscopic activity coefficients, and compare these with simple theoretical models, as outlined in problem 6.

The results of Raman-spectra measurements are given for sulfuric acid solutions by Young et al.[27] From these results, we can calculate directly the

[27]T. F. Young, L. F. Maranville, and H. M. Smith, "Raman Spectral Investigations of Ionic Equilibria in Solutions of Strong Electrolytes," Walter J. Hamer, ed., *The Structure of Electrolytic Solutions* (New York: John Wiley & Sons, Inc., 1959), pp. 35–63.

value of

$$K' = \frac{c_{\mathrm{H}}^* \cdot c_{\mathrm{SO_4^=}}^*}{c_{\mathrm{HSO_4^-}}^*} = \frac{K f_{\mathrm{HSO_4^-}}^*}{f_{\mathrm{H}}^* \cdot f_{\mathrm{SO_4^=}}^*} . \tag{33-4}$$

This quantity is useful for retrieving the values of the microscopic concentrations from the stoichiometric concentration. We see that $K'$ depends on concentration through the activity coefficients. From the development in the preceding sections, we might expect to be able to correlate $K'$ in terms of the microscopic or "true" ionic strength $I_r$:

$$I_r = \tfrac{1}{2} \sum_i z_i^2 c_i^* . \tag{33-5}$$

Such a correlation of Young's data is shown in figure 33-1, where the experimental points are correlated by the equation

$$\ln \frac{K'}{K} = \frac{5.29\sqrt{I_r}}{1 + 0.56\sqrt{I_r}} , \tag{33-6}$$

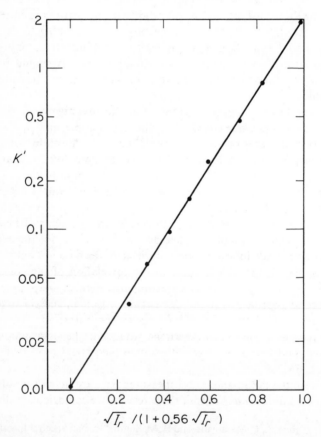

**Figure 33-1.** Correlation of the second dissociation constant of sulfuric acid with the true ionic strength.

where $I_r$ is expressed in mole/liter. The value of $K = 0.0104$ mole/liter, from reference 14, p. 387, is also shown on figure 33-1. The points on figure 33-1 extend to a stoichiometric concentration of sulfuric acid of 2.95 $M$.

Such a correlation can be of value in estimating the microscopic species concentrations in, for example, solutions of copper sulfate and sulfuric acid, where the data are far from complete.

## PROBLEMS

1. Set up the Poisson-Boltzmann equations for a solution containing cations of charge number $z_+$ and anions of charge number $z_-$. Let $x = r/\lambda$; for the cation as the central ion, let $\phi_+ = z_- F\Phi/RT$, and for the anion as the central ion, let $\phi_- = z_+ F\Phi/RT$. Show that $\phi_+ - \phi_-$ satisfies the differential equation

$$\frac{1}{x^2}\frac{d}{dx}\left(x^2\frac{d\phi_+ - \phi_-}{dx}\right) = \frac{z_+}{z_+ - z_-}\left(e^{-\phi_-} - e^{-z_-\phi_-/z_+}\right)$$
$$- \frac{z_-}{z_+ - z_-}\left(e^{-z_+\phi_+/z_-} - e^{-\phi_+}\right)$$

and the boundary conditions

$$\frac{d\phi_+ - \phi_-}{dx} = 0 \quad \text{at} \quad x = a/\lambda,$$

$$\phi_+ - \phi_- \longrightarrow 0 \quad \text{as} \quad x \longrightarrow \infty.$$

The Onsager criterion requires that $\phi_+(x) = \phi_-(x)$ (see equation 29-2). Can this criterion be satisfied exactly when $z_+$ is not equal to $-z_-$, that is, for a nonsymmetric electrolyte?

2. Set up the Poisson-Boltzmann equations as in problem 1. Let $\delta = a/\lambda$ and $S = -z_+z_- Fe/4\pi \epsilon RTa$ so that the boundary condition at $r = a$ becomes

$$\delta \frac{d\phi_+}{dx} = S \quad \text{at} \quad x = \delta.$$

In order to justify the Debye-Hückel limiting law, we wish to solve the problem for small values of $\delta$, that is, for small values of the concentration.

(a) Seek a solution by singular-perturbation expansions, letting the inner variable be $\bar{x} = x/\delta$ and the outer variable be $\tilde{x} = x$. [See John Newman, *Trans. Faraday Soc.*, 61, 2229-2237 (1965) for the application of singular-perturbation techniques to a problem of electrochemical interest.] In the inner region, the potential due to the ion cloud is negligible in the first approximation. In the outer region, the potential is small, and the exponential Boltzmann terms can be linearized.

(b) Use the result of part (a) and the Güntelberg charging process to substantiate the limiting law of Debye and Hückel. If possible, obtain the next term in an expansion for small $\delta$.

(c) Repeat part (b) with the Debye charging process.

3. In section 28, the Helmholtz free energy was used; in section 31, the Gibbs free energy. For a 0.1 molal solution, calculate the difference between $A$ and $G$ and compare with the electrical contribution to the free energy. Obtain an expression for the difference between the derivative of $A$ with respect to $n_i$ when the differentiation is carried out at constant volume or constant pressure. Is the determination of the chemical potential of the solvent according to equation 28-21 completely legitimate? Is the derivation of equation 30-4 from equation 30-3 completely rigorous?

4. Smyrl and Newman [*J. Phys. Chem.*, *72*, 4660-4671 (1968)] express molar activity coefficients of solute species in essentially the following form (compare equation 31-4):

$$\ln f_i - \frac{z_i}{z_n} \ln f_n = -\frac{\alpha' z_i(z_i - z_n)\sqrt{I'}}{1 + B'a\sqrt{I'}} + 2\sum_{j \neq 0} \left( \beta'_{i,j} - \frac{z_i}{z_n} \beta'_{n,j} \right) c_j,$$

where the sum is over solute species, $\beta'_{i,j} = 0$ for a pair of ions of like charge, and $\beta'_{i,j} = \beta'_{j,i}$.

(a) Is this expression consistent with the thermodynamic requirement that $\partial \mu_i / \partial c_j = \partial \mu_j / \partial c_i$?

(b) Show that

$$2\sum_{j \neq 0} \left( \beta'_{i,j} - \frac{z_i}{z_n} \beta'_{n,j} \right) c_j = 2\sum_{j \neq 0} \left( \beta_{i,j} - \frac{z_i}{z_n} \beta_{n,j} \right) \frac{c_j}{\rho_0}$$

$$- \left( 1 - \frac{z_i}{z_n} \right)\left( \frac{c_0 M_0}{\rho_0} - 1 \right) + 0(c^{3/2})$$

if the expression of Smyrl and Newman is to be equivalent to the corresponding one for molal activity coefficients through terms of order $c$ in the solute concentrations. Here, $\rho_0$ is the density of the pure solvent.

(c) Show for a solution of a single electrolyte that

$$\beta'_{+-} = \frac{\beta_{+-}}{\rho_0} + \frac{\nu \bar{V}_e}{4\nu_+ \nu_-},$$

where $\bar{V}_e$ is the partial molar volume of the electrolyte in an infinitely dilute solution. For $\bar{V}_e = 27$ cm³/mole, compare the value of the correction term with typical values of $\beta_{+-}/\rho_0$ for uni-univalent electrolytes from table 30-1.

5. Assume that the Gibbs free energy for a solution of several electrolytes can be expresses as

$$\frac{G}{RT} = \frac{n_0 \mu_0^0}{RT} + \sum_{j \neq 0} n_j [\ln (m_j \lambda_j^\theta) - 1] - \frac{2}{3}\alpha\sqrt{I}\,\tau(Ba\sqrt{I})\sum_j z_j^2 n_j$$

$$- \frac{1}{2}\sum_{i \neq 0}\sum_{j \neq 0} \epsilon_{i,j} z_i z_j n_i m_j + \cdots.$$

Here the last terms, involving $\epsilon_{i,j}$, are supposed to account for short-range specific interactions between pairs of solute ions. The factor $z_i z_j$ is included solely for convenience.

(a) Discuss why this expression is sufficiently general without including terms like $\epsilon_{i,0}$ for interaction of species $i$ with the solvent.

(b) Show that, with no loss of generality, we can require that $\epsilon_{i,j} = \epsilon_{j,i}$.

(c) The molalities of the ions are not independent since they satisfy the electroneutrality relation

$$\sum_i z_i m_i = 0.$$

Therefore, since the $\epsilon_{i,j}$'s cannot be determined separately, define

$$\beta_{i,j} = -\tfrac{1}{4} z_i z_j (2\epsilon_{i,j} - \epsilon_{i,i} - \epsilon_{j,j}),$$

so that $\beta_{i,j} = \beta_{j,i}$ and $\beta_{i,i} = 0$. Show that

$$-\tfrac{1}{2} \sum_{i \neq 0} \sum_{j \neq 0} \epsilon_{i,j} z_i z_j n_i m_j = \sum_{i \neq 0} \sum_{j \neq 0} \beta_{i,j} n_i m_j.$$

Do you think that the $\beta_{i,j}$ values are subject to experimental determination?

(d) Show that, for a solution of a single electrolyte, the expression for $G$ now reduces to

$$\frac{G}{RT} = \frac{n_0 \mu_0^0}{RT} + \sum_{j \neq 0} n_j \left[ \ln (m_j \lambda_j^\theta) - 1 \right] - \frac{2}{3} \alpha \sqrt{I}\, \tau(Ba\sqrt{I}) \sum_j z_j^2 n_j$$
$$+ 2\beta_{+-} M_0 n_0 m_+ m_-.$$

Part (c) reveals why there is no $\beta_{++}$ or $\beta_{--}$ for solutions of a single electrolyte.

(e) For mixtures containing cations of the same charge number $z_+$ and anions of the same charge number $z_-$, Guggenheim[7] expresses the free energy relative to a reference electrolyte for which $\beta_{+-} = \beta_{+-}^0$. Assume that $G^0$, a reference free energy, denotes the expression in part (d) with $\beta_{+-}$ replaced by $\beta_{+-}^0$ and with $m_+$ and $m_-$ having the meaning

$$m_+ = \sum_+ m_i = v_+ m \quad \text{and} \quad m_- = \sum_- m_i = v_- m,$$

where $m$ is the total molality of the solution. Show that the *excess* free energy is given by

$$\frac{G - G^0}{RT} = M_0 n_0 \left[ \sum_{i \neq 0} \sum_{j \neq 0} \beta_{i,j} m_i m_j - 2\beta_{+-}^0 m_+ m_- \right]$$
$$= (\phi - \phi^0) M_0 n_0 v m$$

and that the mean activity coefficient of a cation $k$ and an anion $l$ is

$$v \ln \frac{\gamma_{k,l}}{\gamma_{+-}^0} = 2v_+ \left( \sum_{i \neq 0} \beta_{k,i} m_i - \sum_{i-} \beta_{+-}^0 m_i \right) + 2v_- \left( \sum_{i \neq 0} \beta_{l,i} m_i - \sum_{i+} \beta_{+-}^0 m_i \right),$$

where $\phi^0$ is the osmotic coefficient and $\gamma_{+-}^0$ is the mean molal activity coefficient of the reference electrolyte, both at the molality $m$.

(f) Let

$$\Delta_{i,j} = \beta_{i,j} - \beta_{+-}^0 \quad \text{if} \quad z_i z_j < 0$$
$$\Delta_{i,j} = \beta_{i,j} \quad \text{if} \quad z_i z_j > 0.$$

Show that

$$\frac{G - G^0}{RT} = (\phi - \phi^0)M_0 n_0 vm = M_0 n_0 \sum_{i \neq 0} \sum_{j \neq 0} \Delta_{i,j} m_i m_j$$

and

$$v \ln \frac{\gamma_{k,l}}{\gamma_{+-}^0} = 2v_+ \sum_{i \neq 0} \Delta_{k,i} m_i + 2v_- \sum_{i \neq 0} \Delta_{l,i} m_i.$$

Note that $\beta_{+-}^0$ provides a reference value for $\beta_{i,j}$ for ions with charges of opposite sign, whereas none is available for ions with the same charge.

These exercises indicate that the theory presented in section 31, with $\beta_{i,j}$ not assumed to be zero for ions of opposite charge, is sufficient to account rigorously for the activity coefficients of mixtures of electrolytes through order $m$.

6. For the sulfuric acid solutions considered in section 33,

(a) Express the stoichiometric concentrations of hydrogen and sulfate ions in terms of the microscopic concentrations of hydrogen, bisulfate, and sulfate ions.

(b) Show that the Gibbs-Duhem equation applies to the model system:

$$\sum_i c_i^* \, d\mu_i = 0$$

at constant temperature and pressure.

(c) Show from equation 33-3 that as the ionic strength approaches zero

$$c_{H^+}^* \longrightarrow c_{H^+}, \qquad c_{SO_4^=}^* \longrightarrow c_{SO_4^=}, \qquad c_{HSO_4^-}^* \longrightarrow \frac{c_H \cdot c_{SO_4^=}}{K},$$

where $K$ is as yet undetermined. Assume that equation 14-6 applies to the model system.

(d) Apply equation 14-6 and the result of part (c) to show that

$$(a_{H^+}^\theta)^2 a_{SO_4^=}^\theta = (a_{H^+}^*)^2 a_{SO_4^=}^*.$$

(e) Show from part (d) and (a) that

$$(f_{H^+}^*)^2 f_{SO_4^=}^* = \frac{c_{H^+}^2 \cdot c_{SO_4^=}}{(c_{H^+}^*)^2 c_{SO_4^=}^*} f_{H^+, SO_4^=}^3$$

$$= \frac{c_{H^+}^2 \cdot c_{SO_4^=} f_{H^+, SO_4^=}^3}{(c_{H^+} - c_{HSO_4^-}^*)^2 (c_{SO_4^=} - c_{HSO_4^-}^*)},$$

where $f_{H^+, SO_4^=}$ is the mean molar activity coefficient of hydrogen and sulfate ions and is measurable by thermodynamic means.

(f) If $c_{HSO_4^-}^*$ can be measured at one stoichiometric concentration, then we know, from the above result, the value of $(f_{H^+}^*)^2 f_{SO_4^=}^*$ at the same concentration. Show that

$$K f_{H^+}^* \cdot f_{HSO_4^-}^* = \frac{c_{H^+}^2 \cdot c_{SO_4^=} f_{H^+, SO_4^=}^3}{(c_{H^+} - c_{HSO_4^-}^*) c_{HSO_4^-}^*}$$

and discuss how to get $K$ and $f_{H^+}^* \cdot f_{HSO_4^-}^*$ separately if $c_{HSO_4^-}^*$ is known as a function of stoichiometric concentration.

Note that in this problem we have used only the activity coefficients of neutral combinations of ions, even for the model system.

## NOTATION

$a$        mean diameter of ions, cm

$a_i^*$       property expressing secondary reference state, for microscopic point of view, liter/mole

$A$       Helmholtz free energy, J

$A_n$      coefficients in expression of thermodynamic properties of a solution of a single electrolyte

$B$       Debye-Hückel parameter, $(kg/mole)^{1/2}/Å$

$B'$      Debye-Hückel parameter, $(liter/mole)^{1/2}/Å$

$c_i$      molar concentration of species $i$, mole/liter

$e$       electronic charge, $1.60210 \times 10^{-19}$ C

$f_i$      molar activity coefficient of species $i$

$F$       Faraday's constant, 96,487 C/equiv

$G$       Gibbs free energy, J

$\bar{H}_i$      partial molar enthalpy of species $i$, J/mole

$I$       molal ionic strength, mole/kg

$I'$      molar ionic strength, mole/liter

$I_r$      "true" ionic strength, mole/liter

$K$       dissociation constant, mole/liter

$K'$      dissociation constant, mole/liter

$L$       Avogadro's number, $6.0225 \times 10^{23}$/mole

$m$      molality of a single electrolyte, mole/kg

$m_i$      molality of species $i$, mole/kg

$M_i$      molecular weight of species $i$, g/mole

$n_i$      number of moles of species $i$, mole

$p_i$      partial pressure or fugacity of species $i$, atm

$q$       charge, C

$r$       radial position coordinate, cm

$R$       universal gas constant, 8.3143 J/mole-deg

$t_i^0$      transference number of species $i$ with respect to the velocity of species 0

$T$       absolute temperature, deg K

$U$       open-circuit cell potential, V

$U^\theta$      standard cell potential, V

$U'$      modified open-circuit potential, V

$V$       volume, $cm^3$

$w_1, w_2$   reversible work of electrical charging, J

$z_i$      charge number of species $i$

$\alpha$       Debye-Hückel constant, $(kg/mole)^{1/2}$

$\alpha'$      Debye-Hückel constant, $(liter/mole)^{1/2}$

$\beta_{i,j}$     coefficient for ion-ion specific interactions, kg/mole

| | |
|---|---|
| $\gamma_i$ | molal activity coefficient of species $i$ |
| $\gamma_{+-}$ | mean molal activity coefficient of an electrolyte |
| $\epsilon$ | permittivity, farad/cm |
| $\lambda$ | Debye length, cm |
| $\lambda_i$ | absolute activity of species $i$ |
| $\lambda_i^\theta$ | property expressing secondary reference state, kg/mole |
| $\mu_i$ | chemical or electrochemical potential of species $i$, J/mole |
| $\nu$ | number of ions into which a molecule of electrolyte dissociates |
| $\nu_+, \nu_-$ | numbers of cations and anions into which a molecule of electrolyte dissociates |
| $\xi$ | fraction of charge |
| $\rho_0$ | density of pure solvent, g/cm$^3$ |
| $\rho_e$ | electric charge density, C/cm$^3$ |
| $\sigma$ | see equation 28-24 |
| $\tau$ | see equation 28-5 |
| $\phi$ | osmotic coefficient |
| $\Phi$ | electric potential, V |

subscripts

| | |
|---|---|
| el | electrical |
| 0 | solvent |
| * | from a microscopic point of view |
| $\infty$ | in the bulk solution |

# Reference Electrodes     **5**

In many applications, the ability to assess the potential in the solution is important; this is the primary purpose of reference electrodes. The estimation of potentials across liquid junctions, treated in chapter 6, is directly related to the use of reference electrodes.

Since the absolute potential of a single electrode cannot be measured, all potential measurements in electrochemical systems are performed with a reference electrode. In order to obtain meaningful results, the reference electrode should be reversible, and its potential should remain constant during the measurement. Theoretically, any electrode in an equilibrium state can be used as a reference electrode if its thermodynamic properties are known. However, no real electrode is ideal or has a completely reversible equilibrium potential. Since some electrodes are more reversible and easier to reproduce than others, they are more suitable as reference electrodes.

The purpose of this chapter is to discuss how to select a suitable reference electrode. Some important factors which will cause the electrode potential to deviate from the equilibrium potential will be discussed, and some reference electrodes which are commonly used in electrochemical measurements are introduced.

The material here follows mainly from the book edited by Ives and Janz.[1] A recent review on reference electrodes in nonaqueous solvents has been compiled by Butler.[2]

[1] David J. G. Ives and George J. Janz, eds, *Reference Electrodes* (New York: Academic Press, 1961).

[2] James N. Butler, "Reference Electrodes in Aprotic Organic Solvents," *Advances in Electrochemistry and Electrochemical Engineering*, 7 (1970), 77–175.

## 34. Criteria for reference electrodes

An ideal reference electrode should be reversible and reproducible. In other words, the species which can cross the phase boundary of the reference electrode should exist in equilibrium in both phases of the half cell, and this equilibrium should not be disturbed during the measurement. Practically, this ideal case is impossible to obtain. One can only select a reference electrode for which the deviation from the ideal case is small enough to suit his experimental work. In this section, we discuss the causes of deviation of a reference electrode from the ideal case and how to test a reference electrode.

Even though a reference electrode is carefully selected so that there is no spontaneous reaction between the electrode and the solution, some irreversible reactions still occur during the measurement. Because all potential-detection systems are operated by current, a certain amount of current, even though very small, must be passed through the cell. This current will cause an irreversible reaction to occur at the reference electrode and thus disturbs its equilibrium state. When a current is passed through an electrode, an overpotential representing the deviation from the equilibrium potential is set up and introduces an error in the desired measurement. If the current density is very small, the relation of current density $i$ to the overpotential $\eta_s$ can be represented by the equation (see equation 8-2 and section 56)

$$i = i_0 \frac{(\alpha_a + \alpha_c)F}{RT} \eta_s, \tag{34-1}$$

where $i_0$ is the exchange current density. This equation shows that, if a certain amount of current is passed through an electrode, the overpotential decreases as the exchange current density increases. In other words, electrodes with larger exchange current densities are more suitable for reference electrodes. The larger the surface area of a reference electrode, the smaller the current density for a given amount of current, and the smaller the overpotential will be.

A second source of error can arise from liquid junctions. If liquid junctions exist inside the cell, the measured potential will include the liquid-junction potentials. A cell without liquid junction means that the solution inside the cell is homogeneous. An example is given in equation 15-1. Even in the hydrogen/silver-silver chloride cell 18-12, the solution is not completely homogeneous, since the silver ions should not reach the hydrogen electrode. Normally, the liquid-junction potential in this cell can be neglected, although in dilute solutions it becomes appreciable (see table 18-1).

Other types of cells are even less ideal in this regard. The liquid-junction potentials are established by the activity gradients of species across the cell and can be represented by the integral of equation 43-1. This equation shows that the liquid-junction potential decreases as the difference between the solutions across the junction decreases. Because of the low solubilities of the reactants, the liquid-junction potential of the cell containing the hydrogen electrode, or electrodes of the second kind with a sparingly soluble salt, is small. Therefore, they are more suitable for the purpose of reference electrodes.

For convenience, some cells with liquid junctions are also used for electrochemical measurements. The estimation or minimization of these liquid-junction potentials is treated in chapter 2 and in more detail in chapter 6.

As a third source of error, the equilibrium potential of an electrode can be affected by impurities. The impurities inside the electrode can change the electrode activity, or react with the electrolyte, or react with the electrode.

The impurities can affect the electrode potential in the following ways:

1. They can react corrosively with the electrode, thus disturbing the equilibrium of the electrode and shifting the potential (see section 61). This is a second reason why the exchange current density for the desired electrode reaction should be high, so that the open-circuit potential is determined by the desired reaction and not by chance impurities. To achieve this high exchange current density, the concentration of the reactant should be much larger than the concentration of impurities.

Sometimes, the products of the reaction of impurities are insoluble in the solution and deposit on the electrode covering the electrode surface, thus changing the electrode properties. The most common impurity of this kind is oxygen dissolved in the solution. Some electrodes, such as amalgam electrodes, are extremely sensitive to a trace of oxygen dissolved in the solution.

Some electrodes, such as the hydrogen electrode, need the catalytic action of a metal surface in order to establish equilibrium. If this catalyst is poisoned by impurities, the equilibrium state of the electrode cannot be attained.

2. The impurities in the solution may change the activity of the reacting species. Some impurities have a strong tendency to form complexes with the reacting species in the solution, thus changing the electrode potential. The electrode potential in nonaqueous solutions can be affected seriously by the presence of even a trace of water.

3. Impurities can change the properties of the electrolyte. Some electrodes are very sensitive to the pH of the solution. Impurities such as carbon dioxide in the solution can change the pH of the solution considerably if the solution is near neutral and unbuffered.

## 35. Experimental factors affecting the selection of
## reference electrodes

We mentioned in section 34 that, theoretically, an electrode of the second kind with a large exchange current density is suitable as a reference electrode. Experimentally, a good reference electrode should be reproducible, constant in time, and easy to prepare. The reproducibility and stability of an electrode depend on the purity and sometimes on the surface condition of the metal. Discussion of the purification of chemicals is beyond the scope of this book, but the purity of materials is an important consideration. The common methods of treating a metal surface before it is used as an electrode are:

1. Clean and smooth the metal surface mechanically, either by polishing with sandpaper or by scraping.

2. Smooth the metal surface by electrochemical polishing.

3. Clean the electrode surface by prepolarization.

The last method is very effective in removing any oxide layer.

To eliminate all the impurities from the solution and make the metal surface completely reproducible would require a large amount of work. Electrodes that are much less sensitive to impurities and metal-surface condition are easier to prepare to obtain the same degree of accuracy and should be chosen as reference electrodes.

There is no definite rule about the selection of reference electrodes. A literature search is the best way to find a suitable reference electrode which has been used in the system of interest. In the literature, usually the method of preparation of the reference electrode is given, as well as its reproducibility and stability. If no suitable reference electrode can be found in the literature, the only way to obtain a good reference electrode is by trial and error. Even if the method of preparation of a reference electrode is given in the literature, the reference electrode should be tested before use; the simplest method is to put several electrodes prepared by the same method into the same solution with a different kind of reference electrode. The potential difference between the identical electrodes is a test of reproducibility; the potential difference between them and the different reference electrode as a function of time is a test of stability. The dependence of the behavior of the electrode on the metal surface can be detected in this way by measuring the potential difference between electrodes prepared by different methods.

Generally, a reference electrode should be selected that is reversible to one of the ions in the solution, in order to avoid liquid junctions. However, such an electrode may be difficult to prepare, or unreproducible, or may not exist. Or the solution in question may be of such complex composition that simultaneous reactions are unavoidable. In such a case, a well-behaved reference electrode in its own solution must be connected to the solution in question by means of a liquid junction. The estimation of the resulting liquid-junction potentials then introduces less uncertainty than the use of an unreliable reference electrode.

## 36. The hydrogen electrode

Some reference electrodes commonly used in electrochemistry are introduced in this and the following three sections. It is not the purpose of these sections to present a complete process for preparing a reference electrode. These commonly used reference electrodes are used only as examples to show what important factors should be considered in the selection and preparation of reference electrodes. Therefore, the methods of purification of the chemicals and the structure of the cells, which are very important in experimental work, are not included.

The hydrogen electrode is the best reference electrode in aqueous solutions, not only because its potential is universally adopted as the primary standard with which all other electrodes are compared, but also because it is easy to prepare and capable of the highest degree of reproducibility. Another advantage of the hydrogen electrode is that it has a broad field of application. It can be used over large ranges of temperature, pressure, and pH, and in many nonaqueous or partly aqueous solutions. The disadvantage of the hydrogen electrode is that its equilibrium depends on the catalytic activity of the metal surface. Thus, its reproducibility and stability are affected by the condition and the aging effect of the metal surface.

The reaction mechanism of the hydrogen electrode is still not clear and free from arguments. However, one can see that, before they can undergo the electrochemical reaction, the hydrogen molecules dissolved in the solution must first dissociate into hydrogen atoms:

$$H_2(\text{aq soln}) \rightleftharpoons 2\ H(\text{adsorbed on metal surface}) \rightleftharpoons 2e^- + 2\ H^+(\text{aq soln}).$$
$$(36\text{-}1)$$

Since the dissociation reaction of hydrogen molecules has a very high activation energy comparable to the heat of dissociation (103.2 kcal/mole), the equilibrium can be established only with the aid of a catalyst. Therefore, the metal phase in the hydrogen electrode not only conducts electrons but also acts as a catalyst.

The general requirements for a good metal for the hydrogen electrode are summarized as follows:

1. The metal must be noble and must not itself react or dissolve in the solution.

2. The metal must be a good catalyst for the hydrogen dissociation

reaction; that is, the metal can adsorb hydrogen atoms on the surface but will not react with them to form a stable hydride.

3. The metal should not absorb the hydrogen atoms into its crystal lattice, or the equilibrium of the hydrogen electrode will be disturbed.

4. The metal surface should be made by finely divided deposits. Because the catalytic activity of the metal surface is associated with the crystal imperfections, the metal surface made by finely divided deposits increases not only the real surface area but also the active catalytic sites.

5. In nonaqueous or partly aqueous solutions, the metal must not promote the undesired nonelectrochemical hydrogenation reactions.

Palladium is the best catalyst for the hydrogen dissociation reaction but is not suitable for the hydrogen electrode because a large amount of hydrogen atoms can penetrate into the metal phase; these hydrogen atoms then become inaccessible to the liquid phase, with which they are required to remain in equilibrium. A thin layer of palladium deposited on gold or platinum is satisfactory. Platinized platinum, because of its large surface area, although slightly permeable to hydrogen atoms, is the best metal for the hydrogen electrode. In those cases in which the presence of platinized platinum in the solution will promote some undesirable hydrogenation reactions in nonaqueous or partly aqueous solutions, bright platinum or gold can be used. The surface of bright platinum or gold should be activated by anodic treatment or by chemical treatment with strong oxidizing reagents, such as chromic acid or aqua regia. Transition metals are also suitable as catalysts for the hydrogen dissociation reaction because of their incompletely filled $d$ orbitals.

Some impurities carried by the hydrogen gas are undesirable and should be eliminated. Oxygen will participate in a corrosion reaction on the metal surface. This oxide reacts with the dissolved hydrogen and decreases the concentration of the dissolved hydrogen gas in the neighborhood of the electrode, thus making the potential of the hydrogen electrode shift to the positive side. However, oxygen can activate the catalytic activity of the metal and prolong the life of the electrode. Therefore, a trace of dissolved oxygen in the solution is desirable when a bright platinum or gold electrode is used. Carbon dioxide dissolved in the solution can change the pH of the solution. Other impurities, such as arsenic and sulfur compounds, can act as catalyst poisons and shorten the life of the electrode.

Hydrogen gas directly generated by the electrolytic process is very pure but is unsuitable for hydrogen electrodes because it carries some solution and is not free of oxygen. Pure commercial hydrogen after deoxygenation and passing through a *dust trap* containing potassium hydroxide, which also serves as a carbon dioxide absorbent, is satisfactory. The most common deoxygenation process is to pass the hydrogen gas through a commercial deoxygenating cartridge containing platinum catalysts that are active at room temperature. The other method of deoxygenation is to pass the hydrogen gas through a clean vitreous silica tube containing hot reduced copper (450 to 700°C) or hot palladized or platinum asbestos (200°C).

If rubber tubing is used for the connection, it should be pretreated by boiling in caustic soda solution, thoroughly washed, and then aged for 24 hours in hydrogen gas, because sulfur compounds may come from the rubber tubing.

To maintain a constant concentration of the electrolyte, the hydrogen gas should be presaturated with the solvent at the same vapor pressure as that of the electrolytic solution at the same temperature before it enters the cell.

The stability of hydrogen electrodes can be affected by the presence of impurities in three additional ways:

1. The impurities themselves can be reduced by the dissolved hydrogen gas to form soluble products. This will seriously deplete the concentration of the molecular hydrogen in the solution, and the potential of the hydrogen electrode will be displaced positively. Dissolved oxygen gas, $CrO_4^=$, $Fe^{+++}$, *etc.*, all fall into this category.

2. Some cations of metals—such as silver, mercury, copper, lead, *etc.*—can be reduced and deposit as solids, covering the electrode surface and changing the properties of the electrode.

3. The impurities, such as arsenic and sulfur compounds and some organic compounds, can be adsorbed on the active centers of the metal surface and can paralyze the catalytic activity.

Even in the absence of impurities, the catalytic activity of the metal surface can be destroyed by the *hydrogen poison*, which means that the active centers on the metal surface are *burned out* by the combination reaction of the hydrogen atoms. Since the catalytic activity of the metal surface is associated with the crystal imperfections, the active sites are in a higher energy state than the perfect crystal surface. With the help of a large amount of energy delivered by the combination reaction of hydrogen atoms, the active centers can return to the lower energy state (perfect crystal surface) and lose their catalytic activity. An aged or fatigued hydrogen electrode should be reactivated or replatinized before being used again.

In aqueous solutions, the hydrogen electrode can be used in a wide range of pH. It has been used in alkali hydroxide solutions up to a molality of 4 mole/kg and in sulfuric acid solutions up to a molality of 17.5 mole/kg. However, it fails in neutral solutions in the absence of buffers. The hydrogen electrode potential is very sensitive to the pH. In neutral solutions with even a trace of current passed through the cell, the pH of the solution in the neighborhood of the electrode will change considerably.

The pressure of the hydrogen gas can be measured accurately with a barometer. However, in most cell designs, hydrogen gas is bubbled through the solution. Therefore, the effective pressure of the hydrogen gas should be used in potential calculations. For aqueous solutions, the effective pressure can be calculated from the empirical equation:

$$p_{H_2} = p_{bar} - p_{soln} + \frac{0.4\,h}{13.6},$$ (36-2)

where $p_{bar}$ is the barometric pressure (mm Hg), $p_{soln}$ is the vapor pressure of the solution (mm Hg), and $h$ is the depth of immersion of the bubbler (mm). Recall that fugacities are to be used in potential calculations (see problem 2-14).

The hydrogen electrode can also be used in many nonaqueous and partly aqueous solutions, especially alcoholic solutions.

## 37. The calomel electrode and other mercury-mercurous salt electrodes

Mercury is a noble liquid metal, easy to purify, with a completely reproducible surface. Therefore, it is considered to be the best electrode metal. Many mercurous salts have a very low solubility in water and are suitable for the preparation of an electrode of the second kind. However, these advantages of mercury-mercurous salt electrodes are offset by the fact that mercury has two valence states, and all mercurous salts can disproportionate. The calomel electrode is the most common of all the mercury-mercurous salt electrodes, and is used as an example for the discussion in this section.

The calomel electrode was first introduced by Ostwald in 1890. However, no reproducible potential could be obtained, except in saturated KCl solution, and it was rejected for a period of twenty years. In later studies, it was found that many precautions must be taken in order to obtain a reproducible calomel electrode, as summarized below.

1. The interaction between the calomel and the mercury surface. If calomel is added to the mercury already covered by solution, no satisfactory reversibility can be obtained. If very finely divided calomel contacts the mercury surface in the dry state, it spreads rapidly, almost violently, over the whole surface and forms a pearly skin. This pearly skin is preserved after solution is added and gives a reproducible electrode. It is suggested that a calomel electrode prepared in this way has a monolayer of calomel molecules covering the mercury surface, with their chlorine atoms covalently bonded to the mercury surface. The calomel molecules act as a two-dimensional gas, free to move along the mercury surface and able to sustain fast exchange equilibria.

To prepare a reproducible calomel electrode, the very finely divided calomel should first be mixed with mercury to form *calomel-mercury paste;* then this paste is added to the mercury surface in small amounts until the whole surface is covered with *pearly calomel skin.*

Excess calomel on the mercury surface segregates the solution on the mercury surface from the bulk of the solution. This magnifies the results of any residual nonequilibrium in the system and makes an untidy electrode. The potential is sensitive to movement and is generally unreproducible. A very thin layer of calomel covering the mercury surface is all that is required for a good calomel electrode.

It was found that the coarse size of calomel particles is not suitable for the calomel electrode. It causes a positive deviation of potential, slow to decay, and leads to erratic behavior. Very finely divided (0.1 to 0.5 $\mu$) calomel particles prepared by the chemical precipitation method are satisfactory. The calomel should be stored in a dark place with exclusion of moisture and air before use.

2. The wedge effect. Aqueous solutions can penetrate into the space between the mercury and the glass wall by capillary action, called the *wedge effect*. This will create a large area of thin liquid film contacting the mercury. The properties of this liquid film may not be the same as those in the bulk of the solution. Because of the large area, this thin liquid film can affect the behavior of the electrode seriously.

The wedge effect can be eliminated by rendering the electrode vessel hydrophobic by treatment with a silicone preparation. Any platinum-glass seal should be avoided. The electric connection is formed by filling a capillary tube with mercury and making electric contact with a removable platinum wire in the remote end.

3. Dissolved oxygen. If the solution contains dissolved oxygen, a corrosion reaction will occur:

$$2\,Hg + 2\,HCl + \tfrac{1}{2}O_2 \longrightarrow Hg_2Cl_2 + H_2O, \qquad (37\text{-}1)$$

shifting the potential positively toward that of the oxygen electrode. In addition, the reaction will result in a gradual change in the properties of the electrolytic solution in the neighborhood of the electrode. This reaction had been known for a long time, but it was thought to be of significant effect only in low concentrations of HCl. Later it was found that this reaction can affect the electrode potential seriously, not only in a high concentration of HCl solution but also in KCl solution.

4. Disproportionation reaction. Because of the two valency states of mercury, mercuric ions are produced by a disproportionation reaction. Mercuric ions have a tendency to form complexes. Consequently, two distinct potentials tend to result for the calomel electrode. When a half cell $(Hg/Hg_2Cl_2/HCl)$ is freshly set up, thermal and other equilibria (diffusion, adsorption, *etc.*) will be established at a normal rate; and the potential of the

cell will level off to a value that is constant within 10 $\mu V$ over several hours. This is the potential corresponding to the *metastable calomel electrode*. In this case, the solution of the cell is still free from mercuric ions.

The potential of the *metastable calomel electrode* is not constant. It will increase to a higher value, corresponding to the potential of the *stable calomel electrode*, in which a complete equilibrium among mercury, calomel, and mercuric entities in the solution is established. The difference of the potential between these two calomel electrodes is as much as 0.24 mV at 25°C.

The calomel electrode is best used in acid solution (HCl). The standard potential has been determined over a range of temperatures by a number of workers. Since nearly all potential measurements performed before 1922 were affected by the dissolved oxygen, the standard potential of the calomel electrode is not determined without argument because of the limited amount of reliable experimental data.

Due to the disproportionation reaction, two types of calomel electrodes have been distinguished. The *metastable calomel electrode* is satisfactory for isothermal measurements not extending over a long interval of time, applied to the low ranges of concentration and temperature ($c \leq 0.1$ $N$; $t \leq 25$°C). The *stable calomel electrode* is a very sluggish electrode, but its use is obligatory in the higher ranges of concentration and temperature, in which, however, its useful life is limited.

The calomel electrode is also commonly used as a standard half cell of fixed potential. In this case, concentrated neutral KCl solution is used rather than dilute HCl, because it is very seldom that the fixed-potential half cells are required to show a reproducibility better than 0.1 mV and because the disproportionation reaction of the calomel and the oxidation reaction are much slower in concentrated neutral KCl solution.

Among the mercurous salt electrodes, the mercury-mercurous sulfate electrode is second in popularity to the calomel electrode. It is, of course, reversible to sulfate ions. Because mercurous fluoride can be rapidly and completely hydrolyzed in aqueous solution, it is very seldom used as a reference electrode in aqueous solution. However, it has been used in some nonaqueous solvents, such as liquid hydrogen fluoride.

The behavior of the mercury-mercurous bromide and iodide electrodes is similar to that of the calomel electrode. However, they have two disadvantages compared to the calomel electrode: they are more photosensitive, especially to ultraviolet light; and, while the solubility products decrease markedly in the order from chloride to bromide to iodide, the formation constants of the corresponding complex mercuric halides increase rapidly in the same order and progressively restrict the range of halide concentration in which potential measurements can usefully be made. Therefore, they are used only in special cases.

Mercurous phosphate, iodate, and acetate have also been used as the bases of reference electrodes.

## 38. The mercury-mercuric oxide electrode

As the calomel electrode is commonly used in acid solutions, so the mercury-mercuric oxide electrode (Hg/HgO/OH$^-$) is commonly used in alkaline solutions. Since mercurous oxide does not exist, there is no disturbing effect due to a variable valence of the mercuric oxide. The formal acidic and basic dissociation constants of mercuric oxide have been estimated to be very small. It is more basic than acidic. Therefore, the usefulness of the mercury-mercuric oxide electrode is confined to alkaline solutions.

The mercury-mercuric oxide electrode has a relatively long life, is stable for several days, and is reproducible to better than $\pm0.1$ mV. It is easy to prepare; no special precaution is needed if the chemicals are reasonably pure.

The standard potential of the mercury-mercuric oxide electrode has been investigated intensively by many authors. All data agree within 0.4 mV, which indicates that the mercury-mercuric oxide electrode is well behaved.

## 39. Silver-silver halide electrodes

Due to the low solubility of silver halides, silver-silver halide electrodes are electrodes of the second kind. Among them, the most common is the silver-silver chloride electrode (Ag/AgCl($s$)/Cl$^-$), which is reversible to the chloride ion. The relationship of its standard electrode potential to that of the silver electrode and the solubility product of silver chloride was treated in problem 2-13.

The advantages of the silver-silver chloride electrodes are that they are small and compact, can be used in any orientation, and usually do not significantly contaminate any medium into which they are immersed. The disadvantage is that their thermodynamic properties depend on the physical properties of the solid phases, such as mechanical strain and crystal structure, and thus depend on the method of preparation.

There is still no method to prepare a perfect silver-silver halide electrode. Three methods are commonly used in experimental work:

1. Electrolytic. Platinum metal is generally used as the electrode base on which a layer of silver is electrodeposited from a solution of KAg(CN)$_2$. After thorough washing, the silver-plated electrode is halidized anodically; in 0.1 $N$ HCl solution for a silver chloride electrode and, for silver bromide and silver iodide electrodes, in 0.1 $N$ KBr or KI solutions, sometimes made

weakly acidic by adding the appropriate acid. About 10 percent of the silver is halidized. The reproducibility of silver-silver halide electrodes prepared by this method should be within $\pm 0.02$ mV.

2. Thermal. The thermal electrodes are prepared by the decomposition of a mixture of silver oxide and silver chlorate, bromate, or iodate, the proportion being approximately 90 percent silver oxide by weight. Conductance water is added to the mixture to form a smooth paste. A platinum wire spiral is covered with the paste, heated to the decomposition temperature of 650°C, and then slowly cooled to room temperature.

Silver iodide is sometimes used in the preparation of the silver-silver iodide electrodes (and heated with $Ag_2O$ to 450°C for 10 to 15 minutes), and a mixture of silver oxide and silver perchlorate has also been used in the preparation of silver chloride electrodes.

The thermal silver-silver halide electrodes are less reliable than the electrolytic or thermal-electrolytic electrodes, probably because the surface condition of the latter types is more reproducible.

3. Thermal-electrolytic. By this method, silver is first prepared by decomposing silver oxide and then halidizing by the electrolytic process. The reproducibility of the silver-silver chloride electrode prepared by this method should be within 0.04 mV.

Several less common methods, which we shall not consider here, have also been used in the preparation of silver-silver halide electrodes.

The standard potential of the silver-silver chloride electrode has been thoroughly investigated from the cell

$$H_2 \mid HCl, \ H_2O \mid AgCl(s) \mid Ag. \tag{39-1}$$

In the temperature range from 0 to 95°C, the results can be expressed as

$$U^\theta = 0.23659 - 4.8564 \times 10^{-4} \, t - 3.4205 \times 10^{-6} \, t^2 + 5.869 \times 10^{-9} \, t^3 (V), \tag{39-2}$$

where $t$ is temperature (°C). The standard potentials of the silver bromide and silver iodide electrodes have also been investigated over a range of temperature, although that of the AgI electrode has been more difficult to establish by direct measurement because of experimental difficulties, such as the oxidation of the HI solution.

Certain impurities, such as iodide and sulfide, can form silver salts with solubilities lower than that of silver chloride or silver bromide and can deposit on the electrode surface, thus changing the electrode potential. Even a trace of bromide in the solution will shift the potential of the silver-silver chloride electrode to the positive side. Oxygen dissolved in the electrolytic solutions can affect the behavior of the silver chloride and bromide electrodes if HCl or HBr is used as the electrolyte by means of the slow oxidation reaction:

$$2 \, Ag + 2 \, HCl + \tfrac{1}{2}O_2 \longrightarrow 2 \, AgCl + H_2O. \tag{39-3}$$

For the silver iodide electrode, a marked oxygen effect is noted in both neutral and acidic solutions.

All silver-silver halide electrodes are subject to the aging effect. The potentials of the older electrodes are slightly positive relative to the new electrodes. This effect is always in the same direction and of the same order of magnitude, about 0.05 mV. The potential of a freshly prepared silver-silver halide electrode increases slowly and will reach a stable value. The period of the aging effect varies from a few minutes to one to twenty days. This effect has been attributed to a concentration polarization associated with the electrolytic halidization or the initial immersion of the electrode in the solution being investigated.

Silver-silver halide electrodes are widely used in electrochemical measurements because they are compact and easy to set up. The most important application of the silver-silver chloride electrode is in the investigation of the thermodynamic properties of electrolytes, such as the standard electrode potential and the activity coefficients. Silver-silver halide electrodes can also be applied in nonaqueous solutions. However, this application is limited by the strong tendency of the complex formation of the silver ion, which will greatly increase the solubility of the silver halides. Silver chloride electrodes have important applications in the investigation of the behavior of biological membranes.

## 40. Potentials relative to a given reference electrode

Not infrequently one encounters in the literature a set of potentials of various electrodes, in various solutions, referred to a given reference electrode, for example, to a normal calomel electrode in KCl. It is usually noted that these potentials are *corrected for liquid-junction potentials*, whatever that might mean. Let us inquire into what quantity is tabulated, and why such a tabulation might be useful.

Let $U'$ be "the potential of a given electrode relative to a normal calomel electrode in KCl, corrected for liquid-junction potentials." The half cell of interest is

$$
\begin{array}{c|c|c}
\delta & \beta & \alpha' \\
\text{solution} & \text{electrode} & \text{Pt}(s) \\
(e.g.,\ 0.1\ N\ \text{NaCl}) & (e.g.,\ \text{Hg}) &
\end{array}
\qquad (40\text{-}1)
$$

If $U'$ is to be a thermodynamic quantity of interest, it must assess the electrical state of electrode $\beta$ relative to solution $\delta$, since these are the only relevant phases.

We should want $U'$ to be a thermodynamic quantity, since we should have measured the potential $U$ relative to a well-defined electrode appropriate to the system; for example,

$$
\begin{array}{c|c|c|c|c|c}
\alpha & \phi & \epsilon & \delta & \beta & \alpha' \\
Pt(s) & Hg(l) & Hg_2Cl_2(s) & \begin{array}{c} 0.1\ N\ NaCl \\ in\ H_2O \end{array} & Hg(l) & Pt(s)
\end{array}
\qquad (40\text{-}2)
$$

and the introduction of any liquid-junction potential in the conversion to $U'$ would obviate such a useful measurement. The note, *corrected for liquid-junction potentials*, is a further hint that a thermodynamic quantity is intended, although it may also give the erroneous impression that a liquid junction was involved in the measurement.

By these arguments we are led to the conclusion that $U'$ is given by

$$
FU' = -\mu_{e^-}^{\beta} - F\Phi^{\delta} + \text{const}, \qquad (40\text{-}3)
$$

where $\Phi$ is the quasi-electrostatic potential relative to some ionic species $n$. (The outer potential $\Phi$ might have been intended, but then $U'$ would be a nonthermodynamic quantity.) It seems logical that ion $n$ should be either $Cl^-$ or $K^+$, since these ions are in the statement of the reference electrode (normal calomel electrode in KCl). The chloride ion is the more likely candidate since the calomel electrode responds to this ion.

In seeking the value of the constant in equation 3, we are led next to conclude that $U'$ is the potential of the system

$$
\begin{array}{c|c|c|c|c|c|c}
\alpha & \phi & \epsilon & \lambda & \delta & \beta & \alpha' \\
Pt(s) & Hg(l) & Hg_2Cl_2(s) & \begin{array}{c} 1\ N\ KCl\ \text{-}\text{-}\text{-} \\ in\ H_2O \end{array} & \begin{array}{c} \text{solution} \\ (e.g.,\ 0.1\ N\ NaCl) \end{array} & \begin{array}{c} \text{electrode} \\ (e.g.,\ Hg) \end{array} & Pt(s)
\end{array}
$$
$$
(40\text{-}4)
$$

The dashed line does not denote a junction region. Instead we make the (perhaps imaginary) requirement that the electrical states of solutions $\lambda$ and $\delta$ are related by

$$
\Phi^{\lambda} = \Phi^{\delta}. \qquad (40\text{-}5)
$$

The potential of this system is thus given by

$$
\begin{aligned}
FU' &= -F(\Phi^{\alpha} - \Phi^{\alpha'}) \\
&= -\mu_{e^-}^{\beta} - F\Phi^{\delta} + \mu_{Hg}^{\phi} - \tfrac{1}{2}\mu_{Hg_2Cl_2}^{\epsilon} + RT \ln c_{Cl^-}^{\lambda}. \qquad (40\text{-}6)
\end{aligned}
$$

Since phases $\phi$ and $\epsilon$ are pure phases, we thus determine that the constant in equation 3 is

$$
\text{const} = \mu_{Hg}^{0} - \tfrac{1}{2}\mu_{Hg_2Cl_2}^{0} + RT \ln c_{Cl^-}^{\lambda}, \qquad (40\text{-}7)
$$

where $c_{Cl^-}^{\lambda} = 1$ mole/liter.

Note that the electrode $\beta$ could be a reversible electrode at equilibrium, an ideally polarizable electrode (see section 49), an irreversible electrode, or an electrode undergoing one or more electrode reactions without affecting the above arguments. Similarly, the solution $\delta$ could be of any composition.

Now consider explicitly the system 2, for which the potential $U$ is given by

$$FU = -\mu_{e^-}^\beta + \mu_{Cl^-}^\delta + \mu_{Hg}^0 - \tfrac{1}{2}\mu_{Hg_2Cl_2}^0. \tag{40-8}$$

Hence, the difference is

$$FU - FU' = RT \ln \frac{c_{Cl^-}^\delta}{c_{Cl^-}^\lambda}. \tag{40-9}$$

The conversion from the measured value of $U$ to the desired value of $U'$ is thus very simple in this case.

Now let us take solution $\delta$ to be 0.1 $N$ $Na_2SO_4$ and let us suppose that $U$ is measured relative to a lead-lead sulfate electrode; that is, $U$ is now the potential of the system

$$
\begin{array}{c|c|c|c|c|c}
\alpha & \phi & \epsilon & \delta & \beta & \alpha' \\
Pt(s) & Pb(s) & PbSO_4(s) & 0.1\ N\ Na_2SO_4 & Hg(l) & Pt(s) \\
 & & & \text{in } H_2O & &
\end{array}
\qquad (40\text{-}10)
$$

that is,

$$FU = -\mu_{e^-}^\beta + \tfrac{1}{2}(\mu_{Pb}^\delta + \mu_{SO_4^=}^\delta - \mu_{PbSO_4}^0). \tag{40-11}$$

The difference between $U$ and $U'$ is now given by

$$FU - FU' = \tfrac{1}{2}(\mu_{Pb}^0 + \mu_{SO_4^=}^\delta - \mu_{PbSO_4}^0) + F\Phi^\delta$$
$$- \mu_{Hg}^0 + \tfrac{1}{2}\mu_{Hg_2Cl_2}^0 - RT \ln c_{Cl^-}^\lambda. \tag{40-12}$$

Substitution of equation 26-4 for $\mu_{SO_4^=}^\delta$ gives

$$FU - FU' = \frac{1}{2}(\mu_{Pb}^0 - \mu_{PbSO_4}^0 + RT \ln a_{SO_4^=}^\theta) - \mu_{Hg}^0$$
$$+ \frac{1}{2}\mu_{Hg_2Cl_2}^0 - RT \ln a_{Cl^-}^\theta + \frac{1}{2}RT \ln \frac{c_{SO_4^=}^\delta\, f_{SO_4^=}^\delta}{(c_{Cl^-}^\lambda\, f_{Cl^-}^\delta)^2}. \tag{40-13}$$

For standard electrode potentials referred to the hydrogen electrode, we have for the lead sulfate electrode (table 20-1)

$$FU_{PbSO_4}^\theta = \tfrac{1}{2}(\mu_{H_2}^* + \mu_{PbSO_4}^0 - \mu_{Pb}^0) - \tfrac{1}{2}RT \ln[(\lambda_{H^+}^\theta)^2\, \lambda_{SO_4}^\theta], \tag{40-14}$$
$$U^\theta = -0.356 \text{ V}, \tag{40-15}$$

and for the calomel electrode (table 20-2)

$$FU_{Hg_2Cl_2}^\theta = \tfrac{1}{2}(\mu_{H_2}^* + \mu_{Hg_2Cl_2}^0) - \mu_{Hg}^0 - RT \ln(\lambda_{H^+}^\theta\, \lambda_{Cl^-}^\theta), \tag{40-16}$$
$$U^\theta = 0.2676 \text{ V}. \tag{40-17}$$

Since (see equation 14-7)

$$\lambda_i^\theta = \rho_0 a_i^\theta, \tag{40-18}$$

equation 13 becomes

$$FU - FU' = FU_{Hg_2Cl_2}^\theta - FU_{PbSO_4}^\theta + \frac{1}{2}RT \ln\left[\frac{\rho_0 c_{SO_4^=}^\delta\, f_{SO_4^=}^\delta}{(c_{Cl^-}^\lambda\, f_{Cl^-}^\delta)^2}\right]. \tag{40-19}$$

Because of the conventions that have been adopted in establishing tables 20-1 and 20-2, $\rho_0$, the density of the pure solvent, should be expressed in $g/cm^3$ and the concentrations $c_i$ should be expressed in mole/liter.

We now need the activity coefficient of sulfate ions relative to chloride ions in the 0.1 $N$ $Na_2SO_4$ solution $\delta$. It should be emphasized that this is a thermodynamic quantity despite the fact that no chloride ions exist in this solution. Let us use Guggenheim's expression for multicomponent solutions (see section 31) as written on a concentration scale (see problem 4-4):

$$\ln f_i - \frac{z_i}{z_n} \ln f_n = -\frac{\alpha' z_i (z_i - z_n) \sqrt{I'}}{1 + B'a \sqrt{I'}} + 2 \sum_j \left( \beta'_{ij} - \frac{z_i}{z_n} \beta'_{nj} \right) c_j, \quad (40\text{-}20)$$

where $I'$ is given by equation 28-12 and $\alpha'$ is given by equation 28-15 or table 28-1. See problem 4-4 for the relationship of $\beta'$ to $\beta$. Then

$$\ln \frac{f^\delta_{SO_4^=}}{(f^\delta_{Cl^-})^2} = -\frac{2\alpha' \sqrt{I'}}{1 + B'a \sqrt{I'}} + 2(\beta'_{Na_2SO_4} - 2\beta'_{NaCl}) c^\delta_{Na^+}. \quad (40\text{-}21)$$

The calculation of the potential $U'$ from the measured potential $U$ is somewhat complicated in this case, but it introduces no nonthermodynamic concepts.

Finally, suppose that the potential $U''$ of the cell with liquid junction

$$
\begin{array}{c|c|c|c|c|c|c}
\alpha & \phi & \epsilon & \kappa & \delta & \beta & \alpha' \\
\text{Pt}(s) & \text{Hg}(l) & \text{Hg}_2\text{Cl}_2(s) & 0.1\ N\ \text{KCl} & 0.1\ N\ \text{NaCl} & \text{Hg}(l) & \text{Pt}(s) \\
 & & & \text{in H}_2\text{O} & \text{in H}_2\text{O} & &
\end{array}
\quad (40\text{-}22)
$$

has been measured. How should we obtain $U''$? The potential $U''$ is given by

$$FU'' = -F(\Phi^\alpha - \Phi^{\alpha'})$$
$$= -\mu^\beta_{e^-} - F\Phi^\kappa + \mu^\phi_{Hg} - \tfrac{1}{2}\mu^\epsilon_{Hg_2Cl_2} + RT \ln c^\kappa_{Cl^-}. \quad (40\text{-}23)$$

Hence,

$$FU'' - FU' = F(\Phi^\delta - \Phi^\kappa) + RT \ln \frac{c^\kappa_{Cl^-}}{c^\lambda_{Cl^-}}. \quad (40\text{-}24)$$

Thus, we must obtain an estimate of the liquid-junction potential $\Phi^\delta - \Phi^\kappa$ between solutions $\delta$ and $\kappa$, which is beyond the scope of the present chapter. For the last term, $c^\kappa_{Cl^-} = 0.1\ N$ and $c^\lambda_{Cl^-} = 1\ N$.

If solution $\kappa$ had been a 1 $N$ KCl solution, that is, the left electrode in cell 22 were a normal calomel electrode in KCl, then the last term in equation 24 would be zero, but we would instead have to estimate $\Phi^\delta - \Phi^\kappa$ for the junction between 1 $N$ KCl and 0.1 $N$ NaCl.

The result in equation 24 can be applied to the case where solution $\delta$ is different, say 0.1 $N$ Na$_2$SO$_4$. One still has to estimate the liquid-junction potential $\Phi^\delta - \Phi^\kappa$, referred to the chloride ion.

These examples make it clear that $U'$ represents a thermodynamic quantity. A *correction for liquid-junction potentials* is necessary only if a measurement was made on a cell with liquid junction; and the correction concerns the junction in that cell, which may not involve normal KCl at all. The examples also show that the definition of $U'$ could involve a normal calomel electrode in NaCl without changing the numerical values.

One may well ask why he should use potentials relative to a normal calomel electrode in KCl. Why not use a mercury-mercuric oxide electrode for solutions of KOH, a lead-lead sulfate electrode for solutions of Na$_2$SO$_4$, and a calomel electrode in NaCl for solutions of NaCl, always letting the electrolyte concentration in the reference electrode be the same as in the solution of interest? In at least two cases, some insight into the physical situation is afforded by the use of one given reference electrode. First, electrocapillary curves coincide on the negative branch when plotted against $U'$ (see figure 51-2), indicating that cations are not specifically adsorbed at a mercury-solution interface. Second, anodic current densities in a metal dissolution reaction are relatively independent of solution composition when plotted against $U'$ but not when plotted against, say, the surface overpotential $\eta_s$, which involves a shift of the equilibrium potential with the reactant concentration.

## NOTATION

| | |
|---|---|
| $a$ | mean diameter of ions, cm |
| $a_i^\theta$ | property expressing secondary reference state, liter/mole |
| $B'$ | Debye-Hückel parameter, $(\text{liter/mole})^{1/2}/\text{Å}$ |
| $c_i$ | molar concentration of species $i$, mole/liter |
| $f_i$ | molar activity coefficient of species $i$ |
| $F$ | Faraday's constant, 96,487 C/equiv |
| $h$ | depth of immersion of bubbler, mm |
| $i$ | current density, $\text{A/cm}^2$ |
| $i_0$ | exchange current density, $\text{A/cm}^2$ |
| $I'$ | molar ionic strength, mole/liter |
| $p$ | pressure, mm Hg |
| $p_i$ | partial pressure or fugacity of species $i$, mm Hg |
| $R$ | universal gas constant, 8.3143 J/mole-deg |
| $T$ | absolute temperature, deg K |
| $U$ | open-circuit cell potential, V |
| $U'$ | electrode potential relative to a given reference electrode, V |
| $U''$ | electrode potential measured with a liquid junction present, V |
| $U^\theta$ | standard electrode potential, V |
| $z_i$ | charge number of species $i$ |
| $\alpha_a, \alpha_c$ | transfer coefficients |
| $\alpha'$ | Debye-Hückel constant, $(\text{liter/mole})^{1/2}$ |
| $\beta'_{i,j}$ | coefficient for ion-ion specific interactions, liter/mole |
| $\eta_s$ | surface overpotential, V |
| $\lambda_i^\theta$ | property expressing secondary reference state, kg/mole |
| $\mu_i$ | chemical or electrochemical potential of species $i$, J/mole |
| $\rho_o$ | density of pure solvent, $\text{g/cm}^3$ |
| $\Phi$ | electric potential, V |

# Potentials of Cells with Junctions

**6**

In this chapter, we build on the material developed in chapter 2, on thermodynamics in terms of electrochemical potentials. Numerical values of cell potentials can now be calculated from standard electrode potentials, ionic concentrations, and methods for estimating activity coefficients and liquid-junction potentials.

Taylor[1] showed clearly that the problems of measuring liquid-junction potentials and individual ionic activity coefficients are inexorably tied up with each other. Also of interest here is the work of MacInnes,[2] Wagner,[3] and Smyrl and Newman.[4]

[1] Paul B. Taylor, "Electromotive Force of the Cell with Transference and Theory of Interdiffusion of Electrolytes," *The Journal of Physical Chemistry*, *31* (1927), 1478–1500.

[2] Duncan A. MacInnes, *The Principles of Electrochemistry* (New York: Dover Publications, Inc., 1961).

[3] Carl Wagner, "The Electromotive Force of Galvanic Cells Involving Phases of Locally Variable Composition," *Advances in Electrochemistry and Electrochemical Engineering*, *4* (1966), 1–46.

[4] William H. Smyrl and John Newman, "Potentials of Cells with Liquid Junctions," *Journal of Physical Chemistry*, *72* (1968), 4660–4671.

## 41.  The Nernst equation

The Nernst equation was defined at the end of section 18. We can state here a more definite procedure for arriving at the appropriate form of the Nernst equation for a particular cell. We use the cell 19-4 as an example.

1. Write down the expression for the cell potential using chemical potentials and electrochemical potentials as indicated in chapter 2. For cell 19-4, this is given by equation 19-5:

$$FU = \tfrac{1}{2}\mu_{H_2}^{\alpha} - \mu_{Ag}^{\lambda} + \mu_{AgCl}^{\epsilon} - \mu_{H^+}^{\beta} - \mu_{Cl^-}^{\delta}. \tag{41-1}$$

2. Use equation 26-4 to express the electrochemical potentials of ions in solution. For gaseous components, use the fugacity as outlined in problem 2-14. For the chemical potentials of pure phases, replace the superscript with 0. For alloys, use the expression for the activity, for example, equation 20-4. For cell 19-4, we now have

$$FU = \tfrac{1}{2}\mu_{H_2}^{*} - \mu_{Ag}^{0} + \mu_{AgCl}^{0} - RT \ln (a_{H^+}^{\theta} \cdot a_{Cl^-}^{\theta}) + \tfrac{1}{2}RT \ln p_{H_2}^{\alpha}$$
$$- RT \ln (c_{H^+}^{\beta} \cdot c_{Cl^-}^{\delta}) - RT \ln (f_{H^+}^{\beta} \cdot f_{Cl^-}^{\beta}) + F(\Phi^{\delta} - \Phi^{\beta}), \tag{41-2}$$

where the chloride ion has been chosen as species $n$ is using equation 26-4.

3. Identify the standard cell potential, using equation 14-7 where necessary. Equation 2 becomes

$$FU = FU^{\theta} + \tfrac{1}{2}RT \ln p_{H_2}^{\alpha} - RT \ln \frac{c_{H^+}^{\beta} \cdot c_{Cl^-}^{\delta}}{\rho_0^2}$$
$$- RT \ln (f_{H^+}^{\beta} \cdot f_{Cl^-}^{\beta}) + F(\Phi^{\delta} - \Phi^{\beta}), \tag{41-3}$$

where $FU^{\theta}$ corresponds to entry 7 in table 20-1.

4. Set all ionic activity coefficients equal to one, and neglect any difference in quasi-electrostatic potential between points in the solution. Equation 3 becomes

$$FU = FU^{\theta} + \tfrac{1}{2}RT \ln p_{H_2}^{\alpha} - RT \ln \frac{c_{H^+}^{\beta} \cdot c_{Cl^-}^{\delta}}{\rho_0^2}. \tag{41-4}$$

The potential difference $\Phi^{\delta} - \Phi^{\beta}$ in equations 2 and 3 can be called a liquid-junction potential. These quasi-electrostatic potentials are referred to the chloride ion. The ionic activity coefficients always appear in the equation in a manner that compensates for the arbitrary choice of species $n$. In writing the Nernst equation, both the liquid-junction potential and the ionic activity coefficients are discarded. It would be somewhat inconsistent to retain one but not the other in view of their dependence upon the choice of species $n$.

The rest of this chapter will provide a basis for assessing the error involved in the Nernst equation. We can assert at the outset that we should rather seek other approximations for the cells of section 17, involving a single electrolyte of varying concentration.

## 42. Types of liquid junctions

The potentials of cells with liquid junctions are assessed, using equation 16-3 to evaluate the variation of the electrochemical potential of an ion in the junction region. As already noted in chapter 2, the integration of this equation requires a knowledge of the concentration profiles except in the simple case of a two-component solution, an electrolyte and a solvent or two electrolytes with a common ion in a fused salt. Consequently, we discuss first the popular models of liquid junctions.

1. *Free-diffusion junction.* At time zero, the two solutions are brought into contact to form an intially sharp boundary in a long, vertical tube. The solutions are then allowed to diffuse into each other, and the thickness of the region of varying concentration increases with the square root of time. Even if the transport properties are concentration dependent and the activity coefficients are not unity, the potential of a cell containing such a junction should be independent of time.

2. *Restricted-diffusion junction.* The concentration profiles are allowed to reach a steady state by one-dimensional diffusion in the region between $x = 0$ and $x = L$, in the absence of convection. The composition at $x = 0$ is that of one solution and, at $x = L$, that of the other solution. The potential of a cell containing such a junction is independent of $L$ (as well as time). The condition of no convection is usually not specified (*i.e.*, zero solvent velocity or zero mass-average velocity, *etc.*).

3. *Continuous-mixture junction.* At all points in the junction, the concentrations (excluding, we suppose, that of the solvent) are assumed to be linear combinations of those of the solutions at the ends of the junction. This assumption obviates the problem of calculating the concentration profiles by the laws of diffusion.

4. *Flowing junction.* In some experiments the solutions are brought together and allowed to flow side by side for some distance. It is sometimes supposed that observed potentials should approximate those given by a free-diffusion boundary.

5. *Electrode of the second kind.* To these we add the region of varying composition produced when a sparingly soluble salt is brought into contact with a solution containing a common ion. We might use a model similar to the free-diffusion junction if we imagine the salt to be introduced at the bottom of a vertical tube containing the solution. The sparingly soluble salt will then diffuse up the tube, and the concentration at the bottom will be governed by the solubility product.

## 43. Formulas for liquid-junction potentials

Substitution of equation 26-4 into equation 16-1 yields (see equation 84-3)

$$F\nabla\Phi = -\frac{F}{\kappa}\mathbf{i} - RT\sum_i \frac{t_i^0}{z_i}\nabla\ln c_i - RT\sum_i \frac{t_i^0}{z_i}\nabla\left(\ln f_i - \frac{z_i}{z_n}\ln f_n\right), \quad (43\text{-}1)$$

where $\Phi$ is the quasi-electrostatic potential referred to species $n$. Integration of this equation across the junction region in the absence of current is the basis of the calculation of liquid-junction potentials.

For solutions so dilute that the activity coefficients can be ignored, it becomes immaterial which species is chosen for species $n$. In these dilute solutions, we can use equation 70-5 to express the transference numbers, with the result that

$$F\nabla\Phi = -\frac{F}{\kappa}\mathbf{i} - RT\frac{\sum_i z_i u_i \nabla c_i}{\sum_j z_j^2 u_j c_j}, \quad (43\text{-}2)$$

where $u_i$ is the mobility of species $i$.

It is now a relatively simple matter to perform the integration for the continuous-mixture junction, where the concentrations are given by

$$c_i = c_i^{II} + \xi(c_i^I - c_i^{II}) \quad (43\text{-}3)$$

and where $\xi$ varies from 0 in solution II to 1 in solution I. Equation 2, in the absence of current, becomes

$$F\nabla\Phi = -RT\frac{A\nabla\xi}{B^{II} + (B^I - B^{II})\xi}, \quad (43\text{-}4)$$

where

$$A = \sum_i z_i u_i(c_i^I - c_i^{II}), \qquad B^I = \sum_i z_i^2 u_i c_i^I, \qquad B^{II} = \sum_i z_i^2 u_i c_i^{II}. \quad (43\text{-}5)$$

Integration gives

$$\Phi^I - \Phi^{II} = -\frac{RT}{F}A\frac{\ln\dfrac{B^I}{B^{II}}}{B^I - B^{II}}. \quad (43\text{-}6)$$

This is the Henderson formula[5,6] for the junction potential of a continuous-mixture junction, valid under the conditions cited in its derivation. Because of its simplicity, it is useful for estimating liquid-junction potentials. The ionic mobilities $u_i$ in $A$ and $B$ can be replaced by the ionic diffusion coefficients $D_i$ (see table 75-1).

[5]P. Henderson, "Zur Thermodynamik der Flüssigkeitsketten," *Zeitschrift für physikalische Chemie*, *59* (1907), 118–127.

[6]P. Henderson, "Zur Thermodynamik der Flüssigkeitsketten," *Zeitschrift für physikalische Chemie*, *63* (1908), 325–345.

Planck[7,8] has obtained an implicit expression for the liquid-junction potential for the restricted-diffusion junction for univalent ions where activity coefficients can be ignored. MacInnes[2] has reproduced the derivation of Planck's formula.

The Goldman[9] constant-field equation for liquid-junction potentials is popular among biologists. Although its basis has been justifiably criticized,[10] it predicts values that are in reasonable accord with those obtained by other methods.

## 44. Determination of concentration profiles

The concentration profiles in the junction region are governed by the laws of diffusion in cases 1, 2, and 5 of section 42, the free-diffusion and restricted diffusion junctions and the electrode of the second kind. The transport laws are developed in section 69 for dilute solutions and in section 78 for concentrated solutions. We treat solutions so dilute that we can neglect the interaction of the diffusing species with the other components except the solvent. The appropriate form of the diffusion law is developed in section 77 (see also section 82 and reference 4). However, the activity coefficients are not assumed to be unity. Instead, Guggenheim's expression for dilute solutions of several electrolytes is used (see section 31 and problem 4-4).

To determine the concentration profiles in liquid junctions, then, involves solving the diffusion equation 77-10 or 82-2 in conjunction with the first equation of problem 4-4 and with the material-balance equation 69-3, the electroneutrality equation 69-4, and the condition of zero current.

This problem can be solved numerically for the various models of liquid junction. In the case of restricted diffusion, the equations are already ordinary differential equations. For free diffusion and for an electrode of the second kind, the similarity transformation $Y = y/\sqrt{t}$ reduces the problem to ordinary differential equations. These coupled, nonlinear, ordinary differential equations can be solved readily by the method outlined in appendix C. The equations can be linearized about a trial solution, producing a series of coupled, linear differential equations. In finite difference form, these give coupled, tridiagonal matrices that can be solved on a high-speed, digital computer. The nonlinear problem can then be solved by iteration.

[7] Max Planck, "Ueber die Erregung von Electricität und Wärme in Electrolyten," *Annalen der Physik und Chemie*, NF, *39* (1890), 161–186.

[8] Max Planck, "Ueber die Potentialdifferenz zwischen zwei verdünnten Lösungen binärer Electrolyte," *Annalen der Physik und Chemie*, NF, *40* (1890), 561–576.

[9] David E. Goldman, "Potential, Impedance, and Rectification in Membranes," *The Journal of General Physiology*, 27 (1943), 37–60.

[10] D. Allen Zelman, "An Analysis of the Constant Field Equation," *Journal of Theoretical Biology*, *18* (1968), 396–398.

## 45. Numerical results

We present here calculated[4] values of $\Delta\Phi$ for the several models (section 42) for the junctions between solutions of various compositions. No detailed concentration profiles will be given, since the potentials of cells with liquid junctions can be calculated directly from the tabulated values of $\Delta\Phi$, without further reference to the concentration profiles, as indicated in the next section. The tabulation of the values of $\Delta\Phi$, rather than the potentials of complete cells, is convenient because these values relate to the junction itself, whereas more than one combination of electrodes is possible for a given junction. In addition to $\Delta\Phi$, only thermodynamic data are needed to calculate potentials of complete cells, as the entire effect of the transport phenomena is included in $\Delta\Phi$.

The value of $\Delta\Phi$ depends upon the choice of the reference ion $n$. In each case, this is the last ion for a given junction in the tables. For infinitely dilute solutions, $\Delta\Phi$ becomes independent of this choice and, furthermore, depends only on the ratios of concentrations of the ions in the end solutions. Solutions of zero ionic strength ($f_i = 1$) are indicated by an asterisk, but the concentrations are given nonzero values so that these ratios will be clear. These junctions also provide a basis for comparison with more concentrated solutions, to indicate the effect of the activity coefficients.

Table 1 gives values of $\Delta\Phi$ for the continuous-mixture, restricted-diffusion, and free-diffusion junctions. Table 2 gives values of $\Delta\Phi$ for an electrode of the second kind, where AgCl, with a solubility product* of $10^{-10}$ (mole/liter)$^2$, diffuses into hydrochloric acid solutions of various concentrations. For solutions of zero ionic strength, the values of $\Delta\Phi$ for the continuous-mixture and restricted-diffusion junctions agree with the values calculated from the formulas of Henderson and Planck, respectively (see section 43). In figure 1 are presented the results of more extensive calculations on the HCl-KCl junction. Some of the results in table 2 have already been presented in table 18-1 and discussed in connection with the errors in an electrode of the second kind in very dilute solutions. Table 2 shows that only a small part of the error is a "liquid-junction potential."

The only junction for which our calculations can be compared with other calculations and with experimental results is the 0.1 $M$ HCl–0.1 $M$ KCl junction. MacInnes and Longsworth[11] have made calculations for this junction of the free-diffusion type and reported a value of 28.19 mV to compare with 27.31 mV of the present study. Spiro[12] has discussed cells with liquid

*The solubility product actually is $1.77 \times 10^{-10}$ (mole/kg)$^2$, see problem 2-13.

[11] D. A. MacInnes and L. G. Longsworth, "The Potentials of Galvanic Cells with Liquid Junctions," *Cold Spring Harbor Symposia on Quantitative Biology, 4* (1936), 18–25.

[12] M. Spiro, "The Calculation of Potentials across Liquid Junctions of Uniform Ionic Strength," *Electrochimica Acta, 11* (1966), 569–580.

junctions, including salt bridges, for junctions of constant ionic strength across the junction and of the continuous-mixture type and has included activity-coefficient corrections. For this HCl–KCl junction, Spiro calculated 29.07 mV, and we calculate 27.47 mV. The experimental results are given in table 1. For further discussion of these comparisons, see reference 4.

TABLE 45-1. VALUES OF $\Delta\Phi$ FOR VARIOUS JUNCTIONS AND VARIOUS MODELS AT 25°C. (Values for $f_i = 1$ are indicated by an asterisk. The last ion is the reference ion. Experimental values are given in parentheses.)

| | soln 1 | soln 2 | $\Phi_1 - \Phi_2$, mV | | |
| ion | mole/liter | | free diffusion | restricted diffusion | continuous mixture |
|---|---|---|---|---|---|
| H+ | 0.2 | 0.1 | — | — | −10.31 |
| Cl− | 0.2 | 0.1 | — | — | −11.43* |
| | | | | | |
| K+ | 0.2 | 0.1 | — | — | 1.861 (2.05)[a] |
| Cl− | 0.2 | 0.1 | — | — | 0.335* |
| | | | | | |
| K+ | 0 | 0.01 | −33.50 | −32.65 | −33.75 |
| H+ | 0.02 | 0 | −34.67* | −33.80* | −34.95* |
| Cl− | 0.02 | 0.01 | | | |
| | | | | | |
| K+ | 0 | 0.1 | −27.31 (−27.08)[b] | −27.45 | −27.47 |
| H+ | 0.1 | 0 | (28.25, 18°)[c] | −26.85* | (28.10, 18°)[c] |
| Cl− | 0.1 | 0.1 | (−28.3)[d] −26.69* | | −26.85* |
| | | | | | |
| K+ | 0 | 0.2 | −27.92 | −28.04 | −28.09 |
| H+ | 0.2 | 0 | −26.69* | −26.85* | −26.85* |
| Cl− | 0.2 | 0.2 | | | |
| | | | | | |
| K+ | 0 | 0.2 | −22.58 | −23.03 | −22.31 |
| H+ | 0.1 | 0 | −20.24* | −20.74* | −19.96* |
| Cl− | 0.1 | 0.2 | | | |
| | | | | | |
| K+ | 0 | 0.05 | −20.70 | −21.09 | −20.23 |
| H+ | 0.02 | 0 | −18.50* | −18.97* | −18.02* |
| Cl− | 0.02 | 0.05 | | | |
| | | | | | |
| K+ | 0 | 0.1 | −18.02 | −17.89 | −16.84 |
| H+ | 0.02 | 0 | −14.05* | −14.12* | −12.90* |
| Cl− | 0.02 | 0.1 | | | |
| | | | | | |
| K+ | 0 | 0.1 | −15.91 | −14.99 | −14.04 |
| H+ | 0.01 | 0 | −10.85* | −10.30* | − 9.09* |
| Cl− | 0.01 | 0.1 | | | |
| | | | | | |
| K+ | 0 | 0.1 | −27.24 | −27.38 | −27.40 |
| H+ | 0.09917 | 0 | (−27.98)[e] | −26.77* | −26.76* |
| Cl− | 0.09917 | 0.1 | −26.60* | | |
| | | | | | |
| K+ | 0 | 0.1 | | | |
| H+ | 0.09917 | 0 | −27.39 | −27.48 | −27.55 |
| NO₃⁻ | 0 | 0.05 | −26.53* | −26.62* | −26.70* |
| Cl− | 0.09917 | 0.05 | | | |

TABLE 45-1, continued

| ion | soln 1 | soln 2 | $\Phi_1 - \Phi_2$, mV | | |
|-----|--------|--------|------|------|------|
| | mole/liter | | free diffusion | restricted diffusion | continuous mixture |
| $K^+$ | 0.1 | 0.1 | $-$ 0.157 | $-$ 0.157 | $-$ 0.157 |
| $NO_3^-$ | 0.05 | 0 | $-$ 0.423* | $-$ 0.423* | $-$ 0.423* |
| $Cl^-$ | 0.05 | 0.1 | | | |
| | | | | | |
| $Na^+$ | 0.1 | 0 | | | |
| $H^+$ | 0 | 0.05 | 28.58 | 29.64 | 28.10 |
| $ClO_4^-$ | 0 | 0.05 | 26.72* | 27.90* | 26.22* |
| $Cl^-$ | 0.1 | 0 | | | |
| | | | | | |
| $Na^+$ | 0.1 | 0 | | | |
| $H^+$ | 0 | 0.1 | 32.83 | 33.50 | 33.50 |
| $ClO_4^-$ | 0 | 0.1 | 32.35* | 33.11* | 32.57* |
| $Cl^-$ | 0.1 | 0 | | | |
| | | | | | |
| $Na^+$ | 0.2 | 0 | | | |
| $H^+$ | 0 | 0.2 | 33.29 | 33.88 | 33.53 |
| $ClO_4^-$ | 0 | 0.2 | 32.35* | 33.11* | 32.57* |
| $Cl^-$ | 0.2 | 0 | | | |
| | | | | | |
| $Na^+$ | 0.05 | 0 | | | |
| $H^+$ | 0 | 0.1 | 38.77 | 38.31 | 39.26 |
| $ClO_4^-$ | 0 | 0.1 | 39.96* | 39.58* | 40.48* |
| $Cl^-$ | 0.05 | 0 | | | |
| | | | | | |
| $Cu^{++}$ | 0 | 0.1 | | | |
| $Ag^+$ | 0.2 | 0 | $-$ 6.22* | $-$ 6.22* | $-$ 6.22* |
| $NO_3^-$ | 0.2 | 0 | | | |
| $ClO_4^-$ | 0 | 0.2 | | | |

[a]F. Shedlovsky and D. A. MacInnes, *J. Amer. Chem. Soc.*, 59 (1937), 503.
[b]J. B. Chloupek, V. Z. Kanes, and B. A. Danesova, *Collect. Czech. Chem. Commun.*, 5 (1933), 469, 527.
[c]E. A. Guggenheim and A. Unmack, *Kgl. Danske Videnskab. Selskab, Mat-Fys. Medd.*, 10, #14(1931), 1.
[d]D. C. Grahame and J. I. Cummings, Office of Naval Research Technical Report No. 5, 1950.
[e]N. P. Finkelstein and E. T. Verdier, *Trans. Faraday Soc.*, 53 (1957), 1618.

TABLE 45-2.  VALUES OF $\Delta\Phi$ FOR AN Ag–AgCl ELECTRODE IN HCl SOLUTIONS AT 25°C. (Chloride is the reference ion, and $\beta'$ values are taken to be zero.)

| HCl, bulk mole/liter | $\Phi_0 - \Phi_\infty$, mV | $c_{Cl^-}^0/c_{Cl^-}^\infty$ | $(\mu_{Cl^-}^\infty - \mu_{Cl^-}^0)/F$ mV |
|------|------|------|------|
| $10^{-4}$ | 0.0198 | 1.00961 | $-0.226$ |
| $5 \times 10^{-5}$ | 0.0737 | 1.0392 | $-0.914$ |
| $2 \times 10^{-5}$ | 0.359 | 1.200 | $-4.32$ |
| $10^{-5}$ | 0.915 | 1.604 | $-11.22$ |
| $5 \times 10^{-6}$ | 1.780 | 2.539 | $-22.16$ |
| $2 \times 10^{-6}$ | 3.21 | 5.499 | $-40.58$ |

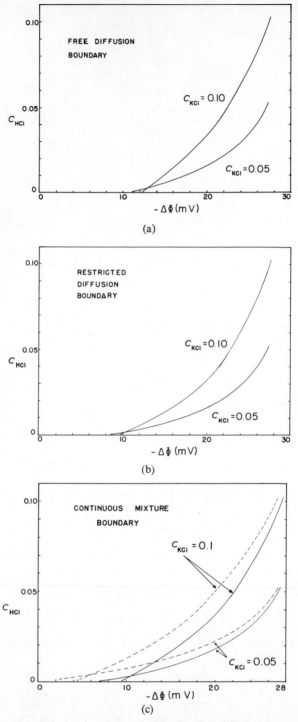

**Figure 45-1.** Calculated values of $\Delta\Phi$ for free-diffusion, restricted-diffusion, and continuous-mixture boundaries between HCl and KCl. The graphs on the left are for given concentrations of KCl on one side of the boundary.

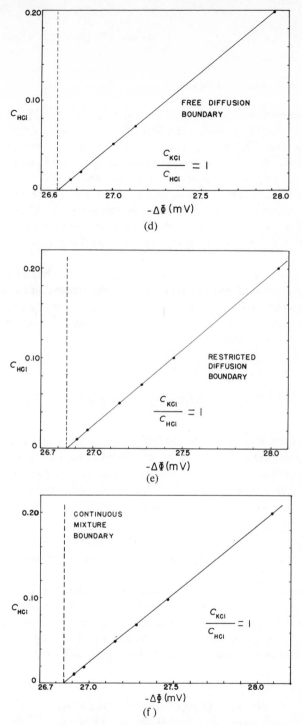

**Figure 45-1.** Continued. The graphs on the right are for a given ratio of concentrations on the two sides of the boundary. The dashed lines represent ideal-solution calculations; the solid lines include activity-coefficient corrections.

133

## 46.  Cells with liquid junction

Once the concentration profiles are known for a liquid-junction region, it is then possible to calculate the effect of the nonuniform composition on the cell potential. The procedure involves first the treatment of electrode equilibria. This allows the expression of the cell potential in terms of a difference in the electrochemical potential of ions in the solutions adjacent to the two electrodes (see chapter 2). The evaluation of this difference involves the integration of equation 16-3 across the junction region. Alternatively, one can proceed as in section 41 to obtain an expression involving the liquid-junction potential, which has been evaluated in the course of determining the concentration profiles.

For the cells of section 17, involving a single electrolyte of varying concentration, one should proceed directly with the evaluation of the integral, since transference numbers and activity coefficients are usually known accurately for solutions of a single electrolyte and the cell potential does not depend on the method of forming the junction.

For cell 19-4, involving a junction between perchloric acid and a sodium chloride solution, the cell potential was expressed in section 41 as

$$FU = FU^\theta + \tfrac{1}{2}RT \ln p_{H_2}^\alpha - RT \ln \frac{c_{H^+}^\beta \cdot c_{Cl^-}^\delta}{p_0^2} - RT \ln (f_{H^+}^\beta \cdot f_{Cl^-}^\beta) + F(\Phi^\delta - \Phi^\beta),$$

(46-1)

where $FU^\theta$ corresponds to entry 7 in table 20-1, and $\Phi$ is referred to the chloride ion. In this case it is particularly convenient to use the quasi-electrostatic potential since it is no longer necessary to select an intermediate solution in the junction (see equation 19-6). Evaluation of the terms in equation 1 for $c_{HClO_4}^\beta = 0.05\ M$ and $c_{NaCl}^\delta = 0.1\ M$ and a fugacity of hydrogen of one atmosphere gives

$$U = 0.222 + 0 + 0.136 + 0.0096 + 0.0281 = 0.396\ \text{V}, \qquad (46\text{-}2)$$

where the term 0.0281 corresponds to the continuous-mixture junction in table 45-1.

A salt bridge is often used to separate two electrolytic solutions, and sometimes the stated purpose is "to eliminate liquid-junction potentials." We should now be in a position to evaluate whether this purpose is achieved, if we could define the liquid-junction potential which is supposed to be eliminated. (Recall that the definition in section 41 depends on the choice of the reference species $n$.)

Such a salt bridge might be

$$\text{HCl (0.1 } M \ \{ \ \text{KCl (0.2 } M \ \} \ \text{HCl (0.2 } M$$
$$\text{in H}_2\text{O)} \quad \text{in H}_2\text{O)} \quad \text{in H}_2\text{O)}$$

(46-3)

It seems clear that the salt bridge does not make the value of $\mu_{Cl^-}$ equal in the two hydrochloric acid solutions. The value of $\mathbf{\Delta\Phi}$ (referred to the chloride ion) for this combination of junctions is 5.78 mV if the junctions are of the continuous-mixture type. This can be compared with the value $\mathbf{\Delta\Phi} = 10.31$ mV for a single, direct junction between 0.1 and 0.2 $M$ HCl solutions.

If the transference numbers of KCl were equal to 0.5 and if departures of activity coefficients from unity could be ignored, the liquid-junction potential of the combination of two junctions of the salt bridge should decrease as the concentration of KCl increases. If one insists on using salt bridges, one might consider as an alternative the series of junctions

$$\text{HCl (0.1 } M \text{ } \{ \text{ KCl (0.1 } M \text{ } \{ \text{ KCl (0.2 } M \text{ } \{ \text{ HCl (0.2 } M}$$
$$\text{in H}_2\text{O)} \quad \{ \quad \text{in H}_2\text{O)} \quad \} \quad \text{in H}_2\text{O)} \quad \{ \quad \text{in H}_2\text{O)} \quad , \tag{46-4}$$

for which $\mathbf{\Delta\Phi} = 1.24$ mV and for which the value of $\mathbf{\Delta\Phi}$ would approach zero as all the concentrations were reduced in proportion if the transference numbers of KCl were 0.5.

## 47.  Error in the Nernst equation

The approximations made in the Nernst equation are to ignore liquid-junction potentials and to ignore the activity coefficients of ionic species in solution. Table 45-1 gives an idea of the range of magnitude of liquid-junction potentials. In the example treated in equation 46-1, the activity-coefficient term amounts to 10 mV, and the liquid-junction potential amounts to 28 mV.

Before one can decide whether it is more serious to neglect activity coefficients or liquid-junction potentials, one should inquire into the effect of using different species for the reference species $n$. The effect, of course, cancels if both activity coefficients and liquid-junction potentials are retained. For the junction between solutions $\delta$ and $\epsilon$, this difference can be expressed as

$$F(\Phi^\delta_{n_*} - \Phi^\epsilon_{n_*}) = F(\Phi^\delta_n - \Phi^\epsilon_n) + \frac{RT}{z_{n_*}} \left( \ln \frac{f^\delta_{n_*}}{f^\epsilon_{n_*}} - \frac{z_{n_*}}{z_n} \ln \frac{f^\delta_n}{f^\epsilon_n} \right), \tag{47-1}$$

where the species chosen for $n$ is denoted by a subscript on the quasi-electrostatic potential. The activity coefficients can be evaluated by the formalism of problem 4-4.

For the junction

$$\begin{array}{cc} \epsilon & \delta \\ 0.1 \ N \ \text{KCl} & 0.1 \ N \ \text{HCl,} \end{array}$$

we have

$$(\Phi^\delta_{K^+} - \Phi^\epsilon_{K^+}) - (\Phi^\delta_{Cl^-} - \Phi^\epsilon_{Cl^-}) = 0.87 \text{ mV} \tag{47-2}$$

and

$$(\Phi^\delta_{H^+} - \Phi^\epsilon_{H^+}) - (\Phi^\delta_{K^+} - \Phi^\epsilon_{K^+}) = 0. \tag{47-3}$$

For the junction

$$\begin{array}{cc} \epsilon & \delta \\ 0.2 \ N \ \text{KCl} & 0.1 \ N \ \text{HCl,} \end{array}$$

we have

$$(\Phi^\delta_{K^+} - \Phi^\epsilon_{K^+}) - (\Phi^\delta_{Cl^-} - \Phi^\epsilon_{Cl^-}) = 4.01 \text{ mV} \tag{47-4}$$

and

$$(\Phi^\delta_{H^+} - \Phi^\epsilon_{H^+}) - (\Phi^\delta_{K^+} - \Phi^\epsilon_{K^+}) = -0.87 \text{ mV}. \tag{47-5}$$

For the junction

$$\begin{array}{ccc} \epsilon & \Big\{ & \delta \\ 0.1 \ N \ \text{KCl} & \Big\{ & 0.05 \ N \ \text{KNO}_3, \ 0.05 \ N \ \text{KCl}, \end{array}$$

we have

$$(\Phi^\delta_{K^+} - \Phi^\epsilon_{K^+}) - (\Phi^\delta_{Cl^-} - \Phi^\epsilon_{Cl^-}) = -0.54 \text{ mV} \tag{47-6}$$

and

$$(\Phi^\delta_{NO_3^-} - \Phi^\epsilon_{NO_3^-}) - (\Phi^\delta_{Cl^-} - \Phi^\epsilon_{Cl^-}) = 0. \tag{47-7}$$

These examples show that liquid-junction potentials can be uncertain by several millivolts, depending on which species is selected for ion $n$, and this uncertainty can be as large as the magnitude of $\Delta\Phi$ itself in some cases.

Next, let us analyze the cell

$$\begin{array}{c|c|c|c|c|c|c} \alpha & \beta & \delta & \text{transition} & \epsilon & \chi & \alpha' \\ \text{Pt}(s) & \text{Li(Hg)} & \begin{array}{c}\text{LiNO}_3, \text{KNO}_3 \\ \text{in } H_2O\end{array} & \begin{array}{c}\text{region} \\ (\text{KNO}_3 \text{ in } H_2O)\end{array} & \begin{array}{c}\text{AgNO}_3, \text{KNO}_3 \\ \text{in } H_2O\end{array} & \text{Ag}(s) & \text{Pt}(s), \end{array}$$

$$\tag{47-8}$$

in which $\text{KNO}_3$ is present throughout the cell at the same concentration. The transition region contains concentration gradients of both $\text{LiNO}_3$ and $\text{AgNO}_3$. (Compare with cell 18-14.) The cell potential can be expressed as

$$FU = -F(\Phi^\alpha - \Phi^{\alpha'}) = \mu^\beta_{Li} - \mu^\chi_{Ag} + \mu^\epsilon_{Ag^+} - \mu^\delta_{Li^+} \tag{47-9}$$

or

$$FU = FU^\theta + RT \ln \frac{a^\beta_{Li} c^\epsilon_{Ag^+}}{c^\delta_{Li^+}} + RT \ln \frac{f^\epsilon_{Ag^+} f^\delta_{K^+}}{f^\delta_{Li^+} f^\epsilon_{K^+}} + F(\Phi^\epsilon - \Phi^\delta), \tag{47-10}$$

where $\Phi$ is referred to the potassium ion and

$$FU^\theta = \mu^0_{Li} - \mu^0_{Ag} + RT \ln \frac{\lambda^\theta_{Ag^+}}{\lambda^\theta_{Li^+}} \tag{47-11}$$

and

$$a^\beta_{Li} = \frac{\lambda^\beta_{Li}}{\lambda^0_{Li}}. \tag{47-12}$$

The activity-coefficient term in equation 10 can be expressed as

$$\ln \frac{f^\epsilon_{Ag^+} f^\delta_{K^+}}{f^\delta_{Li^+} f^\epsilon_{K^+}} = 2(\beta'_{AgNO_3} - \beta'_{KNO_3})c^\epsilon_{NO_3^-} - 2(\beta'_{LiNO_3} - \beta'_{KNO_3})c^\delta_{NO_3^-}. \tag{47-13}$$

Evaluation for $c_{KNO_3} = 0.1 \ M$, $c^\epsilon_{AgNO_3} = 0.01 \ M$, and $c^\delta_{LiNO_3} = 0.01 \ M$ gives a contribution to $U$ of $-2.0 \text{ mV}$ from the activity-coefficient term and a contribution of $-0.47 \text{ mV}$ from the liquid-junction potential. In this case, the Nernst equation is seen to be fairly accurate, the error from the activity coefficients being somewhat larger than the error from the liquid-junction potential.

If we adopt the condition

$$c^\epsilon_{Ag^+} = c^\delta_{Li^+} \ll c_{K^+}, \tag{47-14}$$

then the expression for the cell potential becomes

$$FU = FU^\theta + RT \ln a_{Li}^\beta + 2RT(\beta'_{AgNO_3} - \beta'_{LiNO_3})c_{NO_3^-}. \qquad (47\text{-}15)$$

Thus, the measured cell potential should be a linear function of $c_{NO_3^-}$. As $c_{NO_3^-} \to 0$, the standard cell potential can be determined from the intercept. It would not be necessary to extrapolate to the low concentrations which are necessary for cells without transference. Thus, a cell with a supporting electrolyte throughout, although it is not useful for determining activity coefficients, can be useful for determining standard cell potentials.

## 48.  Potentials across membranes

In many cells of interest, a membrane forms all or part of the junction region. Equation 16-1 can still be used to assess variations of electrochemical potentials across the membrane, although there may be some uncertainty about the values of the transference numbers in the region where the chemical potentials vary.

Membranes can belong to four classes. Some are relatively inert, electrically, such as cellulose acetate membranes used to desalt water by reverse osmosis. A porous glass disk could be in this class. Ion exchange membranes have charged groups bonded to the membrane matrix.[13] Consequently, they tend to exclude co-ions of the same charge as the bound charge. Thus, the transference numbers of anions are small in a cation-exchange resin. Such membranes are used to desalt water by electrodialysis. The third class includes glass, ceramics, and solid electrolytes.[14,15] A glass membrane in which the transference number for hydrogen ions is one in the region where the chemical potentials vary is used to form an electrode which is, in essence, reversible to the hydrogen ion just like the hydrogen electrode. Such electrodes are used in the measurement of pH since they are more convenient than hydrogen electrodes. Biological membranes[16,17] constitute an interesting class which has been the subject of extensive investigation to determine how living cells transport material and operate to create nerve impulses.

[13] G. J. Hills, "Membrane Electrodes," David J. G. Ives and George J. Janz, eds., *Reference Electrodes* (New York: Academic Press, 1961), pp. 411–432.

[14] R. G. Bates, "The Glass Electrode," David J. G. Ives and George J. Janz, eds., *Reference Electrodes* (New York: Academic Press, 1961), pp. 231–269.

[15] George Eisenman, ed., *Glass Electrodes for Hydrogen and Other Cations* (New York: Marcel Dekker, Inc., 1967).

[16] Kenneth S. Cole, *Membranes, Ions and Impulses* (Berkeley: University of California Press, 1968).

[17] J. Walter Woodbury, Stephen H. White, Michael C. Mackey, William L. Hardy, and David B. Chang, "Bioelectrochemistry," Henry Eyring, ed., *Physical Chemistry, An Advanced Treatise*, vol. IXB (New York: Academic Press, 1970), pp. 903–983.

## NOTATION

| | |
|---|---|
| $a_i$ | relative activity of species $i$ |
| $a_i^\theta$ | property expressing secondary reference state, liter/mole |
| $c_i$ | molar concentration of species $i$, mole/liter |
| $D_i$ | diffusion coefficient of species $i$, $cm^2$/sec |
| $f_i$ | molar activity coefficient of species $i$ |
| $F$ | Faraday's constant, 96,487 C/equiv |
| $\mathbf{i}$ | current density, A/$cm^2$ |
| $L$ | thickness of restricted-diffusion junction, cm |
| $p_i$ | partial pressure or fugacity of species $i$, atm |
| $R$ | universal gas constant, 8.3143 J/mole-deg |
| $t_i^0$ | transference number of species $i$ with respect to the velocity of species 0 |
| $T$ | absolute temperature, deg K |
| $u_i$ | mobility of species $i$, $cm^2$-mole/J-sec |
| $U$ | open-circuit cell potential, V |
| $U^\theta$ | standard cell potential, V |
| $x$ | distance, cm |
| $z_i$ | charge number of species $i$ |
| $\kappa$ | conductivity, mho/cm |
| $\lambda_i$ | absolute activity of species $i$ |
| $\lambda_i^\theta$ | property expressing secondary reference state, kg/mole |
| $\mu_i$ | electrochemical potential of species $i$, J/mole |
| $\rho_0$ | density of pure solvent, g/$cm^3$ |
| $\Phi$ | electric potential, V |

# Electrode Kinetics and Other Interfacial Phenomena

The second major area of fundamental electrochemistry necessary in the analysis of electrochemical systems is a knowledge of what goes on at the interface. Part B deals with various aspects of this area of electrochemistry, in particular, with models of the structure of the double layer and with the kinetics of electrode processes. Finally, it deals with electrokinetic and electrocapillary phenomena; although these frequently can be ignored in the analysis of electrochemical systems, they are fundamental parts of electrochemistry and colloid chemistry.

# Structure of the Electric Double Layer

7

Most of our knowledge of the double layer comes from the study of mercury in contact with electrolytic solutions. There are two aspects of this study. Thermodynamics provides a sound basis for expressing relationships among potential, surface tension, and the composition of the bulk solution, and for determining the surface concentrations of various species at the interface. Microscopic models of the diffuse part of the double layer and the inner parts of the double layer offer an explanation for the behavior of macroscopically measurable quantities, such as the surface tension and the double-layer capacity, and provide a useful picture of the detailed structure of the double layer. One is impressed with the forcefulness of the arguments supporting this picture, although he should not expect the properties of the double layer to be exactly predictable on the basis of the model. Rather, experimental measurement of these properties sheds light on the forces acting on species in the interfacial region.

## 49. Qualitative description of double layers

Why is there a double layer? There is a double layer at an interface, first of all, because some species in the solution may have a preference for being near the solid. Let us suppose that we have a solid-solution interface, and let us suppose that there is no charge in the solid itself (see figure 49-1). If the solution is one of potassium iodide in water, then we might suppose that there is a greater tendency for the iodide ions to get very close to the interface than for the potassium ions. This then forms a double charge layer with a diffuse part in the solution, in which there are more potassium ions than iodide ions. The excess of potassium ions in the diffuse part of the double layer balances the excess of iodide ions very close to the interface.

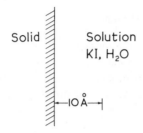

**Figure 49-1.** Solid-solution interface with no charge in the solid.

The iodide ions very near the interface can be regarded as bound by covalent (or specific) forces to the solid itself. The excess potassium ions in the solution are prevented from wandering very far from the interface by the electrical force of attraction to the adsorbed iodide ions. Just how far they wander is determined by a balance of the electric force with the thermal agitation which tries to make ions wander. This distance is characterized by the Debye length (see equation 27-9):

$$\lambda = \sqrt{\frac{\epsilon RT}{2z^2 F^2 c_\infty}} \qquad (49\text{-}1)$$

for a single salt of symmetric ionic valences ($z_+ = -z_- = z$). Figure 49-2 shows the electric charge density $\rho_e$ away from the surface of the solid. The Debye length can amount to perhaps 10 Å.

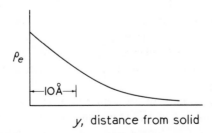

*y*, distance from solid

**Figure 49-2.** Excess electric charge density in the diffuse part of the double layer.

Now consider the interface between a solid (still without charge) and pure water. Suppose that a water molecule looks like that shown in figure 49-3; that is, it has a nonzero electric dipole moment or, in other words, a

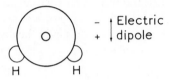

**Figure 49-3.** The dipole moment in a water molecule.

separation of charge within the molecule itself. Now even in this simple case, the oxygen and the hydrogen may have different tendencies to be close to the solid surface, and the water molecules may orient themselves at the surface. This is also a double charge layer at the interface (see figure 49-4). In all cases,

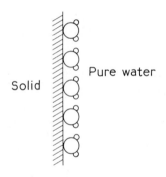

**Figure 49-4.** Oriented water molecules at an interface with no charge in the solid.

the interface as a whole, including all the region in which properties vary from one bulk phase to the other, is electrically neutral, as we can easily see in this particular case since the water molecules are themselves neutral.

A second reason for a double layer to form is that we can vary the charge on the metal side of an interface. Imagine now two metal surfaces exposed to a solution (see figure 49-5). Suppose now that we can apply an appreciable potential difference between these two pieces of metal without there being any appreciable passage of current in the steady state. Where, then, does the potential drop take place? Since it cannot exist in the metal phases or in the solution, due to the absence of an ohmic potential drop, it must occur at the interfaces (see figure 49-6). Thus we have a potential jump at the interface which we can vary by means of our external power supply. In this way we can vary at the same time the charge in the metal at the surface.

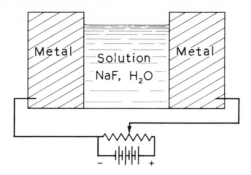

**Figure 49-5.** Metal-solution interfaces arranged so that the charge on the metal can be varied. Now there is a charge in the metal near the interface with the solution.

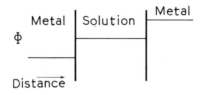

**Figure 49-6.** Steady potential distribution in a system of ideally polarizable electrodes.

The idea that we can vary the charge on the metal side of the surface and also the potential without an electrode reaction occurring is an important one. It perhaps can be only approximated in practice, but it can be approximated sufficiently closely for our purposes. Such an electrode system is called an *ideally polarizable electrode.*

If we superpose these two ways of having a double layer, we get the picture of the double-layer structure shown in figure 49-7. We repeat that the whole of the interfacial region is electrically neutral:

$$q + q_1 + q_2 = 0. \qquad (49\text{-}2)$$

The example concluding section 22 was designed to illustrate that extremely large potentials are required to effect any appreciable separation of charge over any appreciable distance.

If, in figure 5, the two metals are initially uncharged and both are ideally polarizable electrodes, then the application of a current will transfer charge from one metal to the other, leaving them with equal but opposite charges. A current will also flow through the solution, transferring charge from the solution side of one double layer to the solution side of the other double layer so that the charges $q_1 + q_2$ will be equal and opposite in the two double layers. Finally, when a steady state has been attained, the overall system will

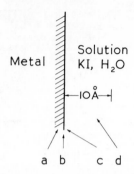

**Figure 49-7.** The structure of the electrical double layer. (a) Charge $q$ in the metal phase, quite close to the surface. (b) The surface of the metal, supposed to be an impenetrable barrier. (c) A layer of adsorbed species which could be oriented dipoles or charged ions, $q_1$. (d) A layer of diffuse charge $q_2$ which is like the bulk of the electrolytic solution except that it is not electrically neutral.

be electrically neutral, the bulk metal and solution phases will be electrically neutral, and the two interfacial regions will each be electrically neutral. We shall have effected, however, a separation of charge within each double layer over a small distance of several angstroms, and the charge $q$ or $q_1 + q_2$ on each side of the interface will not be zero.

Let us consider more closely how we may know the charge in the metal side of the surface. For many of these situations, mercury is a useful electrode material, and many concepts derived from this source are applied to solid electrodes. Let us use mercury dropping from the end of a capillary tube into an electrolytic solution (see figure 49-8). As the mercury drops fall, they

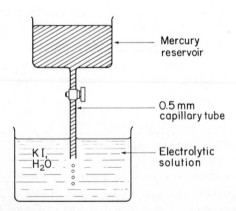

**Figure 49-8.** Apparatus for determining the point of zero charge on mercury in an electrolytic solution.

rapidly deplete the mercury reservoir of excess charge so that soon the mercury drops are uncharged, that is, $q = 0$. If we look at one of these droplets
in the course of its fall, we may find that there is a double layer formed as a
consequence of the desire of iodide ions to be closer to the mercury surface
than the potassium ions. A spherically symmetric shell of adsorbed iodide
ions will not induce any redistribution of charge within the mercury drop,
since the spherical shell of charge cannot exert any electric forces on charges
within the shell. Instead, the adsorbed shell of iodide ions is balanced by an
excess of potassium ions in the diffuse part of the double layer (see the first
two paragraphs of this section).

Now let us add to the system of figure 8 a mechanism for charging the
mercury drops (see figure 49-9). Only the current, the drop size, and the time
between drops are important in determining the charge $q$.

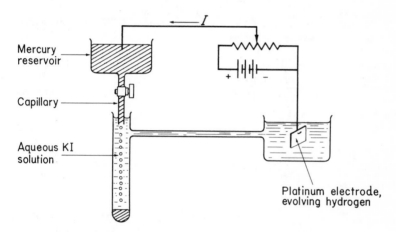

**Figure 49-9.** Apparatus for charging mercury drops in an electrolytic
solution.

The charge-potential relationships for such a system allow one to define
an electric capacity of the double layer, the value of which amounts to about
30 $\mu$F/cm², a fairly large value. For a plane plate capacitor with a relative
dielectric constant of 78.3, this corresponds to a plate separation of about
23 Å and attests to the thinness of the double layer as cited earlier.

In the above discussion we have distinguished between electrical forces
and covalent (or specific) forces. It is really quite difficult to make this distinction precise, even though the concept is useful. The problem has been
discussed in chapter 3. In macroscopic descriptions of interfacial phenomena,
reference electrodes should be used to assess the potential of an electrode
relative to a solution, but for microscopic models one may resort to the concept of the electrostatic potential.

## 50. The Gibbs adsorption isotherm

An interface is the region between two phases, here taken to be homogeneous. There is a transition within the interface from the properties of one phase to those of the other, and the thickness $\tau$ of the interface can range from $10^{-7}$ to $10^{-6}$ cm (see figure 50-1). The thermodynamic treatment of an

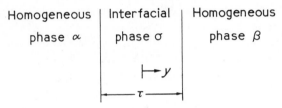

**Figure 50-1.** Interfacial, nonhomogeneous region of thickness $\tau$ between two homogeneous phases.

interface begins generally by considering a system composed of the interface and the two adjacent, homogeneous phases. The extensive properties of the system must be ascribed to these three regions. For example, the number of moles $n_i$ of a species in the system can be written

$$n_i = n_i^\alpha + n_i^\beta + n_i^\sigma. \tag{50-1}$$

Those moles not assigned to the homogeneous phases are assigned to the interface. The surface concentration $\Gamma_i$ is then written as

$$\Gamma_i = \frac{n_i^\sigma}{A}, \tag{50-2}$$

where $A$ is the area of the interface, and is usually expressed in moles/cm². 

One should recognize that there is some ambiguity in the definition of $\Gamma_i$ because the positions of the surfaces bounding the interface have not been specified. Because the detailed structure of the interface is not subject to direct observation, Gibbs took the thickness of the interface to be zero in his classical thermodynamic treatment of the subject. Then the position of only one surface needs to be specified. Let this be at $y = y_{\mathrm{I}}$ on figure 1. Then the definition of the surface concentration can be expressed as

$$\Gamma_i^{\mathrm{I}} = \int_{-\infty}^{y_{\mathrm{I}}} (c_i - c_i^\alpha)\, dy + \int_{y_{\mathrm{I}}}^{\infty} (c_i - c_i^\beta)\, dy. \tag{50-3}$$

The superscript I is added to $\Gamma_i$ in this equation to emphasize that the value obtained for $\Gamma_i$ depends on the position $y_{\mathrm{I}}$ chosen for the Gibbs surface. For example, if we choose the position $y_{\mathrm{II}}$, then the surface concentrations $\Gamma_i^{\mathrm{I}}$ and $\Gamma_i^{\mathrm{II}}$ are related by

$$\Gamma_i^{\mathrm{I}} - \Gamma_i^{\mathrm{II}} = (c_i^\alpha - c_i^\beta)(y_{\mathrm{II}} - y_{\mathrm{I}}). \tag{50-4}$$

Thus, an unambiguous value for $\Gamma_i$ is obtained only if the bulk concentrations of species $i$ are identical in the two adjacent, homogeneous phases. This situation prevails, to a fair approximation, in the case of certain organic compounds which may be adsorbed at an air-solution interface but are essentially insoluble in the adjacent phases.

The surface concentration $\Gamma_i$ as defined above can easily be negative. The ambiguity concerning the choice of the position of the Gibbs surface is usually harmless as long as one is careful to allow for it. The surface concentrations can be fixed by adopting some convention, such as taking $\Gamma_i$ to be zero for a given reference species, usually the solvent, or taking the mass of the interface to be zero. An alternative is to use quantities, called *Gibbs invariants*, which are independent of the position chosen for the Gibbs surface. For example, the quantity

$$\frac{\Gamma_i}{c_i^\alpha - c_i^\beta} - \frac{\Gamma_j}{c_j^\alpha - c_j^\beta}$$

is such an invariant.

Intensive quantities can be assigned to the interface when these quantities have identical values in the adjacent, homogeneous phases. For example, the temperature and the chemical potentials of equilibrated species have meaning for an interface.

The surface tension $\sigma$ is a special intensive property of an interface. It depends on the temperature and composition of the adjacent phases. The surface tension has a mechanical meaning in terms of the forces acting at the interface and a thermodynamic meaning in terms of an energy of the surface per unit area. For example, the variation of the Gibbs function for the system considered in the first paragraph of this section is

$$dG = -S\,dT + V\,dp + \sigma\,dA + \sum_i \mu_i\,dn_i. \tag{50-5}$$

Integration at constant temperature, pressure, and composition, while the area and number of moles are allowed to vary from zero to some nonzero values, gives

$$G = \sigma A + \sum_i \mu_i n_i. \tag{50-6}$$

If we also express this as

$$G = G^\sigma + G^\alpha + G^\beta = G^\sigma + \sum_i \mu_i(n_i^\alpha + n_i^\beta), \tag{50-7}$$

we can show that the surface tension is the excess Gibbs free energy of the surface (per unit area)

$$\sigma = \frac{G^\sigma}{A} - \sum_i \mu_i \Gamma_i. \tag{50-8}$$

Incidentally, one can see from equation 6 that the surface tension is a Gibbs invariant, independent of the choice of the position of the Gibbs surface.

Differentiation of equation 6 and substitution into equation 5 gives

$$A\,d\sigma = -S\,dT + V\,dp - \sum_i n_i\,d\mu_i. \tag{50-9}$$

With the Gibbs-Duhem relations for phases $\alpha$ and $\beta$, for example,

$$0 = -S^\alpha\, dT + V^\alpha\, dp - \sum_i n_i^\alpha\, d\mu_i, \tag{50-10}$$

we obtain

$$A\, d\sigma = -S^\sigma\, dT + V^\sigma\, dp - \sum_i n_i^\sigma\, d\mu_i \tag{50-11}$$

or

$$d\sigma = -s^\sigma\, dT + \tau\, dp - \sum_i \Gamma_i\, d\mu_i, \tag{50-12}$$

where

$$\left.\begin{aligned} S^\sigma &= As^\sigma = S - S^\alpha - S^\beta, \\ V^\sigma &= A\tau = V - V^\alpha - V^\beta. \end{aligned}\right\} \tag{50-13}$$

Equation 12 is the surface analogue of the Gibbs-Duhem relation and is known (for $dT = 0$) as the Gibbs adsorption isotherm. By Gibbs convention, the volume assigned to the interface is zero, and $\tau$ can be set equal to zero. However, Guggenheim prefers to regard the interface to have a nonzero thickness. In either case, equation 12 is applicable, independent of the choice of the position of the surface or surfaces defining the interface.

The Gibbs adsorption equation is useful for determining the surface concentrations $\Gamma_i$ since accurate direct measurement of $\Gamma_i$ is usually more difficult than the determination of variations in surface tension and the use of equation 12.

In applying the Gibbs adsorption equation, one should remember that it applies to the interface between two phases in equilibrium. Consequently, variations must be carried out with the constraint of this phase equilibrium and the consequent loss of a degree of freedom. For a two-component system we can take the temperature and one mole fraction as the independent variables. With the Gibbs-Duhem equations for the homogeneous phases, equation 12 becomes

$$d\sigma = -\left(s^\sigma - \frac{s^\alpha - s^\beta}{c_1^\alpha - c_1^\beta}\Gamma_1\right)dT - \left(\Gamma_2 - \frac{c_2^\alpha - c_2^\beta}{c_1^\alpha - c_1^\beta}\Gamma_1\right)d\mu_2, \tag{50-14}$$

where $s^\alpha$ and $s^\beta$ are the entropies per unit volume of phases $\alpha$ and $\beta$, respectively. Since

$$d\mu_2 = \left[-\bar{S}_2^\alpha + \bar{V}_2^\alpha\left(\frac{\partial p}{\partial T}\right)_{x_2^\alpha,\,\mathrm{sat}}\right]dT + \left[\left(\frac{\partial \mu_2}{\partial x_2^\alpha}\right)_{T,\,p} + \bar{V}_2^\alpha\left(\frac{\partial p}{\partial x_2^\alpha}\right)_{T,\,\mathrm{sat}}\right]dx_2^\alpha, \tag{50-15}$$

we have finally

$$d\sigma = -\left\{s_{(1)}^\sigma + \Gamma_{2(1)}\left[-\bar{S}_2^\alpha + \bar{V}_2^\alpha\left(\frac{\partial p}{\partial T}\right)_{x_2^\alpha,\,\mathrm{sat}}\right]\right\}dT$$

$$- \Gamma_{2(1)}\left[\left(\frac{\partial \mu_2}{\partial x_2^\alpha}\right)_{T,\,p} + \bar{V}_2^\alpha\left(\frac{\partial p}{\partial x_2^\alpha}\right)_{T,\,\mathrm{sat}}\right]dx_\alpha^2, \tag{50-16}$$

where

$$s_{(1)}^\sigma = s^\sigma - \frac{s^\alpha - s^\beta}{c_1^\alpha - c_1^\beta}\Gamma_1 \quad \text{and} \quad \Gamma_{2(1)} = \Gamma_2 - \frac{c_2^\alpha - c_2^\beta}{c_1^\alpha - c_1^\beta}\Gamma_1 \tag{50-17}$$

are the entropy of the interface and the surface concentration of species 2, both evaluated with the Gibbs surface chosen such that $\Gamma_{1(1)} = 0$. We see that a measurement of the variation of the surface tension with $x_2^\alpha$ at constant temperature allows us to determine $\Gamma_{2(1)}$. A subsequent measurement of the variation of surface tension with temperature at constant $x_2^\alpha$ allows us to determine the surface entropy $s_{(1)}^\sigma$.

## 51. The Lippmann equation

We now wish to apply the Gibbs adsorption isotherm to an interface involving an ideally polarizable electrode. We treat the system shown in figure 51-1. Here the counter electrode is used to maintain the potential of

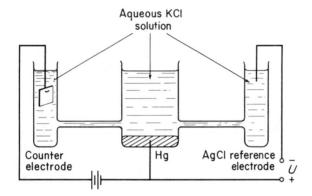

**Figure 51-1.** System for applying a potential to an ideally polarizable electrode.

the mercury, which is measured relative to a silver chloride reference electrode. This latter circuit can be represented by the diagram

$$
\begin{array}{c|c|c|c|c|c}
\alpha & \beta & \delta & \epsilon & \lambda & \alpha' \\
\text{Pt}(s) & \text{Ag}(s) & \text{AgCl}(s) & \text{KCl in} & \text{Hg}(l) & \text{Pt}(s), \\
 & & & \text{H}_2\text{O} & &
\end{array}
\tag{51-1}
$$

for which the potential can be expressed by the methods of chapter 2 as

$$
FU = -F(\Phi^\alpha - \Phi^{\alpha'}) = \mu_{e^-}^\alpha - \mu_{e^-}^{\alpha'}
$$
$$
= \mu_{\text{Cl}^-}^\epsilon - \mu_{e^-}^\lambda + \mu_{\text{Ag}}^\beta - \mu_{\text{AgCl}}^\delta. \tag{51-2}
$$

For variations of the surface tension of the mercury at constant temperature, the Gibbs adsorption isotherm, equation 50-12 becomes

$$
d\sigma = -\Gamma_{e^-}\, d\mu_{e^-}^\lambda - \Gamma_{\text{K}^+}\, d\mu_{\text{K}^+}^\epsilon - \Gamma_{\text{Cl}^-} d\mu_{\text{Cl}^-}^\epsilon. \tag{51-3}
$$

We consider the mercury phase $\lambda$ to be composed of mercury atoms and electrons. The surface concentration of mercury does not appear in equation 3 because we take $d\mu_{\text{Hg}}^\lambda = 0$. The surface concentration of electrons is a Gibbs invariant because the concentration of electrons is zero in the bulk of the homogeneous phases $\lambda$ and $\epsilon$ (see equation 50-4). In fact, $\Gamma_{e^-}$ is related to the surface charge density $q$ discussed in section 49 for an ideally polarizable electrode:

$$q = -F\Gamma_{e^-}. \tag{51-4}$$

We consider the aqueous phase $\epsilon$ to be composed of potassium ions, chloride ions, and water. We choose the Gibbs surface such that $\Gamma_{H_2O} = 0$.

We have emphasized before that the interface as a whole is electrically neutral,

$$\sum_i z_i\Gamma_i = 0. \tag{51-5}$$

If we use equation 5 to eliminate $\Gamma_{Cl^-}$ from equation 3 and use equation 4 to introduce $q$, we obtain

$$d\sigma = -\Gamma_{K^+}\, d\mu^\epsilon_{KCl} - \frac{q}{F}(d\mu^\epsilon_{Cl^-} - d\mu^\lambda_{e^-}). \tag{51-6}$$

Finally, equation 2 can be used to introduce the potential $U$:

$$d\sigma = -\Gamma_{K^+}\, d\mu^\epsilon_{KCl} - q\, dU. \tag{51-7}$$

This important equation is known as the Lippmann equation. It tells us that, if we measure the variation of the surface tension with composition at constant potential, we can obtain the surface concentration of potassium ions and, if we measure the variation with potential at constant composition, we can obtain the surface charge $q$. All this can be done on a firm thermodynamic basis without resort to microscopic models of the interface, although the experiments require considerable effort to obtain accurate results. Bear in mind that $\Gamma_{K^+}$ is relative to the convention that $\Gamma_{H_2O} = 0$.

The above derivation of the Lippmann equation differs from the treatments of reversible electrodes in chapter 2 in that there are no species which are equilibrated between phases $\lambda$ and $\epsilon$. Or, if they are equilibrated, they are assumed to be of negligible concentration in one phase or the other. An alternative treatment assumes that there is an impenetrable barrier through which no species, and hence no current, passes. The surface charge $q$ is then the surface charge density on the electrode side of this barrier, and again no species exists on both sides of the barrier in an appreciable concentration. Both developments lead to the Lippmann equation, and the difference in the bases is of little practical consequence.

The double-layer capacity (per unit area) $C$ is the derivative of the double-layer charge $q$ with respect to potential at constant composition:

$$C = \left(\frac{\partial q}{\partial U}\right)_{\mu,T}, \tag{51-8}$$

where the subscript $\mu$ denotes constant composition. From equation 7 we see that

$$q = -\left(\frac{\partial\sigma}{\partial U}\right)_{\mu,T}. \tag{51-9}$$

Hence,                      $$C = -\left(\frac{\partial^2\sigma}{\partial U^2}\right)_{\mu,T}. \tag{51-10}$$

The double-layer capacity of an ideally polarizable electrode can be measured directly with an alternating current. Because the double layer is thin, it responds rapidly to the alternating current. Consequently, except when the adsorption of long-chain organic compounds is involved, the alternating current capacity does not begin to depart from the static capacity defined by equation 8 until a frequency of about 1 Mc is reached.

Grahame[1] describes an experimental confirmation of the Lippmann equation in which the charge is determined as a function of potential in three independent ways:

1. Differentiation of the surface tension with respect to potential according to equation 9.

2. Integration of the double-layer capacity with respect to potential according to equation 8. The integration constant must be evaluated to give agreement with the other two methods.

3. Direct measurement of $q$ by means of an apparatus such as that sketched in figure 49-8.

Note that only the second method can be applied to solid electrodes and that the determination of the point of zero charge, equivalent to the integration constant, is then uncertain.

The above derivation of the Lippmann equation can be modified to apply to a different reference electrode and to multicomponent solutions, including systems involving the adsorption of neutral organic molecules. The application of thermodynamic principles allows a coherent treatment of a variety of data involving the measurement of surface tension, surface charge, and double-layer capacity as functions of temperature, potential, and solution composition. These data can be manipulated by thermodynamic methods to yield derived quantities of interest, such as the surface concentrations. (See problems 2 and 3 and references 1 and 2.)

The surface tension of mercury in contact with several electrolytic solutions is plotted against potential in figure 51-2. The potential of zero charge is given in table 1 (at 25°C rather than 18°C). From equation 9 we see that zero charge corresponds to the maximum on the surface tension curve. Consequently, this point of zero charge is also referred to as the *electrocapillary maximum*. In figure 51-2, the potentials relative to a normal calomel electrode in KCl have been shifted by $+0.48$ V in order that the electrocapillary maximum for KOH might appear at about zero V. See problem 4.

The surface charge and surface concentrations (the latter being expressed as $z_i F \Gamma_i$) are represented in figures 51-3 and 51-4 for two concentrations of NaCl. Here the potentials are measured relative to a calomel electrode in the same solution as the ideally polarizable electrode, and no questions of

[1]David C. Grahame, "The Electrical Double Layer and the Theory of Electrocapillarity," *Chemical Reviews*, *41* (1947), 441–501.

[2]Paul Delahay, *Double Layer and Electrode Kinetics* (New York: Interscience Publishers, 1965).

TABLE 51-1. POTENTIAL OF ZERO CHARGE FOR MERCURY (RELATIVE TO A
NORMAL CALOMEL ELECTRODE IN KCl) FOR VARIOUS ELECTROLYTIC
SOLUTIONS AT 25°C. (From Grahame et al.[3] with permission of the
American Chemical Society. See also reference 4.)

| Electrolyte | $c$, M | Potential, V | Electrolyte | $c$, M | Potential, V |
|---|---|---|---|---|---|
| LiCl | 1.0 | −0.557 | CsCl | 1.0 | −0.556 |
| | 0.1 | −0.5592 | | 0.1 | −0.5564 |
| NaCl | 1.0 | −0.557 | HCl | 0.1 | −0.558 |
| | 0.1 | −0.5591 | $NH_4Cl$ | 0.1 | −0.5587 |
| KCl | 1.0 | −0.5555 | $CaCl_2$ | 0.1 | −0.5586 |
| | 0.7 | −0.5535 | $SrCl_2$ | 0.1 | −0.5588 |
| | 0.3 | −0.5515 | $BaCl_2$ | 0.1 | −0.5587 |
| | 0.1 | −0.5589 | $MnCl_2$ | 0.1 | −0.5589 |
| | 0.01 | −0.5936 | $CoCl_2$ | 0.1 | −0.5585 |
| | 0.001 | −0.640 | $NiCl_2$ | 0.1 | −0.5588 |
| RbCl | 0.1 | −0.5576 | $AlCl_3$ | 0.1 | −0.5585 |
| NaF | 1.0 | −0.472 | $LaCl_3$ | 0.1 | −0.5588 |
| | 0.1 | −0.474 | $KCH_3COO$ | 0.1 | −0.4884 |
| KF | 0.1 | −0.4714 | $KClO_4$ | 0.1 | −0.5074 |
| $KHCO_3$ | 0.1 | −0.4728 | $KNO_3$ | 0.1 | −0.5166 |
| $K_2CO_3$ | 0.05 | −0.4734 | KBr | 0.1 | −0.5741 |
| $K_2SO_4$ | 0.05 | −0.4705 | KCNS | 0.1 | −0.626 |
| KOH | 0.1 | −0.4767 | KI | 0.1 | −0.732 |

liquid-junction potentials are involved. It is such well-defined potentials
which are used in the Lippmann equation 7.

Figures 51-5 and 51-6 show the double-layer capacity as a function of
potential for NaCl and NaF solutions. More curves of this type can be found
in reference 1.

## 52. The diffuse part of the double layer

The thermodynamics of the double layer was developed for an ideally
polarizable electrode in the preceding two sections. Beyond this one must
resort to microscopic models. These were discussed qualitatively in section 49.

The diffuse part of the double layer is regarded as part of the electrolytic
solution, but here the solution is not electrically neutral. The model used to
treat this region is essentially identical to that of Debye and Hückel, used to
determine the ionic distributions around a central ion and subsequently
to calculate the electrical contribution to the activity coefficients (see sections
27 and 28). The ionic concentrations in the diffuse part of the double layer

[3] D. C. Grahame, E. M. Coffin, J. I. Cummings, and M. A. Poth, "The Potential of
the Electrocapillary Maximum of Mercury. II," *Journal of the American Chemical Society*,
74 (1952), 1207–1211.

[4] Richard S. Perkins and Terrell N. Andersen, "Potentials of Zero Charge of Electrodes," *Modern Aspects of Electrochemistry*, 5 (1969), 203–290.

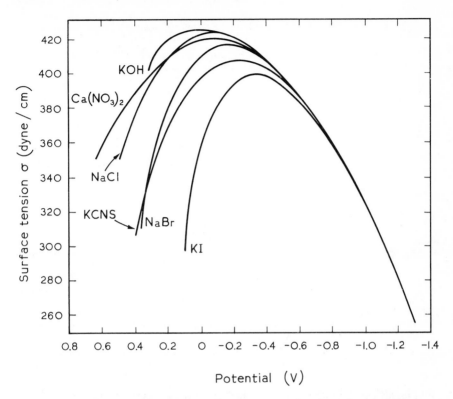

**Figure 51-2.** Interfacial tension of mercury as a function of potential for several electrolytic solutions at 18°C. Potentials relative to a normal calomel electrode are shifted by 0.48 V. (After Grahame,[1] with permission of the American Chemical Society.) These are referred to as electrocapillary curves because the surface tension is often measured with a capillary electrometer.

are assumed to be related to the potential by the Boltzmann distribution (see equation 27-1)

$$c_i = c_{i\infty} \exp\left(-\frac{z_i F \Phi}{RT}\right), \tag{52-1}$$

and Poisson's equation relates the variation of the potential to the charge density (see equation 27-2). For a planar electrode this becomes

$$\frac{d^2\Phi}{dy^2} = -\frac{F}{\epsilon} \sum_i z_i c_{i\infty} \exp\left(-\frac{z_i F \Phi}{RT}\right), \tag{52-2}$$

where $y$ is the distance from the electrode.

Similar limitations apply to the validity of this model as to that of Debye and Hückel (see section 29). For the planar case, in contrast to the spherical case treated in section 27, one can go further without the introduction of the

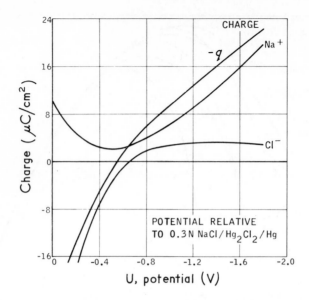

**Figure 51-3.** Charge and adsorption of sodium and chloride ions at a mercury interface in contact with $0.3M$ NaCl at 25°C. (After Grahame,[1] with permission of the American Chemical Society.) The surface concentrations of the ions are expressed as $z_i F \Gamma_i$.

mathematical approximation of Debye and Hückel (see equation 27-7). We should note again that the derivation of the Lippmann equation in the preceding section did not involve the introduction of any model.

The first boundary condition on equation 2 is that

$$\Phi \longrightarrow 0 \quad \text{as} \quad y \longrightarrow \infty. \tag{52-3}$$

From equation 1, we thus see that $c_{i\infty}$ is the concentration of species $i$ approached at large distances from the electrode. Furthermore, since the right side of equation 2 is the charge density divided by $-\epsilon$, integration of this equation allows the potential gradient at $y_2$ to be related to the surface charge density $q_2$ in the diffuse part of the double layer:

$$\frac{d\Phi}{dy} = \frac{q_2}{\epsilon} \quad \text{at} \quad y = y_2. \tag{52-4}$$

This constitutes the second boundary condition for equation 2. Here, $y_2$ is the position of the inner limit of the diffuse layer, that is, the closest distance to which solvated ions can approach the electrode, the same value being applicable to all ionic species. Note the similarity of $y_2$ to the parameter $a$ of the theory of Debye and Hückel.

Let us introduce the electric field $E$:

$$E = -\frac{d\Phi}{dy}. \tag{52-5}$$

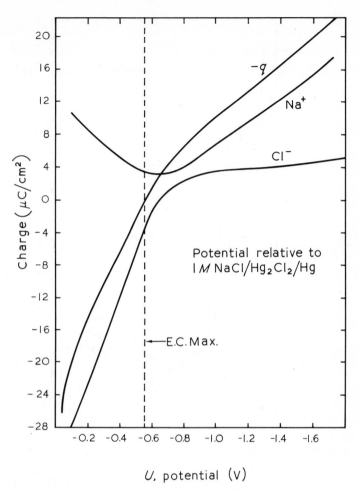

**Figure 51-4.** Charge and adsorption of sodium and chloride ions at a mercury interface in contact with a $1\,M$ NaCl solution at $25°C$. (After Grahame,[1] with permission of the American Chemical Society.) The surface concentrations of the ions are expressed as $z_i F\Gamma_i$.

The electric field can be determined as a function of the potential by rewriting equation 2 as

$$\frac{d^2\Phi}{dy^2} = -\frac{dE}{dy} = -\frac{dE}{d\Phi}\frac{d\Phi}{dy} = E\frac{dE}{d\Phi}$$
$$= -\frac{F}{\epsilon} \sum_i z_i c_{i\infty} \exp\left(-\frac{z_i F\Phi}{RT}\right). \tag{52-6}$$

Integration gives

$$\frac{1}{2}E^2 = \frac{RT}{\epsilon} \sum_i c_{i\infty}\left[\exp\left(-\frac{z_i F\Phi}{RT}\right) - 1\right], \tag{52-7}$$

the integration constant being evaluated from the fact that as $y \rightarrow \infty$, both

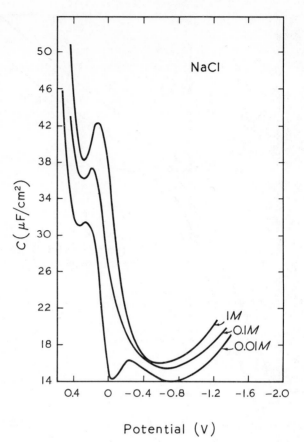

Figure 51-5. Double-layer capacity for mercury in contact with NaCl solutions at 25°C. (After Grahame,[1] with permission of the American Chemical Society.) Potentials are relative to the electrocapillary maximum.

$\Phi$ and $E$ approach zero. The electric field therefore is given in terms of the potential as

$$E = \pm \left\{ \frac{2RT}{\epsilon} \sum_i c_{i\infty} \left[ \exp\left(-\frac{z_i F\Phi}{RT}\right) - 1 \right] \right\}^{1/2}, \qquad (52\text{-}8)$$

the plus sign being used if $\Phi$ is positive and conversely, since $E$ and $\Phi$ must be of the same sign.

Without carrying the problem further, we can now relate the charge in the diffuse layer to the potential at $y_2$ since introduction of the condition 4 gives

$$q_2 = \mp \left\{ 2RT\epsilon \sum_i c_{i\infty} \left[ \exp\left(-\frac{z_i F\Phi_2}{RT}\right) - 1 \right] \right\}^{1/2}, \qquad (52\text{-}9)$$

where $\Phi_2$ is the potential at $y_2$. This relationship has important applications in double-layer theory.

The determination of the potential as a function of distance is straightforward in principle, although it can be complicated in practice. Equation 5 gives

$$y - y_2 = \int_{\Phi}^{\Phi_2} \frac{d\Phi}{E}, \tag{52-10}$$

where $E$ is given as a function of $\Phi$ by equation 8.

Although numerical integration of equation 10 is necessary in general, the analysis can be completed for the special case where the magnitudes of the ionic charges are all the same, $|z_i| = z$. Let us carry out the development in dimensionless form where

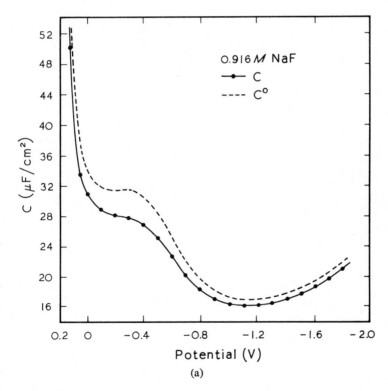

(a)

**Figure 51-6.** Double-layer capacity for mercury in contact with NaF solutions at 25°C. (After Grahame,[5] with permission of the American Chemical Society.) Potentials are relative to the normal calomel electrode.

[5] David C. Grahame, "Differential Capacity of Mercury in Aqueous Sodium Fluoride Solutions. I. Effect of Concentration at 25°," *Journal of the American Chemical Society*, 76 (1954), 4819–4823.

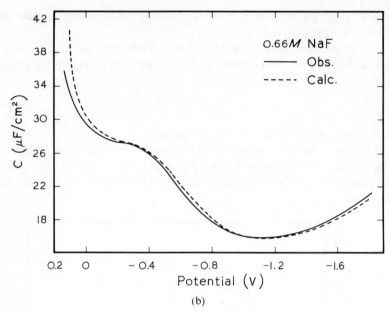

(b)

**Figure 51-6.** Continued.

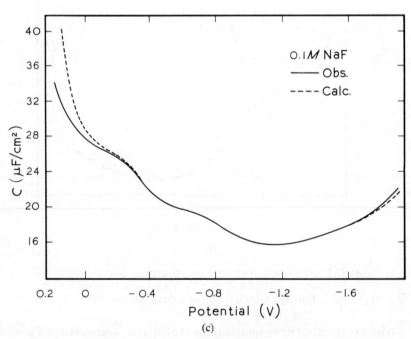

(c)

**Figure 51-6.** Continued.

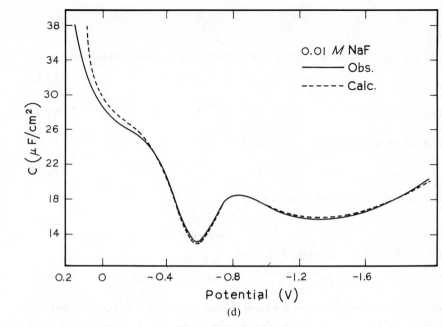

(d)

**Figure 51-6.** Continued.

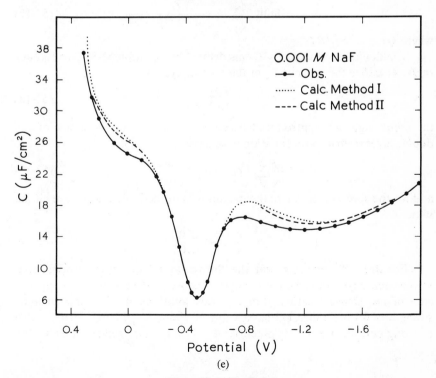

(e)

**Figure 51-6.** Continued.

$$x = \frac{y}{\lambda}, \qquad \phi = \frac{F\Phi}{RT}, \qquad \mathcal{E} = \frac{\lambda FE}{RT}, \qquad C_{i\infty} = \frac{2c_{i\infty}}{\sum_j z_j^2 c_{j\infty}}, \qquad (52\text{-}11)$$

and $\lambda$ is the Debye length given by equation 27-9. Equations 8 and 10 become

$$\mathcal{E} = \pm[\sum_i C_{i\infty}(e^{-z_i\phi} - 1)]^{1/2} \qquad (52\text{-}12)$$

and

$$x - x_2 = \int_\phi^{\phi_2} \frac{d\phi}{\mathcal{E}}. \qquad (52\text{-}13)$$

For the special case of $|z_i| = z$, we have

$$z\mathcal{E} = \pm(e^{z\phi} - 2 + e^{-z\phi})^{1/2} = 2\sinh\frac{z\phi}{2}. \qquad (52\text{-}14)$$

Integration of equation 13 then gives

$$x - x_2 = \ln\frac{\tanh z\phi_2/4}{\tanh z\phi/4}. \qquad (52\text{-}15)$$

This result can be rearranged to yield

$$\phi = \frac{2}{z}\ln\frac{1 - Ke^{-x+x_2}}{1 + Ke^{-x+x_2}}, \qquad z\mathcal{E} = \frac{-4Ke^{-x+x_2}}{1 - (Ke^{-x+x_2})^2}, \qquad (52\text{-}16)$$

where $K$ is a dimensionless constant whose value lies between $-1$ and $+1$ and is related to the potential $\phi_2$ and the charge $q_2$ in the diffuse layer by

$$K = -\tanh\frac{z\phi_2}{4} = \frac{Q_2}{\sqrt{4 + Q_2^2} + 2}, \qquad (52\text{-}17)$$

where $Q_2 = z\lambda Fq_2/RT\epsilon$.

The double-layer capacity $C$ was defined by equation 51-8. Correspondingly, we define the capacity $C_d$ of the diffuse layer as

$$C_d = -\left(\frac{\partial q_2}{\partial \Phi_2}\right)_{\mu,T}, \qquad (52\text{-}18)$$

the minus sign being introduced because $q_2$ is on the opposite side of the double layer from $q$. With equation 9 we have

$$C_d = \frac{\epsilon F}{q_2}\sum_i z_i c_{i\infty}\exp\left(-\frac{z_i F\Phi_2}{RT}\right), \qquad (52\text{-}19)$$

and for the special case where the magnitudes of the ionic charges are all the same

$$C_d = \frac{\epsilon}{\lambda}\cosh\frac{zF\Phi_2}{2RT}. \qquad (52\text{-}20)$$

Equation 20 indicates that the diffuse-layer capacity is proportional to the square root of the ionic strength because of the composition dependence of the Debye length $\lambda$. For aqueous solutions at 25°C and an ionic strength of 0.1 mole/liter, $\epsilon/\lambda$ has a value of about 72 $\mu$F/cm². There is also a strong dependence on the potential $\Phi_2$. For 1–1 electrolytes at 25°C, the diffuse-layer capacity is about 3.6 times higher when $\Phi_2 = 0.1$ V than when $\Phi_2 = 0$.

## 53. Capacity of the double layer in the absence of specific adsorption

The structure of the double layer was discussed qualitatively in section 49, where we indicated that species could be adsorbed by specific forces at the interface. Cations, generally speaking, are not specifically adsorbed, thallous ions being an exception to this rule. Evidence that cations are not specifically adsorbed is given by the fact that the electrocapillary curves of figure 51-2 coincide on the branch at negative electrode potentials. Anions usually *are* specifically adsorbed, exceptions being fluoride, hydroxyl, and sulfate ions. Evidence that chloride ions are specifically adsorbed can be seen in figure 51-3, which shows that sodium ions are again adsorbed toward more positive potentials. This is attributed to adsorption of chloride ions in excess of that dictated solely by the charge $q$ on the electrode.

It is simpler to look first at a system involving an electrolyte, such as NaF, where both ions show little or no tendency for specific adsorption. Then we can say that $q_1 = 0$, and, consequently,

$$q_2 = -q; \qquad (53\text{-}1)$$

that is, the charge in the diffuse layer is given by the charge on the electrode, which can be determined by thermodynamic means. As the subject was developed in the preceding section, all the properties of the diffuse layer depend solely on the charge $q_2$. For example, the potential $\Phi_2$ at the inner limit of the diffuse layer is related to $q_2$ by means of equation 52-9, and the capacity of the diffuse layer $C_d$ is expressed in terms of $\Phi_2$ and $q_2$ in equation 52-19. In the absence of specific adsorption, the ionic surface concentrations $\Gamma_i$ must be accounted for by the diffuse layer, and these can also be related to $q_2$ or $\Phi_2$ (see problem 5). There are thus a variety of ways in which one can test the assumptions that there is no specific adsorption with solutions of NaF and that the diffuse-layer theory is valid.

Grahame[5] looked at the double-layer capacity $C$. Since

$$U = U - \Phi_2 + \Phi_2, \qquad (53\text{-}2)$$

we can write

$$\left(\frac{\partial U}{\partial q}\right)_\mu = \left(\frac{\partial (U - \Phi_2)}{\partial q}\right)_\mu + \left(\frac{\partial \Phi_2}{\partial q}\right)_\mu \qquad (53\text{-}3)$$

or

$$\frac{1}{C} = \frac{1}{C_{M-2}} + \frac{1}{C_d}, \qquad (53\text{-}4)$$

where $C_{M-2}$ is the capacity of the region between the metal and the plane at $y = y_2$. In obtaining equation 4, we have made use of the assumption that $q_2 = -q$ and the definitions 51-5 and 52-18.

In equation 4, we know $C$ by direct measurement, and we know $C_d$ by the assumption of no specific adsorption, as outlined above. Consequently, we can calculate $C_{M-2}$. Instead, we can make various assumptions about $C_{M-2}$ and make predictions of $C$ using diffuse-layer theory to obtain $C_d$. The first plausible assumption that $C_{M-2}$ is constant does not work well at all. Grahame therefore made the second plausible assumption that $C_{M-2}$ depends only on $q$, independent of the bulk concentration. He calculated $C_{M-2}$ as a function of $q$ from equation 4 by using data for $C$ for NaF solutions of about 1 $M$ concentration. At these relatively high concentrations, the contribution of $C_d$ in equation 4 is small. On the basis of this calculated dependence of $C_{M-2}$ on $q$, Grahame then made predictions of $C$ for solution concentrations ranging down to 0.001 $M$ (see figure 51-6). The agreement with experimental values turned out to be quite good, even at the lowest concentration.

In this manner, Grahame has made a substantial case for the relevance of the diffuse-layer theory and the assumption that sodium and fluoride ions are not specifically adsorbed. It remains to explain the charge dependence of $C_{M-2}$, a problem which appears to require a detailed microscopic theory of the region very close to the mercury surface.

## 54. Specific adsorption at an electrode-solution interface

Specific adsorption refers to the attraction of a species toward the mercury surface by forces which are not purely coulombic in nature. Frequently, anions are specifically adsorbed while cations are not. In this case, $\Gamma_+$, which can be obtained by thermodynamic means, can be immediately associated with the surface concentration of cations in the diffuse part of the double layer. The theory in section 52 can then be used to treat the diffuse part of the double layer, all the properties of the diffuse layer being uniquely related by this theory to the surface concentration of cations in that layer. In this manner one can determine the potential $\Phi_2$ at the outer Helmholtz plane, the charge $q_2$ in the diffuse layer, and the surface concentration of anions in the diffuse layer.

From the measured value of $\Gamma_-$, one is now in a position to determine the surface concentration of the specifically adsorbed anions. This quantity is subject to chemical interpretation in terms of adsorption isotherms and the energetics of specific adsorption, with either the electrode charge $q$ or the electrode potential as a correlating variable. A lot of work has been done along these lines, and we must refer to the literature for details.[1,2,6,7] Electro-capillary phenomena and effects of the double layer will be encountered again in chapters 8, 9, and 10.

## PROBLEMS

1. A weighed amount of NaCl solution ($n$ moles) of mole fraction $x^i_{NaCl}$ is added to a highly porous carbon of area $A$. After the carbon has settled, the supernatant solution has a different mole fraction, $x^f_{NaCl}$. The amount of NaCl adsorbed is calculated as $n(x^i_{NaCl} - x^f_{NaCl})/A$. What value of $\Gamma_{NaCl}$ is calculated, that is, relative to what Gibbs convention for the position of the surface?

2. From the Lippmann equation 51-7, derive the Maxwell relations

$$\left(\frac{\partial \Gamma_{K^+}}{\partial U}\right)_\mu = \left(\frac{\partial q}{\partial \mu}\right)_U, \qquad \left(\frac{\partial \mu}{\partial U}\right)_{\Gamma_{K^+}} = -\left(\frac{\partial q}{\partial \Gamma_{K^+}}\right)_U,$$

$$\left(\frac{\partial \Gamma_{K^+}}{\partial q}\right)_\mu = -\left(\frac{\partial U}{\partial \mu}\right)_q, \qquad \left(\frac{\partial \mu}{\partial q}\right)_{\Gamma_{K^+}} = \left(\frac{\partial U}{\partial \Gamma_{K^+}}\right)_q.$$

Show that

$$\left(\frac{\partial q}{\partial \mu}\right)_U = -\left(\frac{\partial q}{\partial U}\right)_\mu \left(\frac{\partial U}{\partial \mu}\right)_q = -C\left(\frac{\partial U}{\partial \mu}\right)_q.$$

3. Show how to obtain the surface concentration $\Gamma_{K^+}$ for the mercury, KCl solution interface from measurements of the double-layer capacity as a function of potential and KCl concentration. In addition, the potential and the surface tension at the point of zero charge can be assumed to be known as functions of concentration.

4. (a) The potential of the point of zero charge for mercury in various electrolytic solutions is given in table 51-1. This is measured relative to a normal calomel electrode in KCl. On the assumption that this is supposed to be a thermodynamic quantity, for example, not involving the uncertainty of liquid-junction potentials, discuss the merit of the suggestion that, for the interface

$$\begin{array}{c|c} \delta & \beta \\ \text{0.1 } N \text{ Na}_2\text{SO}_4 & \text{Hg}(l), \\ \text{in H}_2\text{O} & \end{array}$$

the tabulated value represents (or should represent)

$$\frac{-\mu^\beta_{e^-} - F\Phi^\delta + \mu^0_{Hg} - \frac{1}{2}\mu^0_{Hg_2Cl_2} + RT \ln c^\lambda_{Cl^-}}{F},$$

[6] Roger Parsons, "Equilibrium Properties of Electrified Interphases," *Modern Aspects of Electrochemistry*, *1* (1954), 103–179.

[7] Richard Payne, "The Electrical Double Layer in Nonaqueous Solutions," *Advances in Electrochemistry and Electrochemical Engineering*, 7 (1970), 1–76.

where $\Phi^\delta$ is the quasi-electrostatic potential of phase $\delta$ relative to the chloride ion as species $n$ and $c_{Cl^-}^\lambda$ is the concentration of the chloride ion in the 1 $N$ KCl solution of the reference electrode (see section 40).

(b) If the potential of zero charge for the interface

$$
\begin{array}{c|c}
\delta & \beta \\
\text{0.3 } M \text{ NaCl} & \text{Hg}(l) \\
\text{in H}_2\text{O} &
\end{array}
$$

is measured relative to a calomel electrode in the same solution, how should we calculate the potential of zero charge relative to the normal calomel electrode in KCl?

(c) If the potential of zero charge for the interface of part (b) is measured relative to a calomel electrode in 0.3 $N$ KCl in a system involving a liquid junction, how might we estimate the value relative to the normal calomel electrode in KCl, corrected for the liquid junction? Repeat for the case where the experimental reference electrode is in 1 $N$ KCl.

(d) For the interface of part (a), assume that the potential has been measured relative to a lead sulfate electrode in the same solution. Show how to calculate the potential relative to the normal calomel electrode in KCl.

(e) If the potential for the interface of part (a) has been measured relative to a calomel electrode in 0.1 $N$ KCl in a system involving a liquid junction, show how to estimate the value relative to the normal calomel electrode in KCl, corrected for the liquid junction.

(f) How should we modify the values in table 51-1 in order to obtain tables of potentials of zero charge relative to a hydrogen electrode in 1 $M$ HCl and relative to a hydrogen electrode in 1 $M$ HNO$_3$? Would these two tables be different? Speculate on what we might mean by "potentials relative to a standard hydrogen electrode."

5. Let the surface excess $\Gamma_{i,d}$ of an ionic species in the diffuse layer be defined as

$$
\Gamma_{i,d} = \int_{y_2}^\infty (c_i - c_{i\infty})\, dy
$$

(compare equation 50-3). Show, for the special case where the magnitudes of the ionic charges are all the same, $|z_i| = z$, that diffuse-layer theory yields the expression

$$
\Gamma_{i,d} = 2\lambda c_{i\infty}\, (e^{-z_i\phi_2/2} - 1).
$$

From this result, show that

$$
\left(\frac{\partial \Gamma_{i,d}}{\partial q_2}\right)_\mu = \frac{\dfrac{2}{F}}{1 + e^{z_i\phi_2}} \frac{z_i c_{i\infty}}{\sum\limits_j z_j^2 c_{j\infty}}
$$

and that, consequently, in the absence of specific adsorption, the potential

$U_z$ of zero charge varies with composition as

$$\frac{dU_z}{d\mu} = \frac{1}{zF},$$

where $U_z$ is measured relative to a reference electrode reversible to the anion. Note that for repelling potentials $\Gamma_{i,d}$ shows a limiting amount of exclusion from the double layer, $\Gamma_{i,d} \longrightarrow -2\lambda c_{i\infty}$.

6. (a) Apply the Debye-Hückel approximation, equation 27-7, to the theory of the diffuse layer, and show that the diffuse-layer capacity is given, in this approximation, by

$$C_d = \frac{\epsilon}{\lambda}.$$

(b) Show from equations 52-8 and 52-10 that asymptotically, as $y$ approaches infinity, the potential and electric field in the diffuse layer are given by

$$\Phi = \lambda A e^{-y/\lambda} \quad \text{and} \quad E = A e^{-y/\lambda},$$

where $A$ is a constant independent of position.

7. For an ideally polarizable electrode in a solution of a single salt in the absence of specific adsorption, show how to calculate $C$, $U$, $\Gamma_+$, and $\sigma$ as functions of $q$ and $\mu$ if you are given $C_{M-2}$ as a function of $q$. In addition, the potential and the surface tension at the point of zero charge can be assumed to be known as functions of concentration. Is all of this last information necessary?

## NOTATION

| | |
|---|---|
| $A$ | area, cm² |
| $c_i$ | concentration of species $i$, mole/cm³ |
| $C_i$ | dimensionless concentration |
| $C$ | double-layer capacity, farad/cm² |
| $C_d$ | capacity of the diffuse layer, farad/cm² |
| $C_{M-2}$ | capacity of region between the metal and the inner limit of the diffuse layer, farad/cm² |
| $E$ | electric field, V/cm |
| $\varepsilon$ | dimensionless electric field |
| $F$ | Faraday's constant, 96,487 C/equiv |
| $G$ | Gibbs free energy, J |
| $K$ | see equation 52–17 |
| $n_i$ | number of moles of species $i$, mole |
| $p$ | pressure, dyne/cm² |
| $q$ | surface charge density on the metal side of the double layer, C/cm² |

$q_1$         surface charge density of specifically adsorbed ions, $C/cm^2$
$q_2$         surface charge density in the diffuse layer, $C/cm^2$
$Q_2$        see equation 52-17
$R$          universal gas constant, 8.3143 J/mole-deg
$S$          entropy, J/deg
$T$          absolute temperature, deg K
$U$          electrode potential, V
$V$          volume, $cm^3$
$x$          $y/\lambda$
$x_i$         mole fraction of species $i$
$y$          distance from surface, cm
$y_2$        position of inner limit of diffuse layer, cm
$z_i$         charge number of species $i$
$\Gamma_i$        surface concentration of species $i$, $mole/cm^2$
$\epsilon$          permittivity, farad/cm
$\lambda$          Debye length, cm
$\mu_i$        electrochemical potential of species $i$, J/mole
$\rho_e$        electric charge density, $C/cm^3$
$\sigma$          surface tension, dyne/cm
$\tau$          thickness of surface, cm
$\phi$          dimensionless potential
$\Phi$          electric potential, V

superscripts
$\alpha, \beta$      phases $\alpha$ and $\beta$
$\sigma$          surface

# Electrode Kinetics 8

## 55. Heterogeneous electrode reactions

Current concepts of double-layer structure are based on information obtained from the mercury electrode in the absence of the passage of a faradaic current, that is, from an ideally polarizable electrode. Now we want to turn to the consideration of charge-transfer or faradaic reactions. In electrochemical systems of practical importance, including corrosion, it is the reactions at the electrodes which are of primary importance.

The first thing we want to know about an electrode reaction is its rate. For a single electrode reaction occurring in a steady state, the rate of the reaction is related in a simple manner by Faraday's law to the current density, which is easily measured experimentally. Simultaneous reactions are discussed in section 61. Transient electrode processes involve the double-layer capacity, discussed in chapter 7 and again in section 58. They may also involve transient changes in the nature of the electrode surface.

The rate of the electrode reaction, characterized by the current density, depends first on the nature and previous treatment of the electrode surface. Second, the rate of reaction depends on the composition of the electrolytic solution adjacent to the electrode, just outside the double layer. This may be different from the composition of the bulk solution because of limited rates of mass transfer, treated in parts C and D. However, the diffuse part of the double layer is regarded as part of the interface. It is too thin to probe adequately, and the theory of the diffuse layer is a microscopic model rather than a macroscopic theory. The effect of double-layer structure on electrode kinetics is discussed in section 58.

Finally, the rate of the reaction depends on the electrode potential. This electrode potential is characterized by the surface overpotential $\eta_s$ defined in section 8 as the potential of the working electrode relative to a reference electrode of the same kind placed in the solution adjacent to the surface of the working electrode (just outside the double layer). This is a macroscopically well-defined potential and can be expressed in terms of electrochemical potentials. For the general electrode reaction expressed by equation 12-6, the equilibrium condition 12-7 is

$$\sum_i s_i \mu_i = n \mu_{e^-}. \tag{55-1}$$

The surface overpotential expresses the departure from the equilibrium potential and is given by

$$-F\eta_s = \mu_{e^-} - \sum_i \frac{s_i}{n} \mu_i. \tag{55-2}$$

Superscripts for the appropriate phases in which the species exist should be added; in particular, $\mu_{e^-}$ is the electrochemical potential of the electrons in the metal of the electrode. This definition of $\eta_s$ is equivalent to that given in section 8.

As an example, the surface overpotential for a copper electrode undergoing the reaction

$$Cu \longrightarrow Cu^{++} + 2e^- \tag{55-3}$$

is
$$\eta_s = -\frac{2\mu_{e^-} + \mu_{Cu^{++}} - \mu_{Cu}}{2F}. \tag{55-4}$$

In summary, for analyzing the behavior of electrochemical systems, we seek the macroscopic relationship between the current density and the surface overpotential and the composition adjacent to the electrode surface:

$$i = f(\eta_s, c_i). \tag{55-5}$$

Microscopic models may be useful in correlating these results, although they are not essential. Transient electrode processes can also involve the double-layer capacity and possibly hystereses related to changes in the surface of the electrode.

For sources treating electrode kinetics in general, one should consult references 1, 2, and 3. Vetter[1] has a wealth of experimental information on

[1]Klaus J. Vetter, *Elektrochemische Kinetik* (Berlin: Springer-Verlag, 1961). [English edition: *Electrochemical Kinetics* (New York: Academic Press, 1967).]

[2]Paul Delahay, *Double Layer and Electrode Kinetics* (New York: Interscience Publishers, 1965).

[3]John O'M. Bockris and Amulya K. N. Reddy, *Modern Electrochemistry*, Vol. 2 (New York: Plenum Press, 1970).

specific electrode reactions, and Tanaka and Tamamushi[4] give tables of parameters for a number of reactions.

## 56. Dependence of current density on surface overpotential

We have already indicated in section 8 the simplest type of dependence of the current density on the surface overpotential and the composition adjacent to the electrode surface, that given by the Butler-Volmer equation 8-2:

$$i = i_0 \left[ \exp\left(\frac{\alpha_a F}{RT} \eta_s\right) - \exp\left(-\frac{\alpha_c F}{RT} \eta_s\right) \right], \qquad (56\text{-}1)$$

and we have indicated that this can be regarded as the result of cathodic and anodic reactions proceeding independently, each with an exponential dependence on the surface overpotential $\eta_s$. The exchange current density $i_0$ then depends on the composition of the solution adjacent to the electrode, as well as the temperature and the nature of the electrode surface.

We have also indicated in section 8 the Tafel approximations 8-4 and 8-5 valid for large surface overpotentials. At low surface overpotentials, equation 1 can be approximated by a linear expression:

$$i = i_0 \frac{(\alpha_a + \alpha_c)F}{RT} \eta_s. \qquad (56\text{-}2)$$

Another method of plotting equation 1 is worth noting. Equation 1 can be written

$$i = i_0 \left\{ \exp\left[\frac{(\alpha_a + \alpha_c)F}{RT} \eta_s\right] - 1 \right\} \exp\left(-\frac{\alpha_c F}{RT} \eta_s\right) \qquad (56\text{-}3)$$

or

$$\ln \frac{i}{\exp\left(\frac{\alpha_a + \alpha_c}{RT} F\eta_s\right) - 1} = \ln i_0 - \frac{\alpha_c F}{RT} \eta_s. \qquad (56\text{-}4)$$

If the sum $\alpha_a + \alpha_c$ is known, experimental values of $i$ as a function of $\eta_s$ (at a given composition adjacent to the electrode) yield a straight line when the left side of equation 4 is plotted versus $\eta_s$. Then, the slope gives the value of $\alpha_c$, and the intercept gives the value of $i_0$. As indicated in the next section, there is some reason to expect the sum $\alpha_a + \alpha_c$ to have an integral value.

It should be emphasized again that, for a given composition adjacent to the electrode surface, there are three kinetic parameters in equation 1; these are $i_0$, $\alpha_a$, and $\alpha_c$. Experimental data are needed to determine these constants, in those cases where the experimental data can be adequately represented by equation 1.

[4]N. Tanaka and R. Tamamushi, "Kinetic Parameters of Electrode Reactions," *Electrochimica Acta*, *9*, 963–989 (1964).

The exchange current density $i_0$ depends on the composition of the solution adjacent to the electrode surface. Frequently, this dependence can be represented in terms of the concentrations of reactant and product species raised to some power:

$$i_0 = \left(\frac{c_1}{c_1^\infty}\right)^\gamma \left(\frac{c_2}{c_2^\infty}\right)^\delta i_0(c_i^\infty), \qquad (56\text{-}5)$$

where species 1 and 2 are reactant and product species and $i_0(c_i^\infty)$ is the exchange current density for some conveniently selected values of $c_i^\infty$. The concentrations adjacent to the electrode are denoted by $c_i$.

Many simple electrode reactions follow equation 1, possibly with some allowance for the effect of double-layer structure (see section 58). However, many reactions of technical importance show considerably different behavior. Outstanding among these are anodic dissolution processes showing passivation. A typical curve for such a process is shown in figure 56-1. Here,

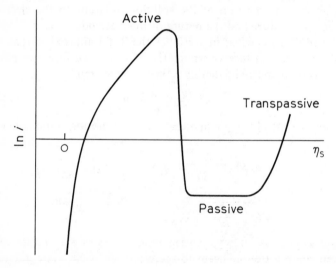

**Figure 56-1.** Illustration of current-potential relation for an electrode exhibiting passivation.

the reaction rate increases for increasing overpotential as indicated by equation 1. However, for sufficiently large overpotentials, a protective anodic oxide film, which may be very thin, forms on the electrode; and the current density drops to a very low value. Eventually, it may increase again either for the anodic dissolution process or, for an electronically conducting film, for anodic evolution of oxygen on the film. This is called the *transpassive region* of the curve. Such passivation phenomena can be reproducible, with very little time required for the formation or removal of the oxide film. Many ferrous alloys show this passivation phenomenon, with the passive current density and the maximum active current density depending on the composition of the alloy as well as the composition of the solution. Such behavior is important in the analysis of corroding systems.

The oxygen reaction generally is sluggish and not reproducible. On noble metals, oxide films can form, and a considerable hysteresis can be present, so that the reaction rate depends strongly on the previous history of the electrode as well as on the present values of the overpotentials and the concentrations adjacent to the electrode.

## 57.  Models for electrode kinetics

We should like to distinguish between surface reactions that are simple reactions and those that are elementary steps. A simple reaction is the simplest that can be observed by analytical methods; that is, the reactants and products can be determined by macroscopic analysis, but intermediates in the reaction cannot be detected or are quite unstable. A simple reaction may be composed of one or more elementary steps. An elementary step is the elementary, mechanistic process by which a reaction occurs. A stable reactant may thus produce an unstable intermediate which immediately enters into reaction in another elementary step.

For example, the copper dissolution and deposition reaction

$$Cu \rightleftharpoons Cu^{++} + 2e^- \tag{57-1}$$

can be regarded as being composed of two elementary steps, each of which involves the transfer of an electron:

$$Cu \rightleftharpoons Cu^+ + e^- \quad (fast), \tag{57-2}$$

$$Cu^+ \rightleftharpoons Cu^{++} + e^- \quad (slow). \tag{57-3}$$

According to Mattsson and Bockris,[5] the second step is inherently much slower than the first step. An elementary step involves the transfer of more than one electron only in exceptional cases.

The rates of the elementary steps comprising a simple reaction should always be proportional to one another. For example, reaction 2 should occur once every time reaction 3 occurs. A reaction is much simpler to analyze if it can be treated as a simple reaction, because then the rates of the elementary steps are simply related to each other. Whether a reaction should be regarded as a simple reaction or two or more simple reactions depends on just how unstable the active intermediate is, on the limits of detection of our analytical methods, and on the accuracy with which we wish to describe the system.

For the copper reaction, the intermediate cuprous ions are not com-

[5]E. Mattsson and J. O'M. Bockris, "Galvanostatic Studies of the Kinetics of Deposition and Dissolution in the Copper + Copper Sulphate System," *Transactions of the Faraday Society*, 55, 1586–1601 (1959).

pletely unstable, and they can diffuse away from the electrode where they decompose by the disproportionation reaction

$$2\,Cu^+ \rightleftharpoons Cu^{++} + Cu. \qquad (57\text{-}4)$$

Furthermore, it makes a difference whether the reaction 1 proceeds in the anodic or the cathodic direction. In the cathodic direction, the slow reaction 3 occurs first and is relatively slow. The cuprous ions thus produced react in reaction 2. Thus, the rate of reaction 2 is limited by the rate of supply of cuprous ions from reaction 3 and is the same as the rate of reaction 3. In the anodic direction, cuprous ions are produced by the relatively fast reaction 2. These can either diffuse away from the electrode or react in reaction 3. Now the rate of reaction 3 is determined in large part by its own kinetic characteristics and may not occur as fast as reaction 2, the difference corresponding to the cuprous ions which diffuse away from the electrode.

The rigorous treatment of the anodic process requires the treatment of reactions 2 and 3 as simultaneous reactions (see section 61) rather than as elementary steps of a simple reaction, and the analysis is complicated by the need to account for the diffusion and convection of cuprous ions, requiring the consideration of transport processes described in part C. On the other hand, if the cuprous ions were a species adsorbed on the electrode, they could not diffuse away. The coverage of cuprous ions would increase so that reactions 2 and 3 again occur at the same rate, and reaction 1 can be regarded as a simple reaction.

In our further treatment of the copper reaction, we shall assume that the cuprous ions do not diffuse away from the electrode, that their concentration reaches a value such that reactions 2 and 3 occur at the same rate, and that reaction 1 can be regarded as a simple reaction.

The distribution of potential in the double layer (see chapter 7) gives rise to a potential difference between the electrode and the solution, which we shall denote by

$$V = \Phi_{met} - \Phi_{soln}, \qquad (57\text{-}5)$$

where $\Phi_{met}$ is the electrostatic potential of the electrode and $\Phi_{soln}$ is the electrostatic potential of the solution just *outside* the double layer. This is usually not a well-defined potential; we can take $V$ to be the potential relative to a given electrode (see section 40).

Consider an elementary step which can be represented by the general equation

$$\sum_i s_i M_i^{z_i} \rightleftharpoons ne^-. \qquad (57\text{-}6)$$

The number of electrons transferred $n$ is zero if the elementary step does not involve charge transfer; it is one if charge transfer is involved, multiple electron transfers being unlikely in an elementary step. The rate of the elementary step can then usually be expressed by the equation

$$r = \frac{i}{nF} = k_a \exp\left[\frac{(1-\beta)nF}{RT} V\right] \prod_i c_i^{p_i}$$
$$- k_c \exp\left(-\frac{\beta nF}{RT} V\right) \prod_i c_i^{q_i}. \tag{57-7}$$

If more than one elementary step is involved in the simple reaction, an additional subscript should be added to $r, i, n, \beta, k_a, k_c, p_i, q_i,$ and $s_i$ to distinguish the elementary steps from each other and from the overall reaction.

In equation 7, $k_a$ and $k_c$ are rate constants for the anodic and cathodic directions, respectively. They would be expected to show an Arrhenius dependence on the temperature. The reaction orders for species $i$ in the anodic and cathodic directions are $p_i$ and $q_i$, respectively. The $k$'s and the exponential factors together represent Arrhenius rate constants with potential dependent activation energies, $\beta$ being a *symmetry factor* representing the fraction of the applied potential $V$ which promotes the cathodic reaction. Similarly, $1 - \beta$ is the fraction of the applied potential which promotes the anodic reaction. Frequently, it is assumed that $\beta$ should have the value $1/2$, although the theoretical justification for this is not completely rigorous.

The meaning of the symmetry factor $\beta$ is usually illustrated by means of a potential energy diagram (see figure 57-1). The potential-energy curve for

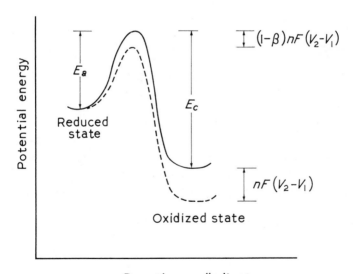

**Figure 57-1.** Potential-energy diagram for an elementary charge-transfer step. The solid curve is for $V = V_1$. The dashed curve is for $V = V_2$, where $V_2$ is greater than $V_1$.

an applied potential $V_1$ is shown by a solid curve, the activation energies $E_a$ and $E_c$ in the anodic and cathodic directions, respectively, being indicated on the figure. A change of the applied potential from $V_1$ to $V_2$ results in a change of the energy of the reduced state relative to the oxidized state by an amount $nF(V_2 - V_1)$, and this tends to drive the reaction anodically if $V_2$ is greater than $V_1$.

However, the potential energy diagram is envisioned to change to the dashed curve. (The zero of potential is not material here; consequently, the reduced state has been sketched at the same energy level. The zero of potential is absorbed into the rate constants $k_a$ and $k_c$ in equation 7.) Then, according to figure 1, the activation energy in the cathodic direction increases by $\beta nF(V_2 - V_1)$, and the activation energy in the anodic direction decreases by $(1 - \beta)nF(V_2 - V_1)$:

$$E_{c2} = E_{c1} + \beta nF(V_2 - V_1),\tag{57-8}$$

$$E_{a2} = E_{a1} - (1 - \beta)nF(V_2 - V_1).\tag{57-9}$$

This corresponds to the exponential terms in equation 7, and, as stated above, $\beta$ is the fraction of the applied potential which favors the cathodic reaction.

Now let us consider a simple reaction involving one elementrary step. Equation 7 describes the current-potential relationship for this reaction. At equilibrium, $i = 0$, and equation 7 yields for the equilibrium potential $V^0$

$$\frac{nFV^0}{RT} = \ln\frac{k_c}{k_a} + \sum_i (q_i - p_i) \ln c_i.\tag{57-10}$$

Comparison with the Nernst equation for this reaction,

$$\frac{nFV^0}{RT} = \text{constant} - \sum_i s_i \ln c_i,\tag{57-11}$$

suggests that $q_i$, $p_i$, and $s_i$ should be related by

$$q_i - p_i = -s_i.\tag{57-12}$$

In fact, for an elementary step, we presume that the reaction orders are given by the elementary reaction 6 itself. If $s_i = 0$, the species is neither a reactant nor a product, and we set $p_i$ and $q_i$ equal to zero. If $s_i > 0$, the species is an anodic reactant, and we set $p_i = s_i$ and $q_i = 0$. If $s_i < 0$, the species is a cathodic reactant; we set $q_i = -s_i$ and $p_i = 0$.

Now, we should recognize that the surface overpotential is the departure of $V$ from the equilibrium value $V^0$:

$$\eta_s = V - V_0.\tag{57-13}$$

This definition of $\eta_s$ is equivalent to those definitions given in section 8 and 55. The exchange current density $i_0$ is, for a simple reaction involving one elementary step, the value of the cathodic term or the anodic term in equation

7 at equilibrium (multiplied by $nF$):

$$\frac{i_0}{nF} = k_a \exp\left[\frac{(1-\beta)nF}{RT} V^0\right] \prod_i c_i^{p_i}$$

$$= k_c \exp\left(-\frac{\beta nF}{RT} V^0\right) \prod_i c_i^{q_i}. \tag{57-14}$$

With these definitions of $\eta_s$ and $i_0$, equation 7 can be written

$$i = i_0 \left\{ \exp\left[\frac{(1-\beta)nF}{RT}\eta_s\right] - \exp\left(-\frac{\beta nF}{RT}\eta_s\right) \right\}. \tag{57-15}$$

Comparison with equation 56-1 shows that for this single elementary step we have $\alpha_a = (1-\beta)n$ and $\alpha_c = \beta n$.

Finally, let us use equation 10 to eliminate $V^0$ from equation 14, with the result

$$i_0 = nFk_c^{1-\beta}k_a^\beta \prod_i c_i^{(q_i+\beta s_i)}. \tag{57-16}$$

This gives an explicit dependence of the exchange current density $i_0$ on the reactant and product concentrations adjacent to the electrode. We see that $\gamma$ and $\delta$ in equation 56-5 correspond to $q_i + \beta s_i$. Because $\beta$ is a fraction, the power on $c_i$ in equation 16 is generally a fraction even though $p_i$, $q_i$, and $s_i$ are all integers. This power is positive if the rules following equation 12 apply. Thus, the exchange current density increases if either the reactant or product concentration is increased.

An increase in the concentration of an anodic reactant for which $p_i = 1$ leads to a proportionate increase in the anodic term in equation 7 at constant $V$. However, $i_0$ is related to this term, not at constant $V$, but at the equilibrium potential. An increase in the concentration of the anodic reactant shifts the equilibrium potential such that the two terms in equation 7 remain equal. Thus, both terms in equation 7 increase, and the exchange current density $i_0$ increases, proportional to a fractional power of the concentration of the anodic reactant.

Let us illustrate the results of this model of electrode kinetics for a redox reaction

$$M_1^{z_1} \longrightarrow M_2^{z_2} + ne^-, \tag{57-17}$$

where there is only one reactant and one product and these are soluble ions. It is assumed that there is only one elementary step, which is the same as the redox reaction 17. Following the rules for reaction orders given below equation 12, we write equation 7 as

$$\frac{i}{nF} = k_a c_1 \exp\left(\frac{(1-\beta)nF}{RT}V\right) - k_c c_2 \exp\left(-\frac{\beta nF}{RT}V\right). \tag{57-18}$$

This implies that the cathodic and anodic reactions are first order in the reactants, but with potential-dependent rate constants.

The equilibrium potential is

$$V^0 = \frac{RT}{nF} \ln\left(\frac{k_c c_2}{k_a c_1}\right),$$  (57-19)

and the exchange current density is given by

$$i_0 = nF k_a^\beta k_c^{1-\beta} c_1^\beta c_2^{1-\beta}$$  (57-20)

(see equation 16). Equation 15 remains valid as the current overpotential relationship.

For more complex electrode reactions, one needs to write down the reaction mechanism in terms of elementary steps and analyze the kinetics of each step, as has been done here for a single elementary step. For the elementary steps 2 and 3 of the copper reaction, we write equation 7 as

$$\frac{i_2}{F} = k_{a2} \exp\left[\frac{(1 - \beta_2)F}{RT} V\right] - k_{c2} c_{Cu^+} \exp\left(-\frac{\beta_2 F}{RT} V\right)$$  (57-21)

and

$$\frac{i_3}{F} = k_{a3} c_{Cu^+} \exp\left[\frac{(1 - \beta_3)F}{RT} V\right] - k_{c3} c_{Cu^{++}} \exp\left(-\frac{\beta_3 F}{RT} V\right).$$  (57-22)

Subscripts 2 and 3 have been added corresponding to reactions 2 and 3. The concentration of the anodic reactant in reaction 2, copper, has not been included in equation 21 since it is constant.

As stated earlier, we shall assume that reactions 2 and 3 occur at the same rate. Hence

$$i = i_2 + i_3 = 2i_2 = 2i_3.$$  (57-23)

Next, we want to introduce the surface overpotential for the overall reaction (see equation 55-4), not those for reactions 2 and 3 individually. We shall also assume that reaction 2 is fast and essentially in equilibrium. (For the more general case, see problem 1.) For large values of $k_{a2}$ and $k_{c2}$, equation 21 yields the potential-dependent equilibrium concentration of cuprous ions:

$$c_{Cu^+} = \frac{k_{a2}}{k_{c2}} \exp\left(\frac{FV}{RT}\right).$$  (57-24)

Substitution into equation 22 gives

$$\frac{i_3}{F} = \frac{i}{2F} = \frac{k_{a3} k_{a2}}{k_{c2}} \exp\left[\frac{(2 - \beta_3)F}{RT} V\right] - k_{c3} c_{Cu^{++}} \exp\left(-\frac{\beta_3 F}{RT} V\right).$$  (57-25)

From equation 25, the equilibrium potential is

$$V^0 = \frac{RT}{2F} \ln\left(\frac{k_{c3} k_{c2}}{k_{a3} k_{a2}} c_{Cu^{++}}\right),$$  (57-26)

and the exchange-current density is

$$i_0 = 2F k_{c3} \left(\frac{k_{a3} k_{a2}}{k_{c3} k_{c2}}\right)^{\beta_3/2} c_{Cu^{++}}^{(2-\beta_3)/2}.$$  (57-27)

With the surface overpotential given by equation 13, equation 25 can now be written as equation 56-1:

$$i = i_0 \left[ \exp\left(\frac{\alpha_a F}{RT} \eta_s\right) - \exp\left(-\frac{\alpha_c F}{RT} \eta_s\right) \right], \qquad (57\text{-}28)$$

where
$$\alpha_a = 2 - \beta_3 \qquad (57\text{-}29)$$
and
$$\alpha_c = \beta_3. \qquad (57\text{-}30)$$

The concentration dependence of the exchange-current density from equation 27 can now be expressed as

$$i_0 = \left(\frac{c_{Cu^{++}}}{c_{Cu^{++}}^\infty}\right)^\gamma i_0(c_{Cu^{++}}^\infty), \qquad (57\text{-}31)$$

where
$$\gamma = \frac{2 - \beta_3}{2}. \qquad (57\text{-}32)$$

Mattsson and Bockris[5] studied the copper deposition and dissolution reaction in 1 $N$ sulfuric acid with various concentrations of copper sulfate. The exchange current density $i_0$ can depend on the concentration of sulfuric acid and the nature of the electrode surface as well as the cupric ion concentration. Mattsson and Bockris studied surfaces prepared by quenching molten copper in purified helium and surfaces prepared by electrodeposition of copper.

Mattsson and Bockris conclude that reaction 2 is inherently fast compared to reaction 3. Within the limits of reproducibility, $\alpha_a$ was 1.5 and $\alpha_c$ was 0.5, indicating that the symmetry factor $\beta_3$ for reaction 3 is equal to $1/2$. A value of $\gamma$ (in equation 31) of 0.6 was obtained for the deposited electrodes, and a value of 0.3 for the helium-prepared electrode. On the other hand, equation 32 gives $\gamma = 0.75$ if $\beta_3 = 0.5$. Our examination of their data suggests a value of 0.42 for $\gamma$ for the deposited electrodes. For later calculations, we shall use this value for $\gamma$ and take $i_0$ equal to 1 mA/cm$^2$ for a cupric ion concentration of 0.1 molar.

The results of Mattsson and Bockris deviate from equation 28 at low overpotentials. They attribute this to slow diffusion of adsorbed ions and atoms to and from lattice sites, indicating that this is not a simple charge-transfer process. This, along with standard deviations of 10 or 20 percent in the exchange-current-density values, indicates that electrode kinetics are, in general, neither predictable nor reproducible on solid electrodes.

The anodic and cathodic contributions to the current density according to equation 25 are plotted in figure 57-2 in order to illustrate the fact that the anodic contribution is independent of cupric ion concentration when plotted against the potential relative to a given reference electrode (see the remarks at the end of section 40).

The models discussed in this section provide a basis for the electrode-kinetic equation 56-1. However, these models do not have the rigor of a

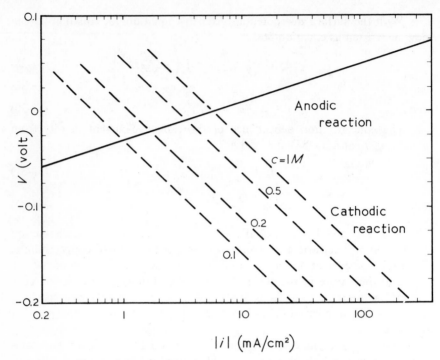

**Figure 57-2.** Anodic and cathodic contributions to the current density
(from equation 57-25 with $\beta_3 = 0.5$) plotted against the
potential relative to a copper electrode in 1 $M$ CuSO$_4$ (see
section 40).

thermodynamic derivation. The expressions in equation 7 are not rigorously
valid, and in equations 10 and 11, it was seen that comparison was made to
the approximate Nernst equation rather than to the exact thermodynamic
expression of the equilibrium potential. Furthermore, the potential $V$
depends on the choice of the given reference electrode. A different reference
electrode would involve, implicitly, different combinations of ionic activity
coefficients.

Equation 56-1 involves an unambiguous potential, the surface overpotential, and is thus superior to equation 7, from which equation 56-1 can
be "derived." In equation 56-1, one can only say that the exchange current
density depends on the composition of the solution. Although not completely
rigorous, the models do provide an explicit idea of the composition dependence of the exchange current density.

The surface overpotential alone is discussed in this chapter. In chapter
20, the concentration overpotential and the total overpotential are developed.

## 58. Effect of double-layer structure

The double-layer structure can have an effect on the overall behavior of the interface, first, by superimposing a capacitive effect on top of the electrode kinetics of the electrode reaction itself. This means that when the potential of the electrode is varied, the current that flows is partly due to charging the double-layer capacity and partly due to a faradaic reaction. The capacity of the double layer in the absence of a faradaic reaction was discussed in chapter 7.

We might now express the current density $i$ for an electrode of constant area as

$$i = f(\eta_s, c_i) + C\frac{d\eta_s}{dt}, \tag{58-1}$$

where $C$ is the double-layer capacity, $\eta_s$ is the surface overpotential, $c_i$ is the concentration of species $i$ just outside the double layer, and $f$ is a function describing the kinetics of the electrode reaction (see equation 55-5).

The double layer can behave like a capacitor that is in parallel with the electrode reactions, so that the current passing from the electrode to the solution either can take part in charge-transfer reactions or can contribute to the charge in the double-layer capacitor. It is this capacitive effect which reduces electrode polarization when alternating current is used for conductivity measurements. The double-layer capacity can also depend on the concentrations $c_i$ and the electrode potential $V$ (see figures 51-5 and 51-6).

Many experiments described in the literature have involved a growing mercury drop, in which the electrode area changes with time. In this case, the last term in equation 1 should be replaced by

$$\frac{1}{A}\frac{dqA}{dt},$$

where $q$ is the surface charge density on the electrode side of the interface and $A$ is the instantaneous electrode area. (Recall that $i$ is the current density flowing *from* the electrode *into* the solution.)

Equation 1 is written for constant concentrations $c_i$ adjacent to the electrode surface. In this case, the equilibrium potential $V^0$ is constant, and

$$\frac{dV}{dt} = \frac{d\eta_s}{dt}. \tag{58-2}$$

When the concentrations vary, the last term in equation 1 should, strictly, be replaced by $dq/dt$, because $q$ depends on $c_i$ as well as $V$.

When more than one reaction occurs, the first term in equation 1 should be replaced by $f(V, c_i)$, where $f$ now describes the faradaic current due to all the reactions at the electrode potential $V$ (see section 61). The last term can also be replaced by $C \, dV/dt$.

The distinction between the contributions of the charging current and the faradaic current to the overall current density $i$ in equation 1 may be subtle since one can measure only the overall current density directly. Ordinarily one would try to measure $f$ by making observations under steady conditions where the second term in equation 1 is zero. With the assumption that $f$ has the same dependence on $\eta_s$ and $c_i$ for unsteady conditions in the interface, the double-layer capacity can then be measured by superimposing a small alternating potential. The situation with a growing electrode area is even more complex and has been the subject of some discussion in the literature.[6,7,8,9]

The second way in which double-layer structure enters into electrode kinetics is in the method of application of the models of section 57 to an elementary step. Long ago, Frumkin[10] proposed that the diffuse part of the double layer should be treated separately from the charge-transfer reaction. Specifically, it was suggested that the concentrations $c_i$ which enter into equation 1 should be the concentrations $c_i^0$ at the inner limit of the diffuse layer and that the potential $\eta_s$ in the first term in equation 1 should be replaced by $\eta_s - \Phi_2$, where $\Phi_2$ is the potential at the inner limit of the diffuse layer.

The value of $\Phi_2$ is to be taken at zero current (in order to eliminate the ohmic potential drop) but at the same electrode potential $V$ as involved during the passage of current. It has been shown that the equilibrium diffuse layer (see section 52) is disturbed to only a minor extent by the passage of current. Thus $\Phi_2$ is still related to the surface charge density $q_2$ in the diffuse layer by equation 52-9 ($c_i$ being represented as $c_{i\infty}$ there), and $c_i^0$ is related to $c_i$ by

$$c_i^0 = c_i \exp\left(-\frac{z_i F \Phi_2}{RT}\right) \qquad (58\text{-}3)$$

(compare equation 52-1).

It is difficult to know *a priori* the magnitude of the surface charge density

[6] Paul Delahay, "Double Layer Studies," *Journal of the Electrochemical Society, 113,* 967–971 (1966).

[7] Karel Holub, Gino Tessari, and Paul Delahay, "Electrode Impedance without *a Priori* Separation of Double-Layer Charging and Faradaic Process," *The Journal of Physical Chemistry, 71,* 2612–2618 (1967).

[8] W. D. Weir, "*A Posteriori* Separation of Faradaic and Double-Layer Charging Processes: Analysis of the Transient Equivalent Network for Electrode Reactions," *The Journal of Physical Chemistry, 71,* 3357–3359 (1967).

[9] Fred C. Anson, "Electrochemistry," *Annual Review of Physical Chemistry, 19,* 83–110 (1968).

[10] A. Frumkin, "Wasserstoffüberspannung und Struktur der Doppelschicht," *Zeitschrift für physikalische Chemie, 164,* 121–133 (1933).

$q_2$ in the diffuse layer since this is determined in large part by interaction with the rest of the interface (metal surface and inner Helmholtz plane), the whole of which is electrically neutral. It is not the value of $q_2$ in the same system at equilibrium (zero current); it depends on the electrode potential in a way that is not readily determined during the passage of current.

The structure of the diffuse layer is studied best with an ideally polarizable electrode in the absence of a faradaic current and, hence, in the absence of any reacting species. In practice, we investigate the double layer with the nonreactive supporting electrolyte alone. Then we add a small amount of the reactant and *assume* that $q_2$ and $\Phi_2$ (at a given electrode potential) are not changed by the small addition or the small current now being passed.

One can also object that the correction is based on the microscopic theory of the diffuse layer and does not have a firm macroscopic basis. For these reasons, we do not expect that the correction can be applied with any certainty to solid electrodes, significant concentrations of reactants, or high ionic strengths. Nevertheless, the Frumkin correction does give an impressive qualitative account of complicated electrode behavior which can be attributed to double-layer structure, as reviewed by Parsons.[11]

## 59. The oxygen electrode

The oxygen electrode is the most complicated electrode commonly encountered. One reason is that the reaction is so irreversible, that is, the exchange current density is so low, that even traces of impurities can successfully compete with it. Consequently, the reversible, equilibrium oxygen potential could not be successfully observed until impurities had been rigorously excluded.[12]

A second reason for the complicated behavior of the oxygen electrode is that the overall reaction

$$O_2 + 4\,H^+ + 4e^- \rightleftharpoons 2\,H_2O \qquad (59\text{-}1)$$

can be regarded as the result of two simple reactions,

$$O_2 + 2\,H^+ + 2e^- \rightleftharpoons H_2O_2 \qquad (59\text{-}2)$$

$$H_2O_2 + 2\,H^+ + 2e^- \rightleftharpoons 2\,H_2O\,, \qquad (59\text{-}3)$$

in which hydrogen peroxide is a relatively stable and detectable intermediate.

This has, first, the consequence that the heterogeneous decomposition of hydrogen peroxide

$$2\,H_2O_2 \rightleftharpoons 2\,H_2O + O_2 \qquad (59\text{-}4)$$

[11] Roger Parsons, "The Structure of the Electrical Double Layer and Its Influence on the Rates of Electrode Reactions," *Advances in Electrochemistry and Electrochemical Engineering*, *1*, 1–64 (1961).

[12] J. O'M. Bockris and A. K. M. Shamshul Huq. "The mechanism of the electrolytic evolution of oxygen on platinum," *Proceedings of the Royal Society*, *A237*, 277–296 (1956).

can thus be regarded as a result of the electrochemical reactions 2 and 3, the first reaction proceeding anodically and the second reaction proceeding cathodically.

Furthermore, the hydrogen peroxide can diffuse away from the electrode at either an anode or a cathode, and, depending on the relative rates of reactions 2 and 3, more or less than four electrons may be required to produce or consume a molecule of oxygen. Since reaction 2 is generally inherently faster than reaction 3, one expects to observe peroxide formation in the cathodic consumption of oxygen, but not in the anodic process. Also, an appreciable time may be needed to get a steady current at a given cathodic potential until a steady bulk concentration of peroxide can build up.

A third complicating factor for the oxygen electrode is that there is an alternative reaction path involving adsorbed species and not the production of hydrogen peroxide. This leads to a fourth complication. The adsorbed layers can become so thick and are so slow to respond to changes in electrode potential that measurements can easily be carried out at the same potential on surfaces of quite different character. Also, the current for the reaction of these layers can be appreciable compared to the current for the primary reaction.

There are still other complications. Some metals, notably platinum, palladium, and rhodium, can dissolve oxygen to an appreciable extent. This can contribute to a hysteresis and is noted particularly in charging curves. Many metals, even gold, tend to corrode near the oxygen potential. Surface oxides are also responsible for the passivation characteristics of, for example, ferrous alloys.

The behavior of the oxygen electrode has been reviewed by several authors.[13,14,15,16]

## 60. Methods of measurement

One wants to determine the composition and overpotential dependence of the current density in electrode kinetics, that is, to find how $f$ in equation 55-5 depends on $c_i$ and $\eta_s$. The electrochemist is interested in elucidating the mechanism of the electrode reaction. The electrochemical engineer is inter-

[13]Klaus J. Vetter. *Electrochemical Kinetics* (New York: Academic Press, 1967), pp. 615–644.

[14]J. P. Hoare, "The Oxygen Electrode on Noble Metals," *Advances in Electrochemistry and Electrochemical Engineering, 6* (1967), 201–288.

[15]James P. Hoare, *The Electrochemistry of Oxygen* (New York: Interscience Publishers, 1968).

[16]A Damjanovic, "Mechanistic Analysis of Oxygen Electrode Reactions," J. O'M. Bockris and B. E. Conway, eds., *Modern Aspects of Electrochemistry*, No. 5 (New York: Plenum Press, 1969), pp. 369–483.

ested in predicting the behavior of practical electrochemical systems. This means that the electrochemist can work with high concentrations of supporting electrolyte on the reproducible surface of liquid mercury in highly purified solutions, whereas the electrochemical engineer needs values of the current density for electrode surfaces and solution compositions (including impurities) that are likely to be encountered.

The electrochemist also wants to study increasingly fast reactions (larger exchange current densities), which becomes increasingly more difficult due to the ohmic potential drop and the concentration variations near the electrode surface. On the other hand, from the point of view of analyzing the overall system behavior, we can assume that the kinetics become unimportant as the electrode reaction becomes too fast to measure conveniently.

For relatively slow reactions, the function $f$ in equation 55-5 can be measured directly under steady conditions by varying the electrode potential and the composition adjacent to the electrode. For faster reactions, there is a preference to use stirred solutions with known hydrodynamic characteristics or to use transient methods. The known hydrodynamic conditions allow us to calculate the composition at the electrode surface, where it may be significantly different from the bulk solution composition (see sections 103, 109, and 110).

Perhaps the simplest transient procedure is to interrupt the current after a steady condition at the interface has been developed. This may occur before the concentrations have changed appreciably from their initial values, in which case the convection is unimportant. Interrupter methods are useful for deposition and dissolution reactions because they allow the character of the surface to be more carefully controlled, since a smaller charge need be passed.

Interruption of the current is also supposed to eliminate the ohmic potential drop from the measurement, while the surface overpotential is maintained for a while by the charge in the double-layer capacitor. Systems used for interruption (and other transient methods) should have a uniform primary current distribution over the electrode of interest (see section 116), since otherwise the current density in the solution may not be zero everywhere after interruption of the current to the electrode.[17]

An ideal geometry in this respect is the sphere, provided that the means of support is constructed so that it does not interfere with the current distribution. Mattsson and Bockris[5] have used such a system to study the copper reaction. A growing mercury drop is another common example of this geometry and has the further advantages of a reproducible surface and known hydrodynamic conditions which further promote a uniform rate of mass transfer to the electrode (see section 110).

[17]John Newman, "Ohmic Potential Measured by Interrupter Techniques," *Journal of the Electrochemical Society, 117* (1970), 507–508.

Rotating cylinders (see section 109) also provide a uniform primary current distribution and uniform, known conditions of mass transfer. With the sphere and cylinder geometries, the ohmic potential drop can be calculated easily, in case interrupter methods are not to be used or one wants a check on the potential change when the current is interrupted.

Methods of studying electrode kinetics are reviewed in references 1, 18, and 19 and include a variety of possibilities among the transient procedures. Mention should be made of the use of a rotating ring-disk system to detect relatively unstable intermediates in the electrode reaction. For example, the presence of hydrogen peroxide produced in carrying out the oxygen reaction on a disk electrode (see section 59) can be quantitatively determined by reacting the hydrogen peroxide back to oxygen on a concentric ring electrode mounted in the same rotating surface as the disk electrode.

## 61. Simultaneous reactions

We have discussed a single electrode reaction, but this idealization is not always achieved. If two or more reactions can occur simultaneously, it is simplest to regard each reaction to occur independently, and the net current density is the sum of the current densities due to the several reactions.[20,21,22,23] Under these conditions, the open-circuit potential is not an equilibrium potential corresponding to any of these reactions but is a *mixed* or *corrosion* potential. At open circuit, equilibrium does not prevail; one reaction proceeds anodically and another cathodically, so that the net current density is zero. In this way, even traces of impurities can obscure the measurement of the equilibrium potential for the oxygen electrode (see section 59).

The same thing occurs in corrosion processes. The anodic process may

[18]John E. B. Randles, "Concentration Polarization and the Study of Electrode Reaction Kinetics," P. Zuman, ed., *Progress in Polarography*, vol. I (New York: Interscience Publishers, 1962), pp. 123–144.

[19]Ernest Yeager and Jaroslav Kuta, "Techniques for the Study of Electrode Processes," Henry Eyring, ed., Vol. IXA, Electrochemistry, pp. 345–461, *Physical Chemistry; An Advanced Treatise* (New York: Academic Press, 1970).

[20]Carl Wagner and Wilhelm Traud, "Über die Deutung von Korrosionvorgängen durch Überlagerung von elektrochemischen Teilvorgängen und über die Potentialbildung an Mischelektroden," *Zeitschrift für Elektrochemie*, 44 (1938), 391–402.

[21]M. Stern and A. L. Geary, "Electrochemical Polarization I. A Theoretical Analysis of the Shape of Polarization Curves," *Journal of the Electrochemical Society*, 104 (1957), 56–63.

[22]Klaus J. Vetter, *Electrochemical Kinetics* (New York: Academic Press, 1967), pp. 732–747.

[23]Mars G. Fontana and Norbert D. Greene, *Corrosion Engineering* (New York: McGraw-Hill Book Company, 1967).

be dissolution of iron

$$Fe \longrightarrow Fe^{++} + 2e^-, \qquad (61\text{-}1)$$

and the cathodic process may be the reduction of oxygen

$$O_2 + 4\,H^+ + 4e^- \longrightarrow 2\,H_2O, \qquad (61\text{-}2)$$

the two processes being coupled so that the electrons produced in equation 1 are consumed in equation 2, leaving a zero net current for the piece of iron. Thus, the rate of corrosion may be determined by the rate of mass transfer of oxygen to the corroding surface. Corrosion in aqueous media is often an electrochemical process.

All corrosion processes do not proceed by simultaneous reactions on the same surface, however. When two dissimilar metals are in contact, an electrochemical cell can easily be established. The anodic dissolution process may occur predominantly on one metal, while the cathodic process of oxygen reduction or hydrogen evolution occurs predominantly on the other metal. In other cases—for example, pitting corrosion—the anodic and cathodic processes may occur on different parts of the same metal. Analysis of these systems requires consideration of the ohmic potential drop and concentration variations in the solution[23] and cannot be confined to the electrochemical reactions at the surface.

Simultaneous reactions are also encountered when, for example, the limiting current for copper deposition is exceeded and hydrogen evolution begins. This is shown in figures 9-1 and 10-2.

The behavior of passivating metals[23] in corroding systems deserves special mention. For these metals, an increase in the severity of the corrosion environment can lead to passivation and a reduction in the corrosion rate. This can be achieved by making the metal more positive, which is the basis of anodic protection. For the iron-oxygen couple, an increase in stirring, which promotes the rate of oxygen transfer to the surface, can passivate the metal and decrease the corrosion rate.

Let us finally treat the oxygen reduction reaction, regarding equations 59-2 and 59-3 as simultaneous reactions. We shall imagine that a stagnant diffusion layer of thickness $\delta$ is adjacent to the metal, and we denote oxygen and hydrogen peroxide as species $A$ and $B$, respectively. Since reaction 59-2 is inherently faster than reaction 59-3, the hydrogen peroxide produced in the first reaction can diffuse away from the surface instead of reacting in the second reaction. The material balances for oxygen and hydrogen peroxide take the form

$$-\frac{i_2}{2F} = D_A \frac{c_A^\infty - c_A^0}{\delta}, \qquad (61\text{-}3)$$

$$\frac{i_3 - i_2}{2F} = -D_B \frac{c_B^\infty - c_B^0}{\delta}. \qquad (61\text{-}4)$$

Equation 3 represents the rate of diffusion of oxygen to the surface, this oxygen being consumed in reaction 59-2. (Subscripts 2 and 3 denote reactions 59-2 and 59-3, respectively.) Equation 4 represents the rate of diffusion of hydrogen peroxide away from the surface, this being equal to the difference between the rates at which it is being produced in reaction 59-2 and consumed in reaction 59-3.

At 25°C in water, the saturation concentration of oxygen is $1.26 \times 10^{-3}$ mole/liter at a partial pressure of one atmosphere (corrected for the vapor pressure of water but not for the fugacity coefficients in the gas phase), and the diffusion coefficient is $1.9 \times 10^{-5}$ cm²/sec. Davis et al.,[24] among others, report the saturation concentration and the diffusion coefficient as functions of the concentration of potassium hydroxide.

In acidic or neutral media, let us assume that reactions 59-2 and 59-3 are pseudo first order in oxygen and peroxide and write

$$\frac{i_2}{2F} = k_{a2} c_B^0 \exp\left(\frac{\alpha_{a2}F}{RT} V\right) - k_{c2} c_A^0 \exp\left(-\frac{\alpha_{c2}F}{RT} V\right), \tag{61-5}$$

$$\frac{i_3}{2F} = k_{a3} \exp\left(\frac{\alpha_{a3}F}{RT} V\right) - k_{c3} c_B^0 \exp\left(-\frac{\alpha_{c3}F}{RT} V\right), \tag{61-6}$$

where the $k$'s may now be pH dependent and where the electrode potential $V$ is measured relative to a hydrogen electrode in the same solution (see section 40).

For sufficiently cathodic potentials, the anodic terms in equations 5 and 6 are negligible, and this seems particularly appropriate for the relatively irreversible oxygen electrode. In addition, we shall take $\alpha_{c2} = \alpha_{c3} = 0.5$ and $c_B^\infty = 0$. With these approximations, equations 3 to 6 can be combined to express the overall current density as

$$-\frac{i_2 + i_3}{2F} \frac{\delta}{c_A^\infty D_A} = \frac{e^{-\phi}}{1 + e^{-\phi}}\left(1 + \frac{1}{1 + Ke^\phi}\right), \tag{61-7}$$

where
$$\phi = \frac{\alpha_{c2}FV'}{RT}, \tag{61-8}$$

$$V' = V - \frac{RT}{\alpha_{c2}F} \ln \frac{k_{c2}\delta}{D_A}, \tag{61-9}$$

and
$$K = \frac{D_B k_{c2}}{D_A k_{c3}}. \tag{61-10}$$

Equation 7 is plotted in figure 61-1.

Figure 61-1 shows limiting currents for the reduction of oxygen and hydrogen peroxide (compare figures 9-1 and 10-2). For small values of $K$, the processes occur simultaneously; for large values of $K$, the processes become clearly distinguishable. For either very small or very large values of $K$, the current reaches half of its plateau value when $V = (RT/\alpha_{c2}F) \ln (k_{c2}\delta/D_A)$,

[24]R. E. Davis, G. L. Horvath, and C. W. Tobias, "The Solubility and Diffusion Coefficient of Oxygen in Potassium Hydroxide Solutions," *Electrochimica Acta*, 12 (1967), 287–297.

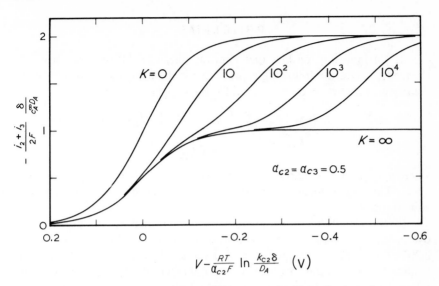

**Figure 61-1.** Theoretical polarographic curves for the reduction of oxygen, with neglect of the anodic reaction terms. In this, $K = D_B k_{c2}/D_A k_{c3}$.

thus providing a method of determining $k_{c2}$. This value of $V$ is known as the *half-wave potential*.

The position of the second wave relative to the first depends on the value of $K$, that is, on the slowness of reaction 59-3 relative to reaction 59-2. Polarographic curves for the reduction of oxygen on mercury[25,26] suggest that $K$ has a value of about $3 \times 10^7$ for this system. A more exact analysis would include the possibility of $\alpha_{c2}$ and $\alpha_{c3}$ being different from each other and different from 0.5 Also, the anodic term in equation 5 should perhaps be included in the analysis.

One can contrast the behavior of cupric ion reduction to oxygen reduction. Since reaction 57-2 is faster than reaction 57-3, cuprous ion reduction should occur simultaneously with reduction of cupric ions to cuprous ions, and only one polarographic wave would be distinguishable. On the other hand, reaction 59-2 is faster than reaction 59-3, and two waves are observed for oxygen reduction.

[25] I. M. Kolthoff and C. S. Miller, "The Reduction of Oxygen at the Dropping Mercury Electrode," *Journal of the American Chemical Society, 63* (1941), 1013–1017.

[26] R. Brdička and K. Wiesner, "Polarographic Determination of the Rate of the Reaction between Ferrohem and Hydrogen Peroxide," *Collection of Czechoslovak Chemical Communications, 12* (1947), 39–63.

## PROBLEMS

**1.** (a) Ignoring the fact that cuprous ions can diffuse away from the electrode, derive the following expression for the dependence of the current density on the surface overpotential for the copper reaction:

$$i = \frac{i_0 \left\{ \exp\left[\frac{(2 - \beta_3)F}{RT} \eta_s\right] - \exp\left(-\frac{\beta_3 F}{RT} \eta_s\right) \right\}}{1 + \frac{k_{a3}}{k_{c2}}\left[\frac{k_{c3}k_{c2}}{k_{a3}k_{a2}} c_{Cu^{+}} \exp\left(\frac{2F}{RT} \eta_s\right)\right]^{(1+\beta_2-\beta_3)/2}},$$

where $\eta_s$ is given by equation 57-13, $V^0$ by equation 57-26, and $i_0$ by equation 57-27. This equation has been written so as to be directly comparable to equation 57-28.

(b) Show that this expression reduces to equation 57-28 for large cathodic overpotentials. Show that the Tafel slope for large anodic overpotentials is $2.303RT/(1 - \beta_2)F$ (see equation 8-4). Reaction 57-2 appears to be rate controlling for large anodic overpotentials, and reaction 57-3 for large cathodic overpotentials. Rationalize this in physical terms.

(c) Show that this expression reduces to equation 57-28 when $k_{c2}$ and $k_{a2}/c_{Cu^{+}}$ are much larger than $k_{c3}$ and $k_{a3}$.

**2.** Develop the analogue of equation 57-15 with consideration of the Frumkin correction,

$$i = i_0 \exp\left[\frac{F\Phi_2}{RT}(\beta n - \sum_i z_i q_i)\right] \times \left\{\exp\left[\frac{(1 - \beta)nF}{RT} \eta_s\right] - \exp\left(-\frac{\beta nF}{RT} \eta_s\right)\right\},$$

where the composition dependence of $i_0$ is given by equation 57-16.

**3.** The standard electrode potential for reaction 59-2 is 0.682 V and that for reaction 59-3 is 1.77 V. Derive the second of these values from the first and entry 15 of table 20-1. Show that the equilibrium concentration of hydrogen peroxide should be about $3.2 \times 10^{-19}$ mole/liter for an oxygen partial pressure of one atmosphere.

## NOTATION

| | |
|---|---|
| $A$ | electrode area, cm² |
| $c_i$ | concentration of species $i$, mole/cm³ |
| $C$ | double-layer capacity, farad/cm² |
| $D_i$ | diffusion coefficient of species $i$, cm²/sec |
| $e^-$ | symbol for the electron |
| $E_a, E_c$ | activation energies in anodic and cathodic directions, J/mole |
| $f$ | function in expression of electrode kinetics |
| $F$ | Faraday's constant, 96,487 C/equiv |

| | |
|---|---|
| $i$ | current density, $A/cm^2$ |
| $i_0$ | exchange current density, $A/cm^2$ |
| $k_a, k_c$ | rate constants in anodic and cathodic directions |
| $M_i$ | symbol for the chemical formula of species $i$ |
| $n$ | number of electrons transferred in electrode reaction |
| $p_i$ | reaction order for anodic reactants |
| $q$ | surface charge density on the metal side of the double layer, $C/cm^2$ |
| $q_2$ | surface charge density in the diffuse layer, $C/cm^2$ |
| $q_i$ | reaction order for cathodic reactants |
| $r$ | reaction rate, $mole/cm^2\text{-}sec$ |
| $R$ | universal gas constant, 8.3143 J/mole-deg |
| $s_i$ | stoichiometric coefficient of species $i$ in electrode reaction |
| $t$ | time, sec |
| $T$ | absolute temperature, deg K |
| $V$ | electrode potential, V |
| $z_i$ | charge number of species $i$ |
| $\alpha_a, \alpha_c$ | transfer coefficients |
| $\beta$ | symmetry factor |
| $\gamma, \delta$ | exponents in composition dependence of the exchange current density |
| $\delta$ | thickness of stagnant diffusion layer, cm |
| $\eta_s$ | surface overpotential, V |
| $\mu_i$ | electrochemical potential of species $i$, J/mole |
| $\Phi$ | electric potential, V |
| $\Phi_2$ | potential at inner limit of diffuse layer, V |

# Electrokinetic Phenomena

# 9

This chapter deals with the effects observed when the diffuse double layer and an external electric field interact in relation to a hydrodynamic flow. A tangential electric field can produce a small change in velocity over the very small thickness of the double layer, and a shear stress or velocity gradient at the surface can produce electrical effects.

This subject is important in the study of colloids. It also yields some information on the structure of the electrical double layer at a solid-solution interface, which cannot be studied with the aid of the interfacial tension. The treatment here is incomplete, particularly with regard to experimental results for specific interfaces, and the reader is referred to the literature.[1,2,3]

## 62. Discontinuous velocity at an interface

Suppose that we have a planar, solid dielectric in contact with an electrolytic solution and that there is a tangential electric field (see figure 62-1).

[1] J. Th. G. Overbeek, "Electrokinetic Phenomena." H. R. Kruyt, ed., *Colloid Science*, vol. I, pp. 194–244 (Amsterdam: Elsevier Publishing Company, 1952).

[2] A. J. Rutgers and M. De Smet, "Electrokinetic Researches in Capillary Systems and in Colloidal Solutions," *Electrochemical Constants*, pp. 263–279, National Bureau of Standards (U.S.) Circular 524 (1953).

[3] A. Klinkenberg and J. L. van der Minne, eds., *Electrostatics in the Petroleum Industry* (Amsterdam: Elsevier Publishing Company, 1958).

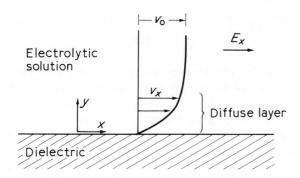

**Figure 62-1.** Velocity produced by a tangential electric field in the diffuse charge layer. A positive charge in the diffuse layer will produce a negative zeta potential and will result in a positive value of $v_0$ if $E_x$ is also positive.

A double layer can exist at the surface due to the specific adsorption of ions, and this means that there will be a counterbalancing charge in a diffuse layer. The structure of this diffuse layer was discussed in section 52. The tangential electric field will exert a force on the charge in the diffuse layer. This layer, being part of the solution, is mobile and can be expected to move relative to the solid as a result of the electric field.

The tangential electric field is taken to be uniform throughout the dielectric and the solution. The structure of the double layer will then not be disturbed from that treated in section 52. The resulting motion of the solution will be described by the Navier-Stokes equation 94-4 with the electrical force included (see equation 93-5). This equation is simplified by the fact that the velocity is only in the $x$-direction and depends only on the distance $y$ from the dielectric. In the steady state and with no significant gradient of the dynamic pressure, we have (see equation B-9)

$$\mu \frac{\partial^2 v_x}{\partial y^2} + \rho_e E_x = 0. \tag{62-1}$$

Substitution of Poisson's equation 22-8 gives

$$\mu \frac{\partial^2 v_x}{\partial y^2} - \epsilon \frac{\partial^2 \Phi}{\partial y^2} E_x = 0, \tag{62-2}$$

and integration gives

$$\mu \frac{\partial v_x}{\partial y} = \epsilon \frac{\partial \Phi}{\partial y} E_x, \tag{62-3}$$

the integration constant being evaluated from the fact that both $\partial v_x/\partial y$ and $\partial \Phi/\partial y$ are zero outside the diffuse layer. A second integration gives

$$\mu(v_x - v_0) = \epsilon(\Phi - \Phi_\infty)E_x, \tag{62-4}$$

where $v_0$ is the value of the velocity outside the diffuse layer.

If we take $\zeta$ to be the value of $\Phi - \Phi_\infty$ at the plane where $v_x = 0$, equation 4 yields

$$v_0 = -\frac{\epsilon \zeta E_x}{\mu}. \qquad (62\text{-}5)$$

The *zeta potential* $\zeta$ can be roughly associated with the potential $\Phi_2$ at the inner limit of the diffuse layer, since this is the plane where we should expect the velocity $v_x$ to become zero. However, we are unlikely to have an independent determination of $\Phi_2$ at a solid-solution interface.

The zeta potential is a property of the dielectric-solution interface and is due to the amount of specific adsorption at that interface. Because $\mu$ and $\epsilon$ are unlikely to be constant through the diffuse layer, $\zeta$ should more be regarded as a macroscopic variable relating the velocity $v_0$ to the tangential electric field $E_x$, and its relationship to the potential $\Phi_2$ thereby becomes more remote.

Because of the thinness of the diffuse layer compared to macroscopic dimensions, equation 5 can be regarded as a relation between the local slip velocity $v_0$ and the local tangential field $E_t$, even though the dielectric-solution interface is not planar, the tangential electric field is not uniform, and the gradient of the dynamic pressure is not zero. We shall attempt to clarify the meaning of this approximation in the context of a straight capillary through a dielectric; and this approximation will be applied to spherical, colloidal particles in the treatment of electrophoretic velocities and sedimentation potentials.

Small metal particles behave much like small dielectric particles if the metal behaves like an ideally polarizable electrode (see section 49). Then, the electric field is essentially zero within the metal, and the local polarization varies because of the tangential electric field. As long as the polarization does not become so great as to violate the condition for ideal polarizability, equation 5 can be applied to relate the local slip velocity $v_0$ to the local tangential electric field $E_t$. With metal particles, one has the possibility to vary the charge density on the metal side of the interface.

Levich[4] prefers to replace the zeta potential by the charge density $q_2$ in the diffuse layer. With the Debye-Hückel approximation we can relate $\zeta$ and $q_2$ by

$$q_2 = -\frac{\epsilon \zeta}{\lambda}, \qquad (62\text{-}6)$$

(see problem 6 of chapter 7) so that equation 5 becomes

$$v_0 = \frac{q_2 \lambda E_x}{\mu}. \qquad (62\text{-}7)$$

A zeta potential can be on the order of 0.1 V. For a relative dielectric constant $\epsilon/\epsilon_0$ of 78.3, a viscosity $\mu$ of $0.89 \times 10^{-2}$ g/cm-sec, and an electric

---

[4]Veniamin G. Levich, *Physicochemical Hydrodynamics*, section 94 (Englewood Cliffs, N. J.: Prentice-Hall, Inc., 1962).

field of 10 V/cm, equation 5 yields

$$v_0 = -7.8 \times 10^{-3} \text{ cm/sec.} \tag{62-8}$$

This is a relatively small value and can be neglected in many applications.

## 63. Electro-osmosis and the streaming potential

Let us consider a capillary of radius $r_0$ in a dielectric material. The capillary is filled with an electrolytic solution, and there is a uniform electric field in the direction $z$ along the axis of the capillary. However, the electric field is not zero in the radial direction. Instead, Poisson's equation is obeyed in the form

$$\frac{1}{r}\frac{\partial}{\partial r}\left(r\frac{\partial \Phi}{\partial r}\right) = -\frac{\rho_e}{\epsilon}. \tag{63-1}$$

The term $\partial^2\Phi/\partial z^2$ is zero since the axial electric field $E_z$ is constant. The momentum equation (see chapter 15) can be written

$$-\frac{dp}{dz} = \frac{\mu}{r}\frac{\partial}{\partial r}\left(r\frac{\partial v_z}{\partial r}\right) + E_z\rho_e = 0, \tag{63-2}$$

the electric force $E_z\rho_e$ appearing in the force balance.

Substitution of equation 2 into equation 1 and integration twice with respect to $r$, subject to the conditions that $v_z = 0$ at $r = r_0$ and $v_z$ and $\Phi$ are finite at $r = 0$, give

$$v_z = E_z\frac{\epsilon}{\mu}(\Phi - \Phi_{r=r_0}) - \frac{dp}{dz}\frac{r_0^2 - r^2}{4\mu}. \tag{63-3}$$

Hence, the volumetric flowrate $Q$ can be expressed as

$$\frac{Q}{2\pi} = \int_0^{r_0} rv_z\, dr = E_z\frac{\epsilon}{\mu}\int_0^{r_0} r(\Phi - \Phi_{r=r_0})\, dr - \frac{dp}{dz}\int_0^{r_0} r\frac{r_0^2 - r^2}{4\mu}\, dr$$

$$= -E_z\frac{\epsilon}{2\mu}\int_0^{r_0} r^2\frac{\partial \Phi}{\partial r}\, dr - \frac{r_0^4}{16\mu}\frac{dp}{dz}. \tag{63-4}$$

Now, let us turn our attention to the electric current. We express the flux of a species as (see equation 69-1)

$$N_{iz} = z_iu_iFc_iE_z + c_iv_z. \tag{63-5}$$

The concentrations do not vary in the $z$-direction for a given value of $r$; hence, there is no diffusion term in equation 5. The current density in the solution becomes

$$i_z = F\sum_i z_iN_{iz} = F^2E_z\sum_i z_i^2u_ic_i + Fv_z\sum_i z_ic_i$$

$$= \kappa E_z + v_z\rho_e, \tag{63-6}$$

where

$$\kappa = F^2\sum_i z_i^2u_ic_i \tag{63-7}$$

is the conductivity (see equation 70-3) and

$$\rho_e = F\sum_i z_ic_i \tag{63-8}$$

is the charge density. The total current $I$ can be expressed as

$$\frac{I}{2\pi} = \int_0^{r_0} r i_z \, dr = E_z \int_0^{r_0} r\kappa \, dr + \int_0^{r_0} r v_z \rho_e \, dr. \qquad (63\text{-}9)$$

Now substitute equation 3 for $v_z$ and equation 1 for $\rho_e$.

$$\frac{I}{2\pi} = E_z \int_0^{r_0} r\kappa \, dr - \frac{E_z \epsilon^2}{\mu} \int_0^{r_0} (\Phi - \Phi_{r=r_0}) \frac{\partial}{\partial r}\left(r\frac{\partial \Phi}{\partial r}\right) dr$$

$$+ \epsilon \frac{dp}{dz} \int_0^{r_0} \frac{r_0^2 - r^2}{4\mu} \frac{\partial}{\partial r}\left(r\frac{\partial \Phi}{\partial r}\right) dr$$

$$= E_z \int_0^{r_0} r\kappa \, dr + E_z \frac{\epsilon^2}{\mu} \int_0^{r_0} r\left(\frac{\partial \Phi}{\partial r}\right)^2 dr + \frac{dp}{dz}\frac{\epsilon}{2\mu} \int_0^{r_0} r^2 \frac{\partial \Phi}{\partial r} \, dr. \qquad (63\text{-}10)$$

Notice that the coefficient of $E_z$ in equation 4 is identical to the coefficient of $-dp/dz$ in equation 10. This is an example of the Onsager reciprocal relation.

In order to obtain explicit expressions for the potential distribution within the capillary, let us use the Debye-Hückel approximation 27-7:

$$c_i = c_i^0 e^{-z_i F\phi/RT} \approx c_i^0\left(1 - \frac{z_i F\phi}{RT}\right), \qquad (63\text{-}11)$$

where $\phi = \Phi - \Phi_{r=0}$ and $c_i^0$ is the concentration of species $i$ on the center line of the capillary.

Now equation 1 becomes

$$\frac{1}{r}\frac{\partial}{\partial r}\left(r\frac{\partial \phi}{\partial r}\right) = -\frac{F}{\epsilon}\sum_i z_i c_i^0 + \frac{F}{\epsilon}\sum_i z_i^2 c_i^0 \frac{F\phi}{RT} \qquad (63\text{-}12)$$

or

$$\frac{1}{x}\frac{d}{dx}\left(x\frac{d\psi}{dx}\right) = -\Gamma + \psi, \qquad (63\text{-}13)$$

where

$$\psi = \frac{F\phi}{RT}, \qquad x = \frac{r}{\lambda}, \qquad \Gamma = \frac{\sum_i z_i c_i^0}{\sum_i z_i^2 c_i^0}, \qquad (63\text{-}14)$$

and

$$\lambda = \left(\frac{\epsilon RT}{F^2 \sum_i z_i^2 c_i^0}\right)^{1/2} \qquad (63\text{-}15)$$

is the Debye length.

The solution to equation 13 is

$$\psi = \Gamma - \Gamma I_0(x), \qquad (63\text{-}16)$$

where $I_0$ is the modified Bessel function of the first kind, of order zero. The coefficient of $I_0$ is evaluated from the condition that $\psi = 0$ at $x = 0$. The other solution, $K_0(x)$, of the homogeneous form of equation 13 is unbounded at $x = 0$ and must be discarded.

The approximate expression for the charge density now is

$$\rho_e = F(\Gamma - \psi)\sum_i z_i^2 c_i^0 = \Gamma I_0(x)F\sum_i z_i^2 c_i^0 = \frac{\Gamma I_0(x)\epsilon RT}{F\lambda^2}. \qquad (63\text{-}17)$$

Note that the charge density on the center line of the capillary is not exactly zero. We should now like to relate the constant $\Gamma$ to the surface charge density $q_2$ per unit of circumferential area of the capillary:

$$\int_0^{r_0} 2\pi r \rho_e \, dr = 2\pi \, r_0 q_2 \tag{63-18}$$

or

$$\frac{\Gamma \epsilon RT}{F} \int_0^{R_0} x I_0(x) \, dx = r_0 q_2 \tag{63-19}$$

or

$$\Gamma = \frac{q_2 F \lambda}{\epsilon RT I_1(R_0)}, \tag{63-20}$$

where $R_0 = r_0/\lambda$ and $I_1$ is the modified Bessel function of the first kind, of order one. Also,

$$\rho_e = \frac{q_2 I_0(x)}{\lambda I_1(R_0)}, \tag{63-21}$$

and

$$\frac{\partial \Phi}{\partial r} = \frac{RT}{\lambda F} \frac{d\psi}{dx} = -\frac{q_2}{\epsilon} \frac{I_1(x)}{I_1(R_0)}. \tag{63-22}$$

We are now in a position to evaluate the integrals in equations 4 and 10:

$$\frac{\epsilon}{2\mu} \int_0^{r_0} r^2 \frac{\partial \Phi}{\partial r} dr = -\frac{\lambda^3 q_2}{2\mu I_1(R_0)} \int_0^{R_0} x^2 I_1(x) \, dx$$
$$= -\frac{\lambda q_2 r_0^2}{2\mu} \frac{I_2(R_0)}{I_1(R_0)} = -\frac{\lambda q_2 r_0^2}{2\mu}\left[\frac{I_0(R_0)}{I_1(R_0)} - \frac{2}{R_0}\right], \tag{63-23}$$

$$\int_0^{r_0} \left(\frac{\partial \Phi}{\partial r}\right)^2 r \, dr = \frac{\lambda^2 q_2^2}{\epsilon^2 I_1^2(R_0)} \int_0^{R_0} I_1^2(x) x \, dx$$
$$= \frac{q_2^2 \lambda r_0}{\epsilon^2}\left\{\frac{I_0(R_0)}{I_1(R_0)} - \frac{R_0}{2}\left[\frac{I_0^2(R_0)}{I_1^2(R_0)} - 1\right]\right\}$$
$$= \frac{q_2^2 r_0^2}{2\epsilon^2}\left[1 - \frac{I_0(R_0)I_2(R_0)}{I_1^2(R_0)}\right], \tag{63-24}$$

$$\int_0^{r_0} r c_i \, dr = \frac{c_i^0 r_0^2}{2}\left\{1 + z_i \frac{q_2 \lambda F}{\epsilon RT}\left[\frac{2}{R_0} - \frac{1}{I_1(R_0)}\right]\right\}, \tag{63-25}$$

and

$$\int_0^{r_0} r\kappa \, dr = \frac{r_0^2}{2}\kappa_{\text{avg}}, \tag{63-26}$$

where

$$\kappa_{\text{avg}} = \kappa^0\left\{1 + \frac{q_2 \lambda F}{\epsilon RT}\left[\frac{2}{R_0} - \frac{1}{I_1(R_0)}\right]\frac{\sum_i z_i^3 u_i c_i^0}{\sum_i z_i^2 u_i c_i^0}\right\} \tag{63-27}$$

and

$$\kappa^0 = F^2 \sum_i z_i^2 u_i c_i^0 \tag{63-28}$$

is the conductivity on the center line of the capillary.

Equation 10 now becomes

$$\frac{I}{\pi r_0^2} = \langle i_z \rangle = E_z\left\{\kappa_{\text{avg}} + \frac{q_2^2}{\mu}\left[1 - \frac{I_0(R_0)I_2(R_0)}{I_1^2(R_0)}\right]\right\} - \frac{dp}{dz}\frac{\lambda q_2}{\mu}\frac{I_2(R_0)}{I_1(R_0)}, \tag{63-29}$$

and equation 4 becomes

$$\frac{Q}{\pi r_0^2} = \langle v_z \rangle = E_z \frac{\lambda q_2}{\mu}\frac{I_2(R_0)}{I_1(R_0)} - \frac{r_0^2}{8\mu}\frac{dp}{dz}. \tag{63-30}$$

Figure 63-1 shows the velocity profile in the capillary when there is no pressure drop, $dp/dz = 0$. This can be obtained from equation 3, which now becomes

$$v_z = \frac{\lambda q_2 E_z}{\mu I_1(R_0)}[I_0(R_0) - I_0(x)] - \frac{dp}{dz}\frac{r_0^2 - r^2}{4\mu}. \tag{63-31}$$

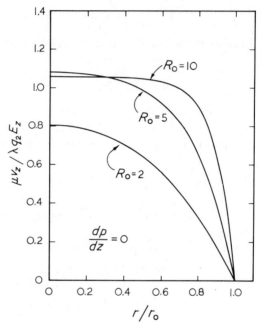

**Figure 63-1.** Velocity profile in the capillary when there is no pressure drop; this is the case of electro-osmosis.

In the absence of a pressure drop, an axial electric field can induce a flow in the capillary, and this is known as *electro-osmosis*. The average velocity or flow rate is then given by equation 30 as

$$\frac{\mu\langle v_z\rangle}{\lambda q_2 E_z} = \frac{I_2(R_0)}{I_1(R_0)} \quad \text{when} \quad \frac{dp}{dz} = 0 \tag{63-32}$$

and is tabulated in table 63-1 as a function of $R_0$, the ratio of the radius of the capillary to the Debye length. For large values of $R_0$, the diffuse layer is relatively thin, and the velocity variation occurs near the wall. This velocity change is given approximately as $\lambda q_2 E_z/\mu$ and is essentially the same as that given by equation 62-7.

The second term on the right in equation 31 corresponds to the usual velocity profile induced by a pressure gradient $dp/dz$. The experiment might be constrained by a condition of no net flow, $Q = 0$, rather than no pressure

TABLE 63-1. DIMENSIONLESS FLOW RATE $\mu\langle v_z\rangle/\lambda q_2 E_z$ AS A FUNCTION OF $R_0$ IN THE ABSENCE OF A PRESSURE DROP. [This also corresponds to the dimensionless pressure drop $(dp/dz)r_0^2/8\lambda q_2 E_z$ generated in the absence of net fluid flow or to the dimensionless streaming potential with no net current or to the dimensionless streaming current at zero potential drop (see equations 33, 34, 36, and 38)].

| $R_0$ | $I_2(R_0)/I_1(R_0)$ | $R_0$ | $I_2(R_0)/I_1(R_0)$ |
|-------|---------------------|-------|---------------------|
| 0     | 0                   | 10    | 0.85419             |
| 0.1   | 0.02499             | 20    | 0.92599             |
| 0.2   | 0.04992             | 50    | 0.97015             |
| 0.5   | 0.12372             | 100   | 0.98504             |
| 1     | 0.24019             | 250   | 0.99401             |
| 2     | 0.43313             | 500   | 0.99700             |
| 5     | 0.71934             | $\infty$ | 1.0              |

drop. In this case, the axial electric field induces a pressure drop rather than a flow, and this pressure drop is given according to equation 30 as

$$\frac{r_0^2}{8\lambda q_2 E_z}\frac{dp}{dz} = \frac{I_2(R_0)}{I_1(R_0)} \quad \text{when} \quad Q = 0, \tag{63-33}$$

also tabulated in table 1. The velocity is not zero throughout the capillary; the second term in equation 31 now assumes a magnitude such that the average velocity is zero. The local velocity profile is shown in figure 63-2. For large values of $R_0$, there appears to be a velocity discontinuity at the wall of magnitude $v_0 = \lambda q_2 E_z/\mu$, in agreement with equation 62-7. This velocity change actually occurs over the thickness of the diffuse layer, which is now very small compared to the radius of the capillary. Superimposed on this is a parabolic Poiseuille velocity profile given by the last term in equation 31.

Equation 4 or 30 shows how an axial electric field can give rise to fluid flow or a pressure drop, that is, how electrical phenomena can produce fluid mechanical phenomena. On the other hand, equation 10 or 29 shows how a pressure drop can produce an electric current or a potential drop. Suppose that the electrolytic solution is forced by a pressure drop to flow through the capillary and that the electrical conditions impose a zero net current. Then the fluid flow generates the so-called *streaming potential*, given by equation 29 as

$$\frac{\mu E_z \kappa_{\text{eff}}}{\lambda q_2 \dfrac{dp}{dz}} = \frac{I_2(R_0)}{I_1(R_0)} \quad \text{with} \quad I = 0, \tag{63-34}$$

where
$$\kappa_{\text{eff}} = \kappa_{\text{avg}} + \frac{q_2^2}{\mu}\left[1 - \frac{I_0(R_0)I_2(R_0)}{I_1^2(R_0)}\right]. \tag{63-35}$$

One may prefer to relate the streaming potential to the flow rate rather than the pressure drop. Elimination of $dp/dz$ between equations 29 and 30 then

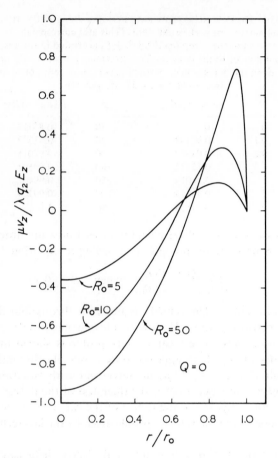

**Figure 63-2.** Velocity profile in the capillary when there is no net fluid flow.

gives

$$-\frac{\pi r_0^4 E_z \kappa'_{\text{eff}}}{8\lambda q_2 Q} = \frac{I_2(R_0)}{I_1(R_0)} \quad \text{with} \quad I = 0, \tag{63-36}$$

where

$$\kappa'_{\text{eff}} = \kappa_{\text{avg}} + \frac{q_2^2}{\mu}\left[1 - \frac{I_0(R_0)I_2(R_0)}{I_1^2(R_0)} - \frac{8}{R_0^2}\frac{I_2^2(R_0)}{I_1^2(R_0)}\right]. \tag{63-37}$$

If the electrical conditions impose a zero potential drop, for example, by having large reversible electrodes at the ends of the capillary and shorting these electrodes together, then the resulting *streaming current* can be expressed according to equations 29 and 30 as

$$\frac{-I\mu}{\pi r_0^2 \lambda q_2 \frac{dp}{dz}} = \frac{I r_0^2}{8\lambda q_2 Q} = \frac{I_2(R_0)}{I_1(R_0)} \quad \text{with} \quad E_z = 0. \tag{63-38}$$

The above development shows how a unified treatment can be given to the phenomena of electro-osmosis and the streaming potential. In the former, electrical effects can give rise to fluid flow, and in the latter fluid flow gives rise to electrical effects. In reality, the two are interrelated by equations 29 and 30, in which there are two driving forces, $E_z$ and $dp/dz$, and two flow quantities, $I$ and $Q$. However, the electrical effects and the fluid flow phenomena are not separated in these equations.

Let us take an example of pure water in a capillary of radius $r_0 = 20\ \mu$ $= 0.002$ cm. We assume the ionic species to be $H^+$ and $OH^-$ ions with concentrations $c_+^0 = c_-^0 = 10^{-7}$ mole/liter at 25°C. Then the Debye length is $\lambda = 0.96\ \mu$, and $R_0$ is about 20. We take $q_2 = 0.01\ \mu C/cm^2$, corresponding to a zeta potential of about $-138$ mV. From these values, we find

$$\frac{q_2 F\lambda}{\epsilon RT} = 5.4, \qquad \kappa^0 = 5.5 \times 10^{-8}\ \text{ohm}^{-1}\text{-cm}^{-1}, \qquad \frac{q_2^2}{\mu\kappa^0} = 2.05,$$

$$\Gamma = 1.3 \times 10^{-7}, \qquad \frac{\kappa_{avg}}{\kappa^0} = 1.15, \qquad \frac{\kappa_{eff}}{\kappa_{avg}} = 1.089, \qquad \frac{\kappa'_{eff}}{\kappa_{avg}} = 1.059.$$

Under these conditions, the application of 100 V across the capillary, with no net fluid flow, should generate a pressure difference equivalent to a column of water 7.3 cm high.

Frequently, the diffuse layer is very thin compared to other geometric lengths, and we want to develop an approximate method of analysis which recognizes this fact. For large values of $R_0$, the diffuse layer is confined to a region close to the wall where the cylindrical geometry is not important. Outside this region, the solution is electrically neutral. We consider first the fluid flow and then the current flow.

For large values of $R_0$, equation 31 reduces to

$$v_z = v_0 - \frac{dp}{dz}\frac{r_0^2 - r^2}{4\mu}, \tag{63-39}$$

where

$$v_0 = \frac{\lambda q_2 E_z}{\mu}. \tag{63-40}$$

Equation 39 applies outside the diffuse layer and corresponds to the solution of equation 2 with zero charge density and with the boundary condition that $v_z = v_0$ at $r = r_0$. This value of $v_0$ comes from the treatment of the thin diffuse layer according to section 62.

Thus, one treats the fluid mechanical problem by solving the momentum equation as though the solution were electrically neutral, but with a slip velocity at a solid wall which is related to the tangential component of the electric field by equation 40, 62-7, or 62-5. Figures 1 and 2 show how this situation is approximated more closely as $R_0$ increases.

Equation 39 yields the flow rate as

$$\frac{Q}{\pi r_0^2} = \langle v_z \rangle = E_z \frac{\lambda q_2}{\mu} - \frac{r_0^2}{8\mu}\frac{dp}{dz}, \tag{63-41}$$

which is also obtained from equation 30 as $R_0$ becomes infinite.

The situation for the current relationships is simple at first sight. As $R_0$ approaches infinity, $\kappa_{avg}$, $\kappa_{eff}$, and $\kappa'_{eff}$ all approach $\kappa^0$, and equation 29 becomes

$$\frac{I}{\pi r_0^2} = \langle i_z \rangle = \kappa^0 E_z - \frac{\lambda q_2}{\mu} \frac{dp}{dz}. \tag{63-42}$$

This yields the correct asymptotic forms for the streaming potential and the streaming current according to equations 34, 36, and 38. Outside the diffuse layer, the solution is taken to be electrically neutral with the conductivity $\kappa^0$, and this region obviously contributes the term $\kappa^0 E_z$ to equation 42. The last term must come then from the diffuse layer.

For the purpose of examining the contribution of the diffuse layer, we want to define a surface current density $j_s$, attributed to the diffuse layer, so that the total current is expressed as

$$I = \pi r_0^2 \kappa^0 E_z + 2\pi r_0 j_s, \tag{63-43}$$

the surface current density being multiplied by the circumference in order to obtain the contribution to the total current. Substitution of equation 43 into equation 29 yields, in the limit of large $R_0$,

$$j_s = \kappa_s E_z + \beta \lambda q_2, \tag{63-44}$$

where
$$\kappa_s = \frac{q_2 \lambda^2 F^3}{\epsilon RT} \sum_i z_i^3 u_i c_i^0 + \frac{\lambda q_2^2}{2\mu} \tag{63-45}$$

is the *surface conductivity* and $\beta$ is the velocity derivative $\partial v_z / \partial y$ just outside the diffuse layer ($y$ being the distance from the surface). In the present case,

$$\beta = -\frac{r_0}{2\mu} \frac{dp}{dz} \tag{63-46}$$

(see equation 39).

The terms in equations 44 and 45 are subject to physical interpretation. The first term on the right in equation 45 might be termed the surface excess conductivity, due to the fact that the ionic concentrations within the diffuse layer differ from their bulk values. This quantity can be positive or negative and is zero for a symmetric electrolyte with equal cation and anion mobilities. In the Debye-Hückel approximation (again for a symmetric electrolyte), $c_+ + c_-$ is uniform within the diffuse layer. Therefore, replacing less mobile ions by more mobile counter ions would increase the local conductivity. If this approximation were relaxed, one should expect the ionic strength to increase in the diffuse layer, and this would lead to more positive values of the surface excess conductivity.

The last term in equation 45 might be called the surface convective conductivity and is due to the motion of the fluid in the diffuse layer as it is induced by the electric field. Since the induced velocity is proportional to $q_2$ and the charge density is proportional to $q_2$, the surface convective conductivity is proportional to the square of $q_2$ and is always positive.

A shear stress in the fluid near the surface induces further fluid motion in the diffuse layer and leads to an additional convective contribution to the surface current density. This term is written in terms of $\beta$ in order to refer to a quantity relevant to the local conditions at the interface. In some cases, erosion corrosion may be related to this last term in equation 44 because the shear stress may vary over the surface.[5,6] This causes the surface current density to vary, and if this current cannot be supplied by the bulk solution because of its low conductivity, it can lead to a corrosion reaction at a metal surface.

If the solid surface itself were moving, it would be necessary to include an additional convective term in equation 44 if the surface current density includes only the current on the mobile side of the double layer. The interface as a whole remains electrically neutral.

Equations 43 and 44 lead to the expression

$$\frac{I}{\pi\,r_0^2} = E_z\left(\kappa^0 + \frac{2\kappa_s}{r_0}\right) - \frac{dp}{dz}\frac{\lambda q_2}{\mu} \tag{63-47}$$

for the total current. Comparison with equation 42 shows that the surface conductivity can be neglected for large values of $r_0/\lambda$.

Equation 44 and equation 40 or 62-5 are vector equations relating to the surface current density, the slip velocity at the surface, the tangential electric field, and the shear stress. They can be applied in other geometric situations where the diffuse layer can be taken to be thin compared to other characteristic lengths. In this approximation, the diffuse layer is essentially planar, and these equations could be refined to account for the fact that the planar diffuse layer can be solved without the Debye-Hückel approximation.

## 64. Electrophoresis

Electrophoresis means the motion of a dielectric particle in an electrolytic solution under the influence of an electric field. The electric field interacts with the diffuse layer to produce a relative motion of the solid and the fluid, as discussed in section 62. This relative motion serves to propel the particle through the fluid. The analysis applies equally well to a metallic particle, if the potential jump across the interface is in a range where the surface is ideally polarizable and the charge in the diffuse layer is essentially uniform over the surface of the particle.

Consider a spherical particle of radius $r_0$, and let the origin of a spherical coordinate system be fixed in the center of the particle. With this conven-

[5] T. R. Beck, D. W. Mahaffey, and J. H. Olsen, "Wear of Small Orifices by Streaming Current Driven Corrosion," *Journal of Basic Engineering*, *92*, 782–788 (1970).

[6] T. R. Beck, D. W. Mahaffey, and J. H. Olsen. "Pitting and Deposits with an Organic Fluid by Electrolysis and by Fluid Flow." *Journal of the Electrochemical Society*, *119*, 155–160 (1972).

tion, the fluid moves past the particle in a steady manner, and, far from the particle, the $z$-component of the velocity is $v_\infty$. Let the $z$-component of the electric field be $E_\infty$ far from the particle. The objective is to relate $v_\infty$ to $E_\infty$. Gravitational forces will be neglected.

Outside the diffuse layer, the solution is electrically neutral, the fluid motion satisfies the Navier-Stokes equation 94-4 and the continuity equation 93-3, and the electric potential satisfies Laplace's equation 71-4. With the assumption that the diffuse layer is thin compared to the radius of the particle, the fluid mechanical equations are to be solved with the boundary conditions that the velocity approaches the uniform velocity at infinity, that the fluid exerts no net force on the particle including the diffuse layer, and that the slip velocity at the surface matches the tangential electric field according to equation 63-40

$$v_\theta = \frac{\lambda q_2 E_\theta}{\mu} \quad \text{at} \quad r = r_0. \tag{64-1}$$

The potential is to satisfy Laplace's equation subject to the conditions that the electric field approaches the uniform field at infinity and there is a charge balance at the interface

$$\kappa \frac{\partial \Phi}{\partial r} = \nabla_s \cdot \mathbf{j}_s \quad \text{at} \quad r = r_0, \tag{64-2}$$

where $\nabla_s \cdot \mathbf{j}_s$ is the surface divergence of the surface current density.

An exact solution of the Navier-Stokes equation satisfying the condition at infinity and the condition of zero net force on the particle plus the diffuse layer is

$$v_r = \left(1 - \frac{r_0^3}{r^3}\right) v_\infty \cos \theta, \tag{64-3}$$

$$v_\theta = -\left(1 + \frac{r_0^3}{2r^3}\right) v_\infty \sin \theta. \tag{64-4}$$

It is not always realized that this is an exact solution of the Navier-Stokes and continuity equations.

From these results,

$$\beta = \frac{3}{2} \frac{v_\infty}{r_0} \sin \theta, \tag{64-5}$$

and the surface current density from equation 63-44 becomes

$$j_s = \frac{3}{2} \frac{v_\infty \lambda q_2}{r_0} \left(1 - \frac{\kappa_s \mu r_0}{\lambda^2 q_2^2}\right) \sin \theta, \tag{64-6}$$

where $E_\theta$ has been eliminated by means of equation 1. The surface divergence of the surface current density therefore is

$$\nabla_s \cdot \mathbf{j}_s = \frac{1}{r_0 \sin \theta} \frac{\partial}{\partial \theta} (j_s \sin \theta) = \frac{3 v_\infty \lambda q_2}{r_0^2} \left(1 - \frac{\kappa_s \mu r_0}{\lambda^2 q_2^2}\right) \cos \theta, \tag{64-7}$$

and boundary condition 2 becomes

$$\frac{\partial \Phi}{\partial r} = \frac{3v_\infty \lambda q_2}{r_0^2 \kappa}\left(1 - \frac{\kappa_s \mu r_0}{\lambda^2 q_2^2}\right)\cos\theta. \tag{64-8}$$

The solution of Laplace's equation satisfying condition 8 and giving the proper electric field at infinity is

$$\Phi = -\left[r + \left(\frac{1}{2} + A\right)\frac{r_0^3}{r^2}\right]E_\infty \cos\theta, \tag{64-9}$$

where

$$A = \frac{3v_\infty \lambda q_2}{2r_0^2 \kappa E_\infty}\left(1 - \frac{\kappa_s \mu r_0}{\lambda^2 q_2^2}\right). \tag{64-10}$$

Equation 9 yields the tangential electric field at the surface of the particle:

$$E_\theta = -\left(\frac{3}{2} + A\right)E_\infty \sin\theta. \tag{64-11}$$

Substitution of this result into equation 1 yields the final expression for the electrophoretic velocity in terms of the applied field:

$$v_\infty = \frac{\dfrac{E_\infty \lambda q_2}{\mu}}{1 - \dfrac{\lambda^2 q_2^2}{r_0^2 \kappa \mu} + \dfrac{\kappa_s}{\kappa r_0}}. \tag{64-12}$$

The second term in the denominator in equation 12 is of order $\lambda^2/r_0^2$ and is small compared to the last term, which is of order $\lambda/r_0$. Both terms should be negligible compared to the first term, and the electrophoretic velocity can be expressed in terms of the zeta potential as

$$v_\infty = -\frac{E_\infty \epsilon \zeta}{\mu}. \tag{64-13}$$

Recall that $v_\infty$ is the velocity of the fluid with respect to the particle. Therefore, a particle with a positive zeta potential moves in the direction of the electric field.

## 65. Sedimentation potential

Small particles will fall through an electrolytic solution in a gravitational field with essentially a Stokes velocity profile. This is not an exact solution of the Navier-Stokes equation; rather, it applies for small values of the Reynolds number $Re = 2v_\infty r_0/\nu$. If the particle has a positive zeta potential, then the charge in the diffuse layer is negative. The shear stress near the particle will cause a surface current density to flow from the back of the particle to the front. This current must then flow through the bulk of the solution from the front to the back. This means that the potential behind a particle with a positive zeta potential will be negative relative to the potential in front of the drop. A number of particles falling through the solution will then establish an electric field whose magnitude is given by

$$E_z = \frac{\dfrac{-6\pi\, nv_\infty r_0 \lambda q_2}{\kappa}}{1 + \dfrac{\kappa_s}{\kappa r_0} - 2\dfrac{\lambda^2 q_2^2}{r_0^2 \kappa \mu}}, \tag{65-1}$$

where $n$ is the number of particles per unit volume of the system. $E_z$ is in the direction opposite to the velocity $v_\infty$ of the particles. The same remarks apply to the terms in the denominator of this equation as to those in equation 64-12.

By means of a force balance, the velocity of fall is given as

$$v_\infty = \frac{2(\rho' - \rho)gr_0^2}{9\mu} \frac{1 + \dfrac{\kappa_s}{\kappa r_0} - 2\dfrac{\lambda^2 q_2^2}{r_0^2 \kappa \mu}}{1 + \dfrac{\kappa_s}{\kappa r_0} - \dfrac{\lambda^2 q_2^2}{r_0^2 \kappa \mu}}, \tag{65-2}$$

which differs insignificantly from the velocity for an uncharged particle

$$v_\infty = \frac{2(\rho' - \rho)gr_0^2}{9\mu} \tag{65-3}$$

given by Stokes law, since $\lambda^2 q_2^2/r_0^2 \kappa \mu$ is of order $\lambda^2/r_0^2$. In equations 2 and 3, $\rho'$ is the density of the particle, $\rho$ is the density of the electrolytic solution, and $g$ is the magnitude of the gravitational acceleration.

The sedimentation potential is analogous to the streaming potential discussed in section 63. In both cases, the relative motion of a solid and an electrolytic solution gives rise to electrical effects. However, the sedimentation potential is not much studied because it is difficult to obtain an appreciable magnitude experimentally.

By adding shorted reversible electrodes to the system, one could maintain a zero electric field. One should then observe a sedimentation current analogous to the streaming current discussed in section 63.

## PROBLEMS

**1.** (a) For a diffuse layer, the electrokinetic slip velocity is given by equation 62-5,

$$v_0 = -\frac{\epsilon \zeta E_x}{\mu},$$

equation 62-7 being valid only when the Debye-Hückel approximation is applicable. The second basic electrokinetic equation is equation 63-44. Show that in general this equation should be replaced by

$$j_s = \kappa_s E_x - \epsilon \beta \zeta,$$

where the surface conductivity is

$$\kappa_s = F^2 \sum_i z_i^2 u_i \Gamma_{i,d} + \frac{\epsilon^2}{\mu} \int_0^\zeta E_y \, d\Phi.$$

Here $E_y$ is given as a function of $\Phi$ by equation 52-8, and $q_2$ can be determined in terms of $\zeta$ from equation 52-9.

(b) For a solution of a binary, symmetric electrolyte, $q_2$ and $\zeta$ are related by

$$q_2 = -\frac{2\epsilon RT}{zF\lambda} \sinh\left(\frac{zF\zeta}{2RT}\right).$$

Show that the expression for the surface conductivity becomes

$$\kappa_s = \frac{\kappa\lambda^2 zFq_2}{\epsilon RT}\left(t_+ - t_- - \tanh\frac{zF\zeta}{4RT}\right) + \frac{\dfrac{\lambda q_2^2}{\mu}}{1 + \sqrt{1 + \left(\dfrac{zF\lambda q_2}{2\epsilon RT}\right)^2}},$$

where $\kappa$ is the conductivity of the bulk solution and

$$t_+ = 1 - t_- = \frac{u_+}{u_+ + u_-}$$

is the cation transference number.

(c) Show that when the Debye-Hückel approximation is valid, that is, when $zF\zeta/RT \ll 1$, the results in part (b) reduce to equations 62-6 and 63-45. Notice that the surface excess conductivity in part (b) is higher than that in equation 63-45, while the surface convective conductivity is lower.

2. Show on the basis of problem 1 that equation 63-47 should be written as

$$\frac{I}{\pi r_0^2} = E_z\left(\kappa^0 + \frac{2\kappa_s}{r_0}\right) + \frac{\epsilon\zeta}{\mu}\frac{dp}{dz}$$

and that consequently, for large values of $r_0/\lambda$, the streaming potential (equations 63-34 and 63-36) should be written

$$-\frac{\mu E_z\kappa^0}{\epsilon\zeta\dfrac{dp}{dz}} = \frac{\pi r_0^4 E_z\kappa^0}{8\epsilon\zeta Q} = 1 \quad \text{with} \quad I = 0,$$

and the streaming current (equation 63-38) should be written

$$\frac{I\mu}{\pi r_0^2\epsilon\zeta\dfrac{dp}{dz}} = \frac{-Ir_0^2}{8\epsilon\zeta Q} = 1 \quad \text{with} \quad E_z = 0.$$

Make the corresponding changes in equation 63-41 and in the expressions for pressure drop at zero flow (equation 63-33) and flow at zero pressure drop (equation 63-32).

3. Show on the basis of problem 1 that the electrophoretic velocity, equation 64-12 should be written

$$v_\infty = \frac{-\dfrac{E_\infty\epsilon\zeta}{\mu}}{1 + \dfrac{\kappa_s}{\kappa r_0} - \dfrac{\epsilon^2\zeta^2}{r_0^2\kappa\mu}}$$

or, for practical purposes, as equation 64-13. The sedimentation potential, equation 65-1, should be written

$$E_z = \frac{6\pi\, nv_\infty r_0 \epsilon \zeta}{\kappa}.$$

These equations are written in the form in the text for easier comparison with the corresponding results in chapter 10.

**4.** (a) Is the normal component of the current density continuous at an interface if the normal component is to be evaluated just outside the interface?

(b) Rationalize or convince yourself of the validity of equation 64-2.

(c) Obtain an estimate of the magnitude of $\nabla_s \cdot \mathbf{j}_s$ from equation 64-7 by using the values $v_\infty = 0.3$ cm/sec, $\lambda = 10$ Å, $r_0 = 20$ $\mu$, $\mu = 8.9 \times 10^{-3}$ g/cm-sec, $\kappa_s/\lambda = 0.01$ mho/cm, and $q_2 = 7$ $\mu$C/cm$^2$.

**5.** Justify the material balance for a species at an interface:

$$\frac{\partial \Gamma_i}{\partial t} + \nabla_s \cdot \mathbf{J}_{is} = N_{iy}^0 - N_{iy}^\infty,$$

where $\mathbf{J}_{is}$ is the surface flux of species $i$, $N_{iy}^0$ is the normal component of the flux at the surface, involved in faradaic electrode reactions, and $N_{iy}^\infty$ is the normal component of the flux evaluated just outside the diffuse layer. Use this equation to derive equation 64-2.

## NOTATION

| | |
|---|---|
| $c_i$ | concentration of species $i$, mole/cm$^3$ |
| $\mathbf{E}$ | electric field, V/cm |
| $E_\infty$ | electric field far from the particle, V/cm |
| $F$ | Faraday's constant, 96,487 C/equiv |
| $g$ | magnitude of the gravitational acceleration, cm/sec$^2$ |
| $\mathbf{i}$ | current density, A/cm$^2$ |
| $I$ | total current, A |
| $j_s$ | surface current density, A/cm |
| $n$ | number of particles per unit volume of the system, cm$^{-3}$ |
| $N_i$ | flux of species $i$, mole/cm$^2$-sec |
| $p$ | pressure, dyne/cm$^2$ |
| $q_2$ | surface charge density in the diffuse layer, C/cm$^2$ |
| $Q$ | volumetric flow rate, cm$^3$/sec |
| $r$ | radial distance in spherical or cylindrical coordinates, cm |
| $r_0$ | radius of particle or capillary, cm |
| $R$ | universal gas constant, 8.3143 J/mole-deg |
| $R_0$ | $r_0/\lambda$ |
| $Re$ | $2v_\infty r_0/\nu$, the Reynolds number |
| $T$ | absolute temperature, deg K |

| | |
|---|---|
| $u_i$ | mobility of species $i$, cm$^2$-mole/J-sec |
| $\mathbf{v}$ | fluid velocity, cm/sec |
| $v_0$ | electrokinetic velocity discontinuity, cm/sec |
| $v_\infty$ | velocity far from the particle, cm/sec |
| $x$ | distance along a surface, cm |
| $x$ | $r/\lambda$ |
| $y$ | distance from surface, cm |
| $z$ | distance along capillary or in direction of particle motion, cm |
| $z_i$ | charge number of species $i$ |
| $\beta$ | velocity derivative outside diffuse layer, sec$^{-1}$ |
| $\Gamma$ | dimensionless charge density on axis of capillary |
| $\epsilon$ | permittivity, farad/cm |
| $\epsilon_0$ | permittivity of free space, $8.8542 \times 10^{-14}$ farad/cm |
| $\zeta$ | zeta potential, V |
| $\theta$ | angle from the axis of particle motion |
| $\kappa$ | solution conductivity, mho/cm |
| $\kappa_s$ | surface conductivity, mho |
| $\lambda$ | Debye length, cm |
| $\mu$ | viscosity, g/cm-sec |
| $\nu$ | kinematic viscosity, cm$^2$/sec |
| $\rho$ | density, g/cm$^3$ |
| $\rho_e$ | electric charge density, C/cm$^3$ |
| $\phi$ | $\Phi - \Phi_{r=0}$ |
| $\Phi$ | electric potential, V |
| $\psi$ | $F\phi/RT$, dimensionless potential |

# Electrocapillary Phenomena

**10**

The motion of charged mercury drops in an electrolytic solution can lead to much larger electrophoretic velocities and sedimentation potentials than with solid particles because the drop can develop an internal circulation. The motion of the surface leads to larger surface current densities and hence larger sedimentation potentials. In electrophoresis, the electric force is not applied close to a solid surface, and larger velocities result.

For mercury drops, we can ignore the usual electrokinetic effects since the surface velocity itself is much larger than the electrokinetic velocity discontinuity discussed in section 62. Instead, we treat the mercury drop as an ideally polarizable electrode, where the surface tension varies with the local electrode potential on the drop. These variations of surface tension serve to propel the drop through the solution in the electrophoretic case. They can also affect the velocity of fall in a gravitational field in the sedimentation case.

The mechanism and observations of electrocapillary motion were described by Christiansen[1] in 1903. Frumkin and Levich[2,3,4] presented a detailed theoretical analysis of the phenomena.

---

[1]C. Christiansen, "Kapillarelektrische Bewegungen," *Annalen der Physik*, ser. 4, *12* (1903), 1072–1079.

[2]A. Frumkin and B. Levich, "The Motion of Solid and Liquid Metallic Bodies in Solutions of Electrolytes. I," *Acta Physicochimica U.R.S.S.*, *20* (1945), 769–808.

[3]A. Frumkin and B. Levich, "The Motion of Solid and Liquid Metallic Bodies in Solutions of Electrolytes. II. Motion in the field of gravity," *Acta Physicochimica U.R.S.S.*, *21* (1946), 193–212.

[4]V. Levich, "Dvizhenie tverdykh i zhidkikh metallicheskikh chastits v rastvorakh elektrolitov. III. Obshchaya teoriya," *Zhurnal Fizicheskoi Khimii*, *21* (1947), 689–701.

## 66. Dynamics of interfaces

The force balance at an interface is treated in section 95. For a spherical drop, equation 95-4 reads

$$\tau'_{r\theta} - \tau_{r\theta} + \frac{1}{r_0} \frac{\partial \sigma}{\partial \theta} = 0, \tag{66-1}$$

where $\tau'_{r\theta}$ is the force in the $\theta$-direction exerted on the interface by the fluid within the drop and $-\tau_{r\theta}$ is the force exerted by the fluid outside the drop. The shear stress is related to the velocity derivatives for a Newtonian fluid

$$\tau_{r\theta} = -\mu \left[ r \frac{\partial}{\partial r} \left( \frac{v_\theta}{r} \right) + \frac{1}{r} \frac{\partial v_r}{\partial \theta} \right]. \tag{66-2}$$

Surface tension driven flows can lead to a variety of interesting phenomena. The surface tension generally depends on the composition of the solution near the interface. In mass-transfer studies, nonuniformities of surface tension can develop, resulting in what is called interfacial turbulence.[5,6] In the Marangoni effect, a nonuniform surface tension arises from differential evaporation of the components of the solution. In the fall of a drop in a solution with surface-active agents, nonuniformities of surface tension can hinder the internal circulation, so that the drop falls like a solid sphere.[7,8] Similar circumstances can hinder the formation of ripples on a falling liquid film.

In the electrocapillary motion of mercury drops, the surface tension can vary due to the nonuniform potential in the solution. This can produce a motion, like electrophoresis, if there is an applied electric field, or it can hinder the fall of drops in a manner similar to the surface-active agents referred to above. Electrocapillary motion can also lead to some maxima observed in polarographic currents with a dropping mercury electrode.[9,10]

If proper account is taken of the variation of surface tension at a fluid-fluid interface, the concept of the *surface viscosity* becomes of dubious value.

[5] Thomas K. Sherwood and James C. Wei, "Interfacial Phenomena in Liquid Extraction," *Industrial and Engineering Chemistry*, 49 (1957), 1030–1034.

[6] C. V. Sternling and L. E. Scriven, "Interfacial Turbulence: Hydrodynamic Stability and Marangoni Effect," *A.I.Ch.E. Journal*, 5 (1959), 514–523.

[7] A. Frumkin and V. Levich, "O vliyanii poverkhnostno-aktivnykh veshchestv na dvizhenie na granitse zhidkikh sred," *Zhurnal Fizicheskoi Khimii*, 21 (1947), 1183–1204.

[8] John Newman, "Retardation of falling drops," *Chemical Engineering Science*, 22 (1967), 83–85.

[9] A. Frumkin and B. Bruns, "Über Maxima der Polarisationskurven von Quecksilberkathoden," *Acta Physicochimica U.R.S.S.*, 1 (1934), 232–246.

[10] A. Frumkin and V. Levich, "Dvizhenie tverdykh i zhedkikh metallicheskikh chastits v rastvorakh elektrolitov. IV. Maksimumy na krivy tok-napryazhenie kapel' nogo elektroda," *Zhurnal Fizicheskoi Khimii*, 21 (1947), 1335–1349.

## 67.  Electrocapillary motion of mercury drops

The electrocapillary motion of mercury drops in an external electric field is very similar to the electrophoretic motion of solid particles treated in section 64. The velocity distribution outside the drop is again given by equations 64-3 and 64-4, so that

$$v_\theta = -\frac{3}{2} v_\infty \sin \theta \quad \text{at} \quad r = r_0 \tag{67-1}$$

and

$$\tau_{r\theta} = -\frac{3\mu \, v_\infty}{r_0} \sin \theta \quad \text{at} \quad r = r_0. \tag{67-2}$$

Since the interface as a whole now moves, the surface current density is

$$j_s = -q \, v_\theta \quad \text{at} \quad r = r_0, \tag{67-3}$$

$-q$ being the charge density on the solution side of the double layer. We take $q$ to be essentially constant over the surface of the drop so that the surface divergence is

$$\nabla_s \cdot \mathbf{j}_s = \frac{3q \, v_\infty}{r_0} \cos \theta. \tag{67-4}$$

For an ideally polarizable drop, the current entering and leaving the surface cannot come from within the drop. Consequently, the current comes from the electrolytic solution, and equation 64-2 applies

$$\frac{\partial \Phi}{\partial r} = \frac{3q \, v_\infty}{r_0 \kappa} \cos \theta \quad \text{at} \quad r = r_0. \tag{67-5}$$

This forms one of the boundary conditions for Laplace's equation, the other being the uniform field far from the drop. Consequently, the potential in the solution outside the drop is again given by equation 64-9, where now

$$A = \frac{3q \, v_\infty}{2r_0 \, \kappa E_\infty}. \tag{67-6}$$

The velocity distribution inside the drop is

$$v'_r = \frac{3}{2}\left(\frac{r^2}{r_0^2} - 1\right) v_\infty \cos \theta, \tag{67-7}$$

$$v'_\theta = -\frac{3}{2}\left(2\frac{r^2}{r_0^2} - 1\right) v_\infty \sin \theta. \tag{67-8}$$

This is not an exact solution of the Navier-Stokes equation. It is a solution of the approximate form of the equation of motion for creeping flow. Equations 7 and 8 yield

$$\tau'_{r\theta} = \frac{9}{2} \frac{\mu' v_\infty}{r_0} \sin \theta \quad \text{at} \quad r = r_0. \tag{67-9}$$

The motion of the surface also creates a surface current density on the metal side of the double layer. However, this current can easily be supplied from within the drop, because of the high conductivity of the metal, and the potential in the drop remains uniform. Consequently, the Lippmann equation 51-7 allows the variation in surface tension to be related to the variation of the potential in the solution near the drop:

$$\frac{\partial \sigma}{\partial \theta} = -q\frac{\partial U}{\partial \theta} = q\frac{\partial \Phi}{\partial \theta} \quad \text{at} \quad r = r_0. \tag{67-10}$$

With equation 64-9, we have

$$\frac{\partial \sigma}{\partial \theta} = qr_0 E_\infty \left(\frac{3}{2} + A\right) \sin \theta. \tag{67-11}$$

Finally, substitution of equations 2, 9, 11, and 6 into the force-balance equation 66-1 allows us to determine the velocity $v_\infty$ in terms of the electric field $E_\infty$

$$v_\infty = \frac{-qE_\infty r_0}{2\mu + 3\mu' + \dfrac{q^2}{\kappa}}. \tag{67-12}$$

As in the case of equations 64-12 and 64-13, a positively charged drop moves in the direction of the electric field, $v_\infty$ being the velocity of the fluid with respect to the particle.

Levich[11] cites from the Russian literature examples of the experimental verification of equation 12. This can be done with some thoroughness, since it is possible to vary the surface charge on the drop (see figure 49-9). One notices that the velocity in equation 12 can be larger than the usual electrophoretic velocities given by equation 64-12 or 64-13 by a factor of the order of $r_0/\lambda$. Again in contrast to equation 64-12, the last term in the denominator of equation 12 generally is not negligible, and in media of low conductivity the velocity of electrocapillary motion can be small.

The charge in the double layer is the origin of the electrocapillary motion and gives rise to the numerator in equation 12. The tangential electric field produces a variation in surface tension which propels the drop (see equation 10). However, if the double layer has too great an ability to carry a surface current compared to the bulk solution, it can reduce the tangential electric field and hence the variation of surface tension around the drop. This will lower the velocity of electrophoretic motion, as represented by the last term in the denominator of equation 12.

## 68.  Sedimentation potentials for falling mercury drops

Mercury drops falling through an electrolytic solution will establish a sedimentation potential in much the same manner as solid particles, as

[11] Veniamin G. Levich, *Physicochemical Hydrodynamics*, section 101 (Englewood Cliffs, N.J.: Prentice-Hall, Inc., 1962).

discussed in section 65. In this case, equation 65-1 is replaced by

$$E_z = \frac{2\pi \, nv_\infty r_0^2 \dfrac{q\mu}{\kappa}}{\mu + \mu' + \dfrac{q^2}{3\kappa}},\qquad(68\text{-}1)$$

where $E_z$ is in the direction opposite to the velocity $v_\infty$ of the particles. The velocity of fall of the drops is now given by

$$v_\infty = \frac{2(\rho' - \rho)gr_0^2}{3\mu} \, \frac{\mu + \mu' + \dfrac{q^2}{3\kappa}}{2\mu + 3\mu' + \dfrac{q^2}{\kappa}}.\qquad(68\text{-}2)$$

Now, both the sedimentation potential and the velocity of fall are subject to experimental verification, as presented by Levich.[12] For large values of $q^2/\kappa$, the drop falls like a solid particle according to Stokes law:

$$v_\infty = \frac{2(\rho' - \rho)gr_0^2}{9\mu}.\qquad(68\text{-}3)$$

The motion of the charge in the double layer then establishes a potential distribution around the drop that determines the variation of surface tension so as to retard strongly the motion of the surface. For a small surface charge, on the other hand, the internal circulation of the drop is not retarded, and it falls with the somewhat larger velocity

$$v_\infty = \frac{2(\rho' - \rho)gr_0^2}{3\mu} \, \frac{\mu + \mu'}{2\mu + 3\mu'}.\qquad(68\text{-}4)$$

For a small viscosity $\mu'$ of the drop compared to the viscosity $\mu$ of the solution, this velocity can be 50 percent larger than the Stokes velocity.

The sedimentation potential for falling drops is much greater than that given by equation 65-1 for solid particles, by a factor of order $r_0/\lambda$. The terms involving $q^2/\kappa$ in the denominator in equations 1 and 2 are, in general, not negligible, in contrast to the corresponding terms in equations 65-1 and 65-2.

## NOTATION

$E_\infty$      electric field far from the drop, V/cm

$g$      magnitude of the gravitational acceleration, cm/sec²

$j_s$      surface current density, A/cm

$n$      number of drops per unit volume of the system, cm⁻³

$q$      surface charge density on the mercury side of the double layer, C/cm²

$r$      radial distance from the center of the drop, cm

[12] *Ibid.*, sections 102 and 103.

| | |
|---|---|
| $r_0$ | radius of the mercury drop, cm |
| $U$ | electrode potential, V |
| $v_r$ | velocity in the r-direction, cm/sec |
| $v_\theta$ | velocity in the $\theta$-direction, cm/sec |
| $v_\infty$ | velocity far from the drop, cm/sec |
| $\theta$ | angle from the axis of drop motion |
| $\kappa$ | solution conductivity, mho/cm |
| $\lambda$ | Debye length, cm |
| $\mu$ | viscosity, g/cm-sec |
| $\rho$ | density, g/cm³ |
| $\sigma$ | surface tension, dyne/cm |
| $\tau_{r\theta}$ | shear stress, dyne/cm² |
| $\Phi$ | electric potential in the solution, V |

superscript

       in the drop

*Part* **C**

# Transport Processes in Electrolytic Solutions

Frequently, the rate of an electrochemical process is governed by the transport of a reactant species to the electrode surface by diffusion and convection. In other processes, the ohmic potential drop in the solution is decisive. This part of the book treats the transport processes, *migration* and *diffusion*, in electrolytic solutions from a descriptive point of view. For example, it is recognized that electric conduction is a manifestation of the movement of charged species, but the quantitative characterization of conduction in terms of the molecular properties of the species is not considered vital. Engineering applications do not require values of transport properties predicted from molecular theory if measured values are available.

Basic diffusion laws for dilute solutions are presented in chapter 11 and are modified for concentrated solutions in chapter 12. The dilute-solution theory has been applied fruitfully to many electrochemical problems; it is adequate for approximate analysis. It is more or less familiar to all electrochemists. Nevertheless, a careful definition of transport properties requires modifications of that theory except at infinite dilution. Furthermore, there are questions for which the dilute-solution theory promotes circular or incorrect reasoning. A classic example is the question of liquid-junction potentials and individual ionic activity coefficients. Such questions can be clarified or avoided in the theory for concentrated solutions.

One should be aware that the consequences of dilute-solution theory developed in chapter 11 are subject to qualification or reinterpretation as a

215

result of the theory of concentrated solutions. It is not always indicated in the text whether a particular result has a strong analogy or is of little meaning in the concentrated-solution theory.

Thermal effects and transport properties are developed in chapters 13 and 14. The fluid mechanics, necessary to calculate the convective velocity, is introduced in chapter 15.

Transport equations are given in vector notation for generality and for brevity. A short statement of the information needed to comprehend such equations will be found in appendix B.

Chapters 11 and 12 are taken largely from J. Newman, "Transport Processes in Electrolytic Solutions," in C. W. Tobias, *Advances in Electrochemical Engineering*, Vol. 5; copyright © 1967 by John Wiley & Sons, Inc., and reprinted by permission.

# *Infinitely Dilute Solutions*

**11**

## 69. Transport laws

Mass transfer in an electrolytic solution requires a description of the movement of mobile ionic species, material balances, current flow, electroneutrality, and fluid mechanics. Equations for the first four of these will be presented in this section and will be elaborated upon in the following sections. The medium which we wish to describe consists of an un-ionized solvent, ionized electrolytes, and uncharged minor components. This description should be restricted to dilute solutions.

The flux of each dissolved species is given by

$$\mathbf{N}_i = \underset{\text{migration}}{-z_i u_i F c_i \, \nabla \Phi} - \underset{\text{diffusion}}{D_i \, \nabla c_i} + \underset{\text{convection}}{c_i \mathbf{v}} \qquad (69\text{-}1)$$
$\underset{\text{flux}}{\mathbf{N}_i}$

The flux $\mathbf{N}_i$ of species $i$, expressed in moles/cm$^2$-sec, is a vector quantity indicating the direction in which the species is moving and the number of moles going per unit time across a plane of 1 cm$^2$, oriented perpendicular to the flow of the species. This movement is due first of all to the motion of the fluid with the bulk velocity $\mathbf{v}$. However, the movement of the species can deviate from this average velocity by diffusion if there is a concentration gradient $\nabla c_i$ or by migration if there is an electric field $-\nabla \Phi$ and if the species is charged ($z_i$ is the number of proton charges carried by an ion).

217

The migration term is peculiar to electrochemical systems or systems containing charged species. Here $\Phi$ is the electrostatic potential whose gradient is the negative of the electric field. These are not quantities which can be measured easily and directly in a liquid solution. The quantity $u_i$ is called the *mobility* and denotes the average velocity of a species in the solution when acted upon by a force of 1 newton/mole, independent of the origin of the force. Thus, $z_i F$ is the charge per mole on a species. Multiplication by the electric field $-\nabla\Phi$ gives the force per mole. Multiplication by the mobility $u_i$ gives the migration velocity, and finally multiplication by the concentration $c_i$ gives the contribution to the net flux $\mathbf{N}_i$ due to migration in an electric field.

The second and third terms on the right side of equation 1 are the usual terms required to describe nonelectrolytic systems. The species will diffuse from regions of high concentration to regions of lower concentration. The three terms on the right in equation 1 thus represent three mechanisms of mass transfer: migration of a charged species in an electric field, molecular diffusion due to a concentration gradient, and convection due to the bulk motion of the medium. Equation 1 thus serves to define two *transport properties*, the diffusion coefficient $D_i$ and the mobility $u_i$.

The current in an electrolytic solution is, of course, due to the motion of charged particles, and we can easily express this quantitatively:

$$\mathbf{i} = F \sum_i z_i \mathbf{N}_i. \qquad (69\text{-}2)$$

Here, $\mathbf{i}$ is the current density expressed in amperes per square centimeter, and $z_i F$ is again the charge per mole.

Next we need to state a material balance for a minor component:

$$\underset{\text{accumulation}}{\frac{\partial c_i}{\partial t}} = \underset{\text{net input}}{-\nabla\cdot\mathbf{N}_i} + \underset{\substack{\text{production (in homogeneous} \\ \text{chemical reactions)}}}{R_i} \qquad (69\text{-}3)$$

In engineering parlance, accumulation is equal to input minus output plus production. For a differential volume element, accumulation is simply the time rate of change of concentration.

For the net input, it is necessary to compute the net amount of material brought in by the different fluxes on the various faces of the volume element (see figure 69-1). The difference in fluxes contributes to accumulation or depletion.

$$\lim_{\Delta x \to 0} \frac{N_{ix}|_x - N_{ix}|_{x+\Delta x}}{\Delta x} = -\frac{\partial N_{ix}}{\partial x}.$$

The $\Delta x$ in the denominator comes from dividing by the volume of the element.

The production per unit volume $R_i$ involves homogeneous chemical reactions in the bulk of the solution, but not any electrode reactions, which

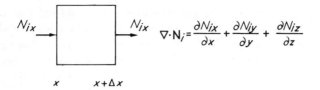

**Figure 69-1.** Accumulation due to differences in the fluxes at the faces of a volume element.

occur at the boundaries of the solution. In electrochemical systems, the reaction is frequently restricted to electrode surfaces, in which case $R_i$ is zero.

Finally, we can say that the solution is electrically neutral.

$$\sum_i z_i c_i = 0. \tag{69-4}$$

Such electroneutrality is observed in all solutions except in a thin double charge layer near electrodes and other boundaries. This double layer may be of the order of 10 to 100 Å in thickness. The phenomena related to the double layer at electrodes can usually be taken into account by the boundary conditions. Hence, it is reasonable to adopt equation 4 in a description of the bulk of a solution. The validity of this equation will be considered again in section 76.

These four equations provide a consistent description of transport processes in electrolytic solutions, and their physical significance is worth repeating. The first states that species in the solution can move by migration, diffusion, and convection. The second equation merely says that the flux of a charged species constitutes an electric current. The third is a material balance for a species, and the fourth is the condition of electroneutrality. Although the specific description may be refined, any theory of electrolytic solutions will need to consider these physical phenomena.

Note that in order to solve a mass-transfer problem it is necessary to know the convective velocity **v**. This requires the equations of fluid mechanics, which are discussed in chapter 15. The analysis of electrochemical systems by means of the above differential equations requires in addition a statement of the geometry of the system and of conditions existing at boundaries of the system. Important among these are the electrode kinetics treated in part B. Boundary conditions will be discussed as they arise, mostly in part D. (See also equations 72-9 to 72-11.)

We can also obtain physical insight by considering the validity of the above four equations. The validity of the electroneutrality equation 4 will be discussed separately in section 76, where we come to the conclusion that electroneutrality is an *accurate approximation*. Equations 2 and 3 can be regarded as expressions of basic physical laws, stating that current is due to the motion of charged particles and that individual species either are conserved or take part in homogeneous chemical reactions. However, the rate processes in the expression of the production rate and the flux introduce uncertainties. The production rate involves chemical kinetics, for which rate expressions are neither predictable nor general. The flux has been expressed by equation 1, but even this breaks down in concentrated solutions.

It is always possible to write mathematical expressions for the basic physical laws of conservation of mass, energy, and momentum in terms of the fluxes of these quantities, but the difficult part is to find correct expressions for these fluxes in terms of the appropriate driving forces in the system. We are not speaking of the microscopic, theoretical explanation of transport properties, but rather of the macroscopic definition of the appropriate transport properties.

The flux equation 1 breaks down, first of all, because migration and diffusion fluxes must be defined with respect to some average velocity of the fluid ($v$ in equation 1), and the flux relations so defined must be consistent with this choice. We have not been careful to specify the fluid velocity. In a concentrated solution, it is not just the solvent velocity which contributes to the average velocity. This difficulty is avoided here by not applying equation 1 to the solvent and by restricting ourselves to dilute solutions where $v$ is essentially the same as the velocity of the solvent.

Furthermore, the flux equation 1 incorrectly defines the transport properties; in fact, it defines an incorrect number of transport properties. This situation arises because equation 1 considers the interaction or friction force of a solute species with the solvent and essentially neglects interactions with the other solutes.

Finally, the driving force for diffusion should be an activity gradient, and activity gradients are identical to concentration gradients only in extremely dilute solutions. However, in a generalization of equation 1, one should avoid the use of single ionic activity coefficients, which are not physically measurable. Furthermore, care is needed in the definition of potentials in media of varying composition (see chapter 3). One concludes that the correct driving force for both diffusion and migration is the gradient of an electrochemical potential (discussed in chapter 2), and any decomposition of this into $\nabla c_i$ and $c_i \nabla \Phi$ is unnecessary.

The multicomponent-diffusion equation, presented in section 78, avoids these difficulties. Nevertheless, equation 1 is recommended for general use because it is prevalent, both explicitly and implicitly, in the electrochemical literature and because it gives a good account of the physical processes involved without excessive complication. One should remember that it is strictly valid only in dilute solutions.

The remaining sections of this chapter are designed to illustrate further the meaning, the application, and the limitations of the basic transport laws discussed here.

## 70. Conductivity, diffusion potentials, and transference numbers

Let us expand the expression for the current density in the solution, equation 69-2, in terms of the species fluxes, equation 69-1.

$$\mathbf{i} = -F^2\, \boldsymbol{\nabla}\Phi \sum_i z_i^2 u_i c_i - F \sum_i z_i D_i\, \boldsymbol{\nabla} c_i + F\mathbf{v} \sum_i z_i c_i. \tag{70-1}$$

By virtue of electroneutrality, the last term on the right is zero, which is equivalent to saying that bulk motion of a fluid with no charge density can contribute nothing to the current density. When there are no concentration variations in the solution, this equation reduces to the common concept of electrolytic conductance.

$$\mathbf{i} = -\kappa\, \boldsymbol{\nabla}\Phi, \tag{70-2}$$

where
$$\kappa = F^2 \sum_i z_i^2 u_i c_i \tag{70-3}$$

is the conductivity of the solution. This is an expression of Ohm's law, valid for electrolytes in the absence of concentration gradients.

Still with no concentration variations, we can say that the current carried by species $j$ is

$$t_j\mathbf{i} = -F^2 z_j^2 u_j c_j\, \boldsymbol{\nabla}\Phi = \frac{z_j^2 u_j c_j}{\sum_i z_i^2 u_i c_i}\, \mathbf{i}, \tag{70-4}$$

where
$$t_j = \frac{z_j^2 u_j c_j}{\sum_i z_i^2 u_i c_i} \tag{70-5}$$

is the fraction of the current carried by species $j$ and is also known as the transference number. In such a case it is convenient and proper to identify a migration flux of species $i$

$$\mathbf{N}_i^{\text{migr}} = -z_i u_i F c_i\, \boldsymbol{\nabla}\Phi = \frac{t_i}{z_i F}\, \mathbf{i}. \tag{70-6}$$

When there are concentration gradients, the current density is not proportional to the electric field, and Ohm's law does not hold. Due to the diffusion current represented by the second term in equation 1, the current density could even have a different direction from the electric field. One can turn equation 1 around:

$$\boldsymbol{\nabla}\Phi = -\frac{\mathbf{i}}{\kappa} - \frac{F}{\kappa} \sum_i z_i D_i\, \boldsymbol{\nabla} c_i, \tag{70-7}$$

and say the same thing backwards. Even in the absence of current, there may be a gradient of potential. The second term in this equation gives rise to what is known as the diffusion potential. If all the diffusion coefficients were equal, this would be zero by electroneutrality. In conductivity measurements an alternating current is used so that concentration differences will not build up (and so as to reduce polarization at the electrodes).

The conductivity and the transference number are additional transport properties, defined in equations 3 and 5 in terms of the ionic mobilities introduced earlier. These transport properties have relevance in solutions of varying composition, but they do not retain their same physical significance. Ohm's law is valid and the transference number has the physical meaning of the fraction of current carried by an ionic species only in the absence of concentration gradients.

When there are concentration gradients, one can identify contributions to the species flux $\mathbf{N}_i$ due to migration, molecular diffusion, and convection, according to equation 69-1. However, the current density in equation 1 is composed of portions due to migration and to diffusion, and it is no longer proper to identify the migration flux according to the last expression in equation 6, although one finds in the literature such deceptively simple statements as

$$\mathbf{N}_i^{\text{diff}} = \frac{1 - t_i}{z_i F} \mathbf{i}. \tag{70-8}$$

It should be apparent that the transference number and the expression of the migration flux in terms of current density should be used with caution in cases where concentration gradients exist.

## 71. Conservation of charge

It is a physical law of nature that electric charge is conserved. This fact is already built into the basic transport relations. Multiplication of equation 69-3 by $z_i F$ and addition over species yield

$$\frac{\partial}{\partial t} F \sum_i z_i c_i = -\nabla \cdot F \sum_i z_i \mathbf{N}_i + F \sum_i z_i R_i. \tag{71-1}$$

The last term will be zero as long as all the homogeneous reactions giving rise to the $R_i$ are electrically balanced. Then the term on the left is the time rate of change of the charge density; the first term on the right is minus the divergence of the current density; and the equation describes conservation of charge. In view of the assumption of electroneutrality, the equation reduces to

$$\nabla \cdot \mathbf{i} = 0. \tag{71-2}$$

In physical terms, our line of reasoning has been that charge is carried by particles of matter and that conservation (or electrically balanced reaction) of these particles implies conservation of charge.

Insertion of equation 70-1 into equation 2 yields

$$\nabla \cdot (\kappa \, \nabla \Phi) + F \sum_i z_i \nabla \cdot (D_i \, \nabla c_i) = 0. \tag{71-3}$$

In the absence of concentration gradients and with a uniform value of the conductivity $\kappa$, this reduces to

$$\nabla^2\Phi = 0, \tag{71-4}$$

that is, the potential satisfies Laplace's equation in a region of uniform composition.

## 72. The binary electrolyte

By a binary electrolyte we mean the solution of a single salt composed of one kind of cation and one kind of anion. At times the term has been known to denote a symmetric electrolyte, which dissociates into equal numbers of anions and cations. Let the positive species be denoted by the subscript $+$, and the negative species by the subscript $-$. The mobilities and the diffusion coefficients will be assumed to be constant.

Let $v_+$ and $v_-$ be the numbers of cations and anions produced by the dissociation of one molecule of electrolyte. The concentration of the electrolyte is then defined by

$$c = \frac{c_+}{v_+} = \frac{c_-}{v_-}, \tag{72-1}$$

so that the electroneutrality equation 69-4 is satisfied. Substitution of the flux equation 69-1 into the material balance 69-3 with $R_i = 0$ yields equations for each of the ionic species:

$$\frac{\partial c}{\partial t} + \mathbf{v}\cdot\nabla c = z_+ u_+ F\,\nabla\cdot(c\,\nabla\Phi) + D_+\,\nabla^2 c. \tag{72-2}$$

$$\frac{\partial c}{\partial t} + \mathbf{v}\cdot\nabla c = z_- u_- F\,\nabla\cdot(c\,\nabla\Phi) + D_-\,\nabla^2 c. \tag{72-3}$$

Subtraction gives

$$(z_+ u_+ - z_- u_-)F\,\nabla\cdot(c\,\nabla\Phi) + (D_+ - D_-)\,\nabla^2 c = 0. \tag{72-4}$$

This can be used to eliminate the potential from either of equations 2 and 3, with the result

$$\frac{\partial c}{\partial t} + \mathbf{v}\cdot\nabla c = D\,\nabla^2 c, \tag{72-5}$$

where

$$D = \frac{z_+ u_+ D_- - z_- u_- D_+}{z_+ u_+ - z_- u_-}. \tag{72-6}$$

Equation 5 is called the equation of convective diffusion. This equation or its analogue applies to heat transfer or nonelectrolytic mass transfer, and its solutions have been extensively studied in the literature. Consequently, it is possible to apply many of these results to electrochemical systems with only minor changes in notation. This will be taken up in part D.

Equation 5 shows that in the absence of current a salt, such as copper sulfate in water, will behave like one species because of the requirement of electroneutrality. The observed diffusion coefficient $D$ represents a compromise between the diffusion coefficient of the anion and the cation.[1] If these diffusion coefficients are different, the species will tend to separate, thereby creating a minute charge density which prevents further separation. The charge density creates a nonuniform potential which acts to speed up the ion with the smaller diffusion coefficient and slow down the ion with the larger diffusion coefficient.

But equation 5 was derived without assuming that the current density is zero. The interesting and useful conclusion is that the concentration distribution in a solution of a single salt is governed by the same equation as the concentration distribution of a neutral species, even when a current is being passed.

The potential distribution in a solution of a single salt is to be determined from equation 4. An integrated form of this equation can be obtained from the expression 69-2 for the current density

$$-\frac{\mathbf{i}}{z_+v_+F} = (z_+u_+ - z_-u_-)Fc\,\nabla\Phi + (D_+ - D_-)\,\nabla c. \qquad (72\text{-}7)$$

This equation shows directly how the potential gradient is related to the concentration gradient and the difference in diffusion coefficients for the diffusion of a salt in the absence of current

$$F\,\nabla\Phi = -\frac{D_+ - D_-}{z_+u_+ - z_-u_-}\,\nabla \ln c. \qquad (72\text{-}8)$$

This is the diffusion potential, discussed in connection with equation 70-7, which prevents any substantial separation of charge in a diffusing system. Equation 7 is analogous to equation 70-1, while equation 4 is analogous to equation 71-3.

In situations where the boundary conditions permit, equation 5 can be solved first for the concentration distribution. If the current density distribution is known, the potential distribution can then be readily determined from equation 7. If the current density distribution is not known, the potential distribution must be determined from equation 4 and the current distribution subsequently from equation 7.

However, it is frequently possible to determine the current density distribution at an electrode from the concentration distribution but without the potential distribution. The normal component of the velocity will be zero at the electrode. Let us also assume that only the cation reacts at the electrode, a common situation for a binary electrolyte. Then the normal components of the cation and anion fluxes at the electrode are

$$N_{+y} = \frac{i_y}{z_+F} = -z_+u_+Fv_+c\,\frac{\partial\Phi}{\partial y} - D_+v_+\frac{\partial c}{\partial y} \qquad (72\text{-}9)$$

[1]W. Nernst, "Zur Kinetik der in Lösung befindlichen Körper," *Zeitschrift für physikalische Chemie*, 2 (1888), 613–637.

and
$$N_{-y} = 0 = -z_-u_-Fv_-c\frac{\partial\Phi}{\partial y} - D_-v_-\frac{\partial c}{\partial y}, \tag{72-10}$$

where $y$ is the distance from the electrode. Elimination of the potential gradient gives

$$\frac{i_y}{z_+v_+F} = -\frac{z_-u_-D_+ - z_+u_+D_-}{z_-u_-}\frac{\partial c}{\partial y} = -\frac{D}{1 - t_+}\frac{\partial c}{\partial y} \quad \text{at} \quad y = 0. \tag{72-11}$$

Here $D$ is given by equation 6, and $t_+$ is the cation transference number given by equation 70-5, which reduces to

$$t_+ = 1 - t_- = \frac{z_+u_+}{z_+u_+ - z_-u_-} \tag{72-12}$$

for a binary electrolyte. Equation 11 shows that the current density is directly related to the concentration derivative at the electrode.

In this discussion, we have treated the diffusion coefficients and mobilities (but not the conductivity) as constants. Usually these quantities depend upon the concentration. However, restriction to a constant-property fluid is common in the literature and has advantages of simplicity and generality. If we relax the assumption of constant properties, equation 7 still stands, but equation 5 is to be replaced by

$$\frac{\partial c}{\partial t} + \mathbf{v}\cdot\nabla c = \nabla\cdot(D\,\nabla c) - \frac{\mathbf{i}\cdot\nabla t_+}{z_+v_+F}, \tag{72-13}$$

where $D$ is given by equation 6 and $t_+$ by equation 12. The first term on the right is the expected generalization of the diffusion term in the convective-diffusion equation for a varying diffusion coefficient.


## 73. Supporting electrolyte

When the flux equation 69-1 is inserted into the material balance equation 69-3, one obtains

$$\frac{\partial c_i}{\partial t} + \mathbf{v}\cdot\nabla c_i = z_iF\,\nabla\cdot(u_ic_i\,\nabla\Phi) + \nabla\cdot(D_i\,\nabla c_i) + R_i. \tag{73-1}$$

The fact that the fluid is incompressible ($\nabla\cdot\mathbf{v} = 0$) has been used in obtaining this equation. This equation is useful for describing the medium, since the flux has been eliminated. The equation could therefore be used to determine the concentration distribution when the velocity and potential distributions are known. The flux equation 69-1 is still useful for formulating boundary conditions.

For a mass-transfer problem in forced convection, the velocity distribution can be assumed to be known, but usually the potential distribution needs to be determined. This means that equation 1 for each ionic species must be solved simultaneously, the electroneutrality equation 69-4 providing the additional relation needed to determine the potential. That is to say, all the equations are coupled through the potential. The problem thus posed is quite complicated.

We have already seen the simplification possible when only two ionic species are present. Then the requirement of electroneutrality allows the potential to be eliminated, and the concentration of the electrolyte satisfies the equation of convective diffusion. A similar simplification applies when migration and reactions in the bulk of the solution can be neglected. Then equation 1 becomes

$$\frac{\partial c_i}{\partial t} + \mathbf{v} \cdot \nabla c_i = D_i \, \nabla^2 c_i, \tag{73-2}$$

for a constant diffusion coefficient. This is again the equation of convective diffusion.

In mass-transfer studies in electrolytic systems, in studies of electrode kinetics, and in some commercial electrochemical cells, a *supporting* or *indifferent* electrolyte is frequently added in order to increase the conductivity of the solution and thereby reduce the electric field. The mass transfer of minor species then will be primarily due to diffusion and convection, and the effect of migration can be qualitatively dismissed. The concentration distribution is then governed by equation 2.

Levich[2] has given a more formal statement of this procedure, one which also allows investigation of the concentration distribution of the major species. Let us develop this for three ionic components, the third of which is present in small amount. We do not consider the possibility of reaction in the bulk of the solution, that is, $R_i = 0$. In the zero approximation, one assumes that the minor constituent is absent and solves for the potential and the concentration of the major species by using the method for binary electrolytes (see section 72). Let this solution be denoted by $c_1^0$, $c_2^0$, and $\Phi^0$.

Then we can write

$$c_1 = c_1^0 + c_1^{(1)}, \qquad c_2 = c_2^0 + c_2^{(1)},$$
$$c_3 = c_3^{(1)}, \qquad \Phi = \Phi^0 + \Phi^{(1)}. \tag{73-3}$$

These are substituted into the basic equations; and, in the first approximation, terms of degree greater than one in $c_1^{(1)}$, $c_2^{(1)}$, $c_3^{(1)}$, and $\Phi^{(1)}$ are dropped. The equations for the first approximation are then linear.

In many cases of importance, the minor constituent is the only one taking part in electrode reactions, and the zero solution yields constants for $c_1^0$, $c_2^0$, and $\Phi^0$. This applies to mass-transfer studies, where the system is selected so that the behavior of the minor component is of interest. For commercial

[2]V. Levich, "The Theory of Concentration Polarization," *Acta Physicochimica U.R.S.S.*, *17* (1942), 257–307.

cells, a loss of current efficiency would result if the supporting electrolyte were to participate in electrode reactions. Hence the name *indifferent* electrolyte. For this case, the equations for the first approximation reduce to

$$\frac{\partial c_3^{(1)}}{\partial t} + \mathbf{v} \cdot \nabla c_3^{(1)} = D_3 \, \nabla^2 c_3^{(1)}, \tag{73-4}$$

$$\frac{\partial c_1^{(1)}}{\partial t} + \mathbf{v} \cdot \nabla c_1^{(1)} = D_e \, \nabla^2 c_1^{(1)} + \frac{z_3 u_1 (D_2 - D_3)}{z_1 u_1 - z_2 u_2} \, \nabla^2 c_3^{(1)}, \tag{73-5}$$

and
$$-\frac{\mathbf{i}}{z_1 F} = (z_1 u_1 - z_2 u_2) F c_1^0 \, \nabla \Phi^{(1)}$$

$$+ (D_1 - D_2) \nabla c_1^{(1)} + \frac{z_3}{z_1} (D_3 - D_2) \nabla c_3^{(1)}, \tag{73-6}$$

where
$$D_e = \frac{z_1 u_1 D_2 - z_2 u_2 D_1}{z_1 u_1 - z_2 u_2} \tag{73-7}$$

is the diffusion coefficient of the supporting electrolyte. Here $c_2$ has been eliminated by means of the electroneutrality equation, and the mobilities and diffusion coefficients have been assumed to be constant.

The minor species obeys the equation of convective diffusion with its ionic diffusion coefficient; equation 4 is the same as equation 2. The supporting electrolyte obeys the equation of convective diffusion with the diffusion coefficient of the salt, but with an additional term of interaction with the minor species. The equations are to be solved in the order given: first, for the concentration of the minor component, second, for the concentration of the supporting electrolyte, and finally, for the potential from equation 6. In case the current is not known at this point, one can take the divergence of this equation (see equation 71-2) and solve a second-order differential equation for the potential.

It is not difficult to extend the development to a case where two minor constituents are involved in the electrode reaction but the major species are not involved. An example would be an oxidation-reduction reaction with a supporting electrolyte. Whether the above treatment applies to the reaction of a nonelectrolyte, such as oxygen, will be considered in the problems. (See also section 121.)

The treatment of a supporting electrolyte can be considered to be the beginning of a perturbation expansion of the problem. The expansion parameter would be a characteristic concentration of the minor species divided by a characteristic concentration of the supporting electrolyte. The procedure is, of course, valid only when this ratio is small. In practice, one is usually content to solve equation 4 for the minor component.

The concept of supporting electrolytes raises a number of interesting and paradoxical questions. Some of these will be considered in chapter 19.

## 74. Multicomponent diffusion by elimination of the electric field

Multicomponent diffusion in nonelectrolytic solutions has been treated in the literature. In concentrated solutions, the diffusing species interact with each other; but in dilute solutions, each species diffuses independently according to its own concentration gradient and diffusion coefficient. However, in a dilute electrolytic solution even in the absence of current, the solute species do not diffuse independently. A diffusion potential will be established, and the diffusing ions will interact with it.

Substitution of equation 70-7 into the flux equation 69-1 yields

$$N_i = \frac{t_i}{z_i F} i - D_i \nabla c_i + v c_i + \frac{t_i}{z_i} \sum_j z_j D_j \nabla c_j. \tag{74-1}$$

In order to satisfy the condition of electroneutrality, the concentration of an ionic species $n$ can be eliminated:

$$z_n c_n = -\sum_{j \neq n} z_j c_j, \tag{74-2}$$

with the result

$$N_i = \frac{t_i}{z_i F} i - D_i \nabla c_i + v c_i + \frac{t_i}{z_i} \sum_j z_j (D_j - D_n) \nabla c_j. \tag{74-3}$$

These equations for all minor species except species $n$ can be substituted into the appropriate material balances (equation 69-3) to give

$$\frac{\partial c_i}{\partial t} + v \cdot \nabla c_i = D_i \nabla^2 c_i + R_i - \frac{i \cdot \nabla t_i}{z_i F}$$

$$- \sum_j \frac{z_j}{z_i} (D_j - D_n) \nabla \cdot (t_i \nabla c_j). \tag{74-4}$$

Even in the absence of current and homogeneous chemical reactions, diffusion of species in an electrolytic solution is coupled in much the same way as diffusion in concentrated, multicomponent, nonelectrolytic solutions. One may also note that the migration flux $-z_i u_i F c_i \nabla \Phi$ is not the same as $t_i i / z_i F$ when there are concentration gradients. This was discussed before in connection with equation 70-6.

## 75. Mobilities and diffusion coefficients

We mentioned in section 69 that a single driving force, the gradient of the electrochemical potential of a species, is appropriate for both diffusion and migration. We are thus led to expect that the ionic mobility and diffusion coefficient are related. This relationship is provided by the Nernst-

Einstein equation

$$D_i = RTu_i. \tag{75-1}$$

This equation is strictly applicable only at infinite dilution, although its failure is related to the approximate nature of the flux equation 69-1. The quantities $D_i$ and $u_i$ in equation 1 are not adequately defined at nonzero concentrations, and further inquiry into the nature of this equation should await the consideration of concentrated electrolytes in chapter 12.

With the Nernst-Einstein relation, the expression 72-6 for the diffusion coefficient of a binary electrolyte becomes

$$D = \frac{D_+ D_- (z_+ - z_-)}{z_+ D_+ - z_- D_-}. \tag{75-2}$$

One commonly encounters the statement that "a salt bridge used to eliminate liquid junction potentials should contain a salt with equal cation and anion transference numbers." (Liquid junction potentials are diffusion potentials which arise when one connects two electrolytic solutions of different composition, see chapter 6.) This can be interpreted with the aid of the Nernst-Einstein equation. For the solution of a single salt, the transference numbers are nominally independent of concentration (due to electroneutrality) and are given by equation 72-12 .With equation 1 we have

$$t_+ = \frac{z_+ D_+}{z_+ D_+ - z_- D_-}, \qquad t_- = \frac{-z_- D_-}{z_+ D_+ - z_- D_-}. \tag{75-3}$$

Equality of the transference numbers, coupled with the Nernst-Einstein equation, implies that the diffusion coefficients are equal for symmetric salts $(z_+ = -z_-)$. Then the concentration can vary without giving rise to diffusion potentials (see equation 72-8 or 70-7). (We still do not have a satisfactory answer to the question of what happens at the junctions of the salt bridge with the two solutions we were trying to connect.)

Alternatively, equation 70-7 can now be written, by means of the Nernst-Einstein relation, as

$$F \nabla \Phi = -\frac{F}{\kappa} \mathbf{i} - RT \sum_i \frac{t_i}{z_i} \nabla \ln c_i. \tag{75-4}$$

Table 1 gives an indication of the magnitudes of ionic diffusion coefficients and mobilities. Ionic mobilities are usually not found directly in the literature; instead values of *ionic equivalent conductances* are reported. These are related to ionic mobilities by

$$\lambda_i = |z_i| F^2 u_i. \tag{75-5}$$

Ionic diffusion coefficients can then be calculated with the aid of the Nernst-Einstein relation:

$$D_i = \frac{RT \lambda_i}{|z_i| F^2}. \tag{75-6}$$

TABLE 75-1. VALUES OF EQUIVALENT CONDUCTANCES AND DIFFUSION
COEFFICIENTS OF SELECTED IONS AT INFINITE DILUTION IN WATER AT 25°C.

| ion | $z_i$ | $\lambda_i^0$ mho-cm² equiv | $D_i \times 10^5$ cm² sec | ion | $z_i$ | $\lambda_i^0$ mho-cm² equiv | $D_i \times 10^5$ cm² sec |
|-----|-------|------|------|-----|-------|------|------|
| H⁺ | 1 | 349.8 | 9.312 | OH⁻ | −1 | 197.6 | 5.260 |
| Li⁺ | 1 | 38.69 | 1.030 | Cl⁻ | −1 | 76.34 | 2.032 |
| Na⁺ | 1 | 50.11 | 1.334 | Br⁻ | −1 | 78.3 | 2.084 |
| K⁺ | 1 | 73.52 | 1.957 | I⁻ | −1 | 76.8 | 2.044 |
| NH₄⁺ | 1 | 73.4 | 1.954 | NO₃⁻ | −1 | 71.44 | 1.902 |
| Ag⁺ | 1 | 61.92 | 1.648 | HCO₃⁻ | −1 | 41.5 | 1.105 |
| Tl⁺ | 1 | 74.7 | 1.989 | HCO₂⁻ | −1 | 54.6 | 1.454 |
| Mg⁺⁺ | 2 | 53.06 | 0.7063 | CH₃CO₂⁻ | −1 | 40.9 | 1.089 |
| Ca⁺⁺ | 2 | 59.50 | 0.7920 | SO₄⁼ | −2 | 80 | 1.065 |
| Sr⁺⁺ | 2 | 59.46 | 0.7914 | Fe(CN)₆³⁻ | −3 | 101 | 0.896 |
| Ba⁺⁺ | 2 | 63.64 | 0.8471 | Fe(CN)₆⁴⁻ | −4 | 111 | 0.739 |
| Cu⁺⁺ | 2 | 54 | 0.72 | IO₄⁻ | −1 | 54.38 | 1.448 |
| Zn⁺⁺ | 2 | 53 | 0.71 | ClO₄⁻ | −1 | 67.32 | 1.792 |
| La⁺⁺⁺ | 3 | 69.5 | 0.617 | BrO₃⁻ | −1 | 55.78 | 1.485 |
| Co(NH₃)₆⁺⁺⁺ | 3 | 102.3 | 0.908 | HSO₄⁻ | −1 | 50 | 1.33 |

Table 1 shows that most ionic diffusion coefficients are about 1 or $2 \times 10^{-5}$ cm²/sec. Exceptions are hydrogen ions and hydroxyl ions, for which $D_i$ values are 9.3 and $5.3 \times 10^{-5}$ cm²/sec.

The equivalent conductance $\Lambda$ of a single salt is the sum of the values for the two ions

$$\Lambda = \lambda_+ + \lambda_- \tag{75-7}$$

and is related to the conductivity of the solution by

$$\Lambda = \frac{\kappa}{z_+ \nu_+ c}. \tag{75-8}$$

The value of $\Lambda$ will thus be about 100 mho-cm²/equiv except for acids and bases. The conductivity of the solution is obtained by multiplying $\Lambda$ by the equivalent concentration $z_+ \nu_+ c$, but this should be in equiv/cm³ in order for $\kappa$ to be in mho/cm. Thus, the conductivity of 0.6 $M$ NaCl solution (roughly sea water) will be about 0.04 ohm⁻¹–cm⁻¹ when due allowance is made for the concentration dependence of $\Lambda$.

The transference number of an ion in a binary salt solution will be

$$t_+ = 1 - t_- = \frac{\lambda_+}{\lambda_+ + \lambda_-} \tag{75-9}$$

and will be close to 0.5 except for acids and bases, where $t_+$ can be as high as 0.8 or as low as 0.2. For solutions with an excess of inert electrolyte, the transference number of a minor ionic species will be proportional to its concentration and inversely proportional to the concentration of the supporting electrolyte and hence will be small.

Ionic equivalent conductances, such as those in table 1, are ordinarily determined by measuring the equivalent conductance $\Lambda$ and the transference number $t_+$ for solutions of single salts and extrapolating the values so obtained to infinite dilution. Equations 7 and 9 then yield $\lambda_+$ and $\lambda_-$. Good agreement is usually obtained for, say, $\lambda_i$ for chloride ions determined from solutions of NaCl and separately from solutions of KCl. Diffusion coefficients calculated from equation 2 are also in good agreement with values measured at high dilution.

An approximate guide to the temperature dependence of ionic diffusion coefficients is provided by the relationship

$$\frac{D_i \mu}{T} = \text{constant},\tag{75-10}$$

where $\mu$ is the viscosity of the solution. Thus, ionic diffusion coefficients and equivalent conductances can vary by 2 to 3 percent per degree Celsius. This is a fairly strong temperature dependence. Equation 10 can also be used to estimate the concentration dependence of ionic diffusion coefficients.

## 76. Electroneutrality and Laplace's equation

The electroneutrality equation 69-4 is not a fundamental law of nature. Perhaps a more nearly correct relationship would be Poisson's equation, which, for a medium of uniform dielectric constant, reads (see equation 22-8)

$$\nabla^2 \Phi = -\frac{F}{\epsilon} \sum_i z_i c_i,\tag{76-1}$$

and relates the charge density to the Laplacian of the electric potential. The proportionality constant in this equation is Faraday's constant $F$ divided by the permittivity or dielectric constant $\epsilon$. The value of this proportionality constant is quite large ($1.392 \times 10^{16}$ V-cm/equiv for a relative dielectric constant of 78.303), so that what, in terms of concentrations, would be a negligible deviation from electroneutrality amounts to a considerable deviation from Laplace's equation for the potential.

Another way of saying the same thing is that $F/\epsilon$ is so large that an appreciable separation of charge would require prohibitively large electric forces. Still another way is that the conductivity is so large that any initial charge density would be neutralized very rapidly or would rapidly flow to the boundaries of the solution.

The equations given in section 69, with appropriately stated boundary conditions, are sufficient to describe transport processes in electrolytic solutions. Therefore, the use of both Poisson's equation and electroneutrality would be inconsistent. The proper thing to do is to *replace* Poisson's equation in the analysis by the electroneutrality condition 69-4 on the basis of the large value of $F/\epsilon$. Thus, electroneutrality does *not* imply Laplace's equation for the potential

$$\nabla^2\Phi = 0; \qquad\qquad (76\text{-}2)$$

this would be inconsistent. Of course, one could retain Poisson's equation and discard the assumption of electroneutrality in the description of electrochemical systems. However, the close adherence of electrolytic solutions to the condition of electroneutrality, as well as the consequent mathematical simplification in the treatment of specific problems, justifies the approach taken here. For the perturbation analysis of phenomena near an electrode with the use of Poisson's equation, see references 3 and 4.

Electroneutrality and Laplace's equation are firmly entrenched in electrochemistry, but the assumption of electroneutrality does not imply that Laplace's equation holds for the potential. In many cases, the distribution of potential and current in cells of various configurations is determined from Laplace's equation for the potential and Ohm's law

$$\mathbf{i} = -\kappa\,\nabla\Phi \qquad\qquad (76\text{-}3)$$

for the current. This procedure is valid when the current is not appreciably limited by mass transfer of reactants to the electrodes. Then the concentrations are fairly constant, and equation 3 applies with a fairly constant conductivity. Conservation of charge then yields Laplace's equation for the potential (see equations 71-2, -3, and -4). This justification of Laplace's equation is considerably different from the statement that electroneutrality implies Laplace's equation for the potential. The procedure outlined here can be expected to lead to inconsistencies if one subsequently attempts to investigate the detailed behavior of each species in the solution near electrodes.

It should be pointed out that it is not permissible to neglect the charge density in the electrode double layer, since the electric field is indeed very large in this region. This region may be 10 to 100 Å in thickness and was treated in section 52. The double layer can legitimately be regarded as part of the interface and not part of the solution. In extremely dilute solutions, the charge density may also be appreciable compared to the total ionic concentration.

We illustrate next the validity of the assumption of electroneutrality by means of an example. Let us consider a cell in which a binary electrolyte is used to deposit the cation on the cathode while the anode dissolves and replenishes the solution. Let us further simplify the problem by assuming a steady state with no convection and with variations in only one dimension. This is not supposed to represent a common system; it is merely a test of the electroneutrality assumption. Consider a uni-univalent electrolyte, such as

---

[3] John Newman, "The Polarized Diffuse Double Layer," *Transactions of the Faraday Society, 61* (1965), 2229–2237.

[4] William H. Smyrl and John Newman, "Double Layer Structure at the Limiting Current," *Transactions of the Faraday Society, 63* (1967), 207–216.

silver nitrate, and use the Nernst-Einstein relation (equation 75-1) throughout.

The procedure is to solve the problem using the electroneutrality equation, and, in the end, the deviation from electroneutrality can be assessed by means of Poisson's equation 1. Since the flux of the anion is zero, equation 69-1 yields

$$Fc\frac{d\Phi}{dx} = RT\frac{dc}{dx}. \tag{76-4}$$

Equation 72-7 becomes

$$-\frac{i}{F} = (u_+ + u_-)Fc\frac{d\Phi}{dx} + (D_+ - D_-)\frac{dc}{dx} = 2D_+\frac{dc}{dx}. \tag{76-5}$$

Integration gives the steady-state concentration in terms of the current density

$$c = c_{avg} - \frac{i}{2D_+F}\left(x - \frac{1}{2}L\right), \tag{76-6}$$

where $c_{avg}$ is the average concentration in the cell, $x$ is the distance measured from one electrode, and $L$ is the distance between the electrodes.

Now for

$$i = \pm\frac{4D_+Fc_{avg}}{L} \tag{76-7}$$

the concentration at one electrode will drop to zero, and the *limiting current* is attained. A higher current can be passed only if another electrode reaction occurs. Let us operate at half of the limiting current, so that

$$c = \frac{c_{avg}}{2}\left(1 + \frac{2x}{L}\right). \tag{76-8}$$

Equation 4 gives

$$\frac{d^2\Phi}{dx^2} = -\frac{4RT}{F}\frac{1}{(L+2x)^2}. \tag{76-9}$$

If, at the same time, both electroneutrality and Poisson's equation were exact, this second derivative would be equal to zero, but it is not. Thus, one can see the incompatibility of these relationships.

Let us test the electroneutrality assumption by calculating the charge density and the difference in concentration of anions and cations required to produce this value of $d^2\Phi/dx^2$.

$$\frac{d^2\Phi}{dx^2} = -\frac{4RT}{F}\frac{1}{(L+2x)^2} = -\frac{F}{\epsilon}(c_+ - c_-). \tag{76-10}$$

At 25°C and for a relative dielectric constant of 78.303, $RT/F = 25.692$ mV, and $\epsilon RT/F^2 = 1.846 \times 10^{-18}$ equiv/cm. For $L = 0.1$ mm and $x = 0.05$ mm, this gives

$$\frac{d^2\Phi}{dx^2} = -256.92 \text{ V/cm}^2 \tag{76-11}$$

and

$$c_+ - c_- = 1.846 \times 10^{-14} \text{ equiv/cm}^3 = 1.846 \times 10^{-11} \text{ equiv/liter.} \quad (76\text{-}12)$$

This indicates that the assumption of electroneutrality is very good in electrochemical systems.

## 77. Moderately dilute solutions

When the gradient of the electrochemical potential is used as the driving force for diffusion and migration, the flux equation 69-1 for an ionic component becomes

$$\mathbf{N}_i = -u_i c_i \nabla \mu_i + c_i \mathbf{v}. \quad (77\text{-}1)$$

The driving force per mole is $-\nabla \mu_i$. Multiplication by the mobility $u_i$ gives the velocity for diffusion and migration, and multiplication by the concentration $c_i$ gives the contribution to the net flux $\mathbf{N}_i$. With the use of the Nernst-Einstein relation 75-1, equation 1 becomes

$$\mathbf{N}_i = -\frac{D_i c_i}{RT} \nabla \mu_i + c_i \mathbf{v}. \quad (77\text{-}2)$$

The use of the electrochemical potential avoids the problem of defining the electric potential in a medium of varying composition. However, substitution of equation 1 into the material balance 69-3 yields

$$\frac{\partial c_i}{\partial t} + \mathbf{v} \cdot \nabla c_i = \nabla \cdot (u_i c_i \nabla \mu_i) + R_i. \quad (77\text{-}3)$$

Here one does not have a simple equation for the concentration, as in the case of the equation of convective diffusion 73-2, since $\mu_i$ depends on the local electrical state as well as the local composition.

One way to proceed is to define an electric potential on the basis of a chosen ionic species $n$ (the *quasi-electrostatic potential*, see equation 26-3)

$$\mu_n = RT \ln c_n + z_n F \Phi. \quad (77\text{-}4)$$

Then we can write for the gradient of the electrochemical potential of any species

$$\nabla \mu_i = \nabla \left( \mu_i - \frac{z_i}{z_n} \mu_n \right) + \frac{z_i}{z_n} \nabla \mu_n. \quad (77\text{-}5)$$

The term in parentheses now corresponds to a neutral combination of ions and can be expressed as

$$\mu_i - \frac{z_i}{z_n} \mu_n = RT \left[ \ln (a_i^\theta c_i f_i) - \frac{z_i}{z_n} \ln (a_n^\theta c_n f_n) \right]$$
$$= RT \left( \ln a_i^\theta - \frac{z_i}{z_n} \ln a_n^\theta \right) + RT \left( \ln c_i - \frac{z_i}{z_n} \ln c_n \right)$$
$$+ RT \left( \ln f_i - \frac{z_i}{z_n} \ln f_n \right). \quad (77\text{-}6)$$

These combinations of $a_i^\theta$'s and ionic activity coefficients $f_i$'s are well defined according to the considerations of section 14, and there need be no hesitation in their use.

At uniform temperature, equation 5 now becomes

$$\nabla \mu_i = RT \, \nabla \ln c_i + z_i F \, \nabla \Phi + RT \, \nabla \left( \ln f_i - \frac{z_i}{z_n} \ln f_n \right), \qquad (77\text{-}7)$$

and equation 2 becomes

$$\mathbf{N}_i = -\frac{z_i D_i F}{RT} c_i \, \nabla \Phi - D_i \, \nabla c_i - D_i c_i \, \nabla \left( \ln f_i - \frac{z_i}{z_n} \ln f_n \right) + c_i \mathbf{v}. \qquad (77\text{-}8)$$

Let

$$f_{i,n} = \frac{f_i}{f_n^{z_i/z_n}}, \qquad (77\text{-}9)$$

so that equation 8 can be written as

$$\mathbf{N}_i = -\frac{z_i D_i F}{RT} c_i \, \nabla \Phi - D_i \, \nabla c_i - D_i c_i \, \nabla \ln f_{i,n} + c_i \mathbf{v}. \qquad (77\text{-}10)$$

The electric potential $\Phi$ is introduced because we need a means of assessing the electrical state of the solution. Its arbitrariness is indicated by the necessity of choosing a particular ionic species $n$ in equation 4. The advantage of this procedure is that the structure of the equations is now essentially the same as that of the dilute-solution theory of section 69. There are flux equations 10 and material balance equations 69-3 for each species, and these correspond to the unknown fluxes $\mathbf{N}_i$ and concentrations $c_i$. In addition there is the electroneutrality equation 69-4, corresponding to the unknown potential $\Phi$. Calculation procedures worked out for the dilute-solution theory can still be applied here.

The equations are now more complicated than before because the activity coefficients $f_{i,n}$ relative to species $n$ depend on the local composition of the solution. Here the thermodynamic properties of multicomponent solutions, discussed in section 31, can be applied to express these activity coefficients in terms of the concentrations. This procedure also shows how activity coefficients can be introduced into the dilute-solution theory without using activity coefficients of individual ions. The arbitrariness in the potential $\Phi$ and the reference of ionic activities to the species $n$ reflect in a complementary manner the arbitrariness in selecting species $n$ in equation 4. However, the potential $\Phi$ is well defined, though arbitrary, and can be used to determine relationships of the electrical state at phase boundaries.

At the same time, this procedure illuminates the limitations of the dilute-solution theory. For sufficiently dilute solutions, $f_{i,n} \rightarrow 1$ (see equation 14-6). The use of an electric potential $\Phi$ in the dilute-solution theory is therefore not vague; which species $n$ is chosen becomes immaterial when the solution is so dilute that $f_{i,n} \rightarrow 1$. One also sees that variations in $f_{i,n}$ are neglected in the dilute-solution theory. This theory works fairly well in moderately concentrated solutions, not so much because $f_{i,n}$ is close to 1 as because variations in $f_{i,n}$ can be neglected.

The use of electrochemical potentials and the considerations of activity-coefficient variations might appear to be the most important first correction to dilute-solution theory and have been treated as such in references 5 and 6 and in chapter 6 on the calculation of the potentials of cells with liquid junctions. However, the variation of ionic diffusion coefficients with concentration may be equally important. It should also be recalled that interactions between a diffusing species and species other than the solvent are not included in equation 2 or 10 and that the fluid velocity $\mathbf{v}$ has not been carefully defined. These are considered in the next chapter.

## PROBLEMS

**1.** Write down expressions for the diffusion coefficient $D$ of the electrolyte, the cation transference number $t_+$, and the conductivity $\kappa$ for solutions of sulfuric acid when the electrolyte is assumed to dissociate either as

$$H_2SO_4 \rightleftharpoons H^+ + HSO_4^-$$

or as

$$H_2SO_4 \rightleftharpoons 2 H^+ + SO_4^=.$$

Make numerical comparisons for these quantities on the basis of the information given in section 75.

**2.** At the negative electrode in a lead-acid battery, the reaction is

$$Pb(s) + SO_4^= \rightleftharpoons PbSO_4(s) + 2e^-.$$

Regard the solution as a binary electrolyte of $H_2SO_4$ dissociated into $H^+$ and $SO_4^=$ ions and show that the current density at the electrode surface is related to the concentration gradient by

$$\frac{i_y}{z_-v_-F} = -\frac{D}{1-t_-}\frac{\partial c}{\partial y} \quad \text{at} \quad y = 0,$$

analogous to equation 72-11. Discuss any difficulties presented by the presence of the solid $PbSO_4$ at the electrode.

**3.** At the positive electrode in a lead-acid battery, the reaction is

$$PbO_2(s) + SO_4^= + 4 H^+ + 2e^- \rightleftharpoons PbSO_4(s) + 2 H_2O.$$

Regard the solution as a binary electrolyte of $H_2SO_4$ dissociated into $H^+$ and $SO_4^=$ ions, and show that the current density at the electrode surface is related to the concentration gradient by

$$\frac{i_y}{F} = -\frac{2D}{2-t_+}\frac{\partial c}{\partial y} \quad \text{at} \quad y = 0,$$

[5]William H. Smyrl and John Newman, "Potentials of Cells with Liquid Junctions," *Journal of Physical Chemistry*, 72 (1968), 4660–4671.

[6]John Newman and Limin Hsueh, "Currents Limited by Gas Solubility," *Industrial and Engineering Chemistry Fundamentals*, 9 (1970), 677–679.

analogous to equation 72-11. Discuss any difficulties presented by the presence of the solid $PbSO_4$ at the electrode.

**4.** The treatment of supporting electrolyte in section 73 should be applicable to the reaction of a dissolved, neutral species such as oxygen:

$$O_2 + 2\,H_2O + 4e^- \rightleftharpoons 4\,OH^-.$$

Would the concentration of supporting electrolyte change at all near the electrode surface? Equation 73-5 suggests that it would not since $z_3 = 0$ in this case. Sketch the concentration profiles for the various species when the supporting electrolyte is

(a) NaOH    (b) NaCl    (c) HCl.

Rationalize the shape of each profile in terms of the net flux of the species determined by the electrode reaction and the contributions of diffusion, migration, and convection to this flux. Remember that the condition of electroneutrality must be satisfied.

For NaCl as a supporting electrolyte, there must be two minor species, $O_2$ and $OH^-$. For HCl as a supporting electrolyte, it should be convenient to write the electrode reaction as

$$O_2 + 4\,H^+ + 4e^- \rightleftharpoons 2\,H_2O.$$

## NOTATION

| | |
|---|---|
| $a_i^\theta$ | property expressing secondary reference state, liter/mole |
| $c$ | molar concentration of a single electrolyte, mole/cm³ |
| $c_i$ | concentration of species $i$, mole/cm³ |
| $D$ | diffusion coefficient of electrolyte, cm²/sec |
| $D_i$ | diffusion coefficient of species $i$, cm²/sec |
| $f_i$ | molar activity coefficient of species $i$ |
| $f_{i,n}$ | molar activity coefficient of species $i$ relative to the ionic species $n$ |
| $F$ | Faraday's constant, 96,487 C/equiv |
| $i$ | current density, A/cm² |
| $L$ | distance between electrodes, cm |
| $N_i$ | flux of species $i$, mole/cm²-sec |
| $R$ | universal gas constant, 8.3143 J/mole-deg |
| $R_i$ | rate of homogeneous production of species $i$, mole/cm³-sec |
| $t$ | time, sec |
| $t_i$ | transference number of species $i$ |
| $T$ | absolute temperature, deg K |
| $u_i$ | mobility of species $i$, cm²-mole/J-sec |
| $v$ | fluid velocity, cm/sec |
| $z_i$ | charge number of species $i$ |
| $\epsilon$ | permittivity, farad/cm |

$\kappa$         conductivity, mho/cm

$\lambda_i$        ionic equivalent conductance, mho-cm$^2$/equiv

$\Lambda$        equivalent conductance of binary electrolyte, mho-cm$^2$/equiv

$\mu$         viscosity, g/cm-sec

$\mu_i$        electrochemical potential of species $i$, J/mole

$\nu_+, \nu_-$    numbers of cations and anions into which a molecule of electrolyte
               dissociates

$\Phi$         electric potential, V

# Concentrated Solutions 12

Although the use of the material of chapter 11 has been quite successful in the analysis of electrochemical problems, in this chapter we want to develop a description of transport processes which is more generally valid.

## 78. Transport laws

Mass transfer in electrolytic solutions requires a description of the movement of mobile ionic species (equation 69-1 or 77-2), material balances (equation 69-3), current flow (equation 69-2), electroneutrality (equation 69-4), and fluid mechanics (see chapter 15). The equations for material balances, current flow, and electroneutrality given in section 69 remain valid for concentrated solutions, but the flux equation requires modification.

The flux equations treated earlier fail even in ternary solutions of nonelectrolytes, since in such solutions there are two independent concentration gradients and the diffusion flux of each species can be affected by both concentration gradients.

In order to avoid the difficulties mentioned in section 69, equation 69-1 can be replaced by the multicomponent diffusion equation

$$c_i \, \nabla \mu_i = \sum_j K_{ij}(\mathbf{v}_j - \mathbf{v}_i) = RT \sum_j \frac{c_i c_j}{c_T \mathfrak{D}_{ij}} (\mathbf{v}_j - \mathbf{v}_i), \qquad (78\text{-}1)$$

where $\mu_i$ is the *electro*chemical potential of species $i$ and $K_{ij}$ are friction coefficients or interaction coefficients. $\mathbf{v}_i$ is the velocity of species $i$, an average

239

velocity for the species but not the velocity of individual molecules. Thus, the flux of species $i$ is $\mathbf{N}_i = c_i\mathbf{v}_i$. The total concentration is

$$c_T = \sum_i c_i, \tag{78-2}$$

where the sum includes the solvent, and $\mathfrak{D}_{ij}$ is a *diffusion coefficient* describing the interaction of species $i$ and $j$. These diffusion coefficients are, for the moment, simply parameters that can replace the drag coefficients $K_{ij}$:

$$K_{ij} = \frac{RTc_ic_j}{c_T\mathfrak{D}_{ij}}. \tag{78-3}$$

The term $-c_i\,\nabla\mu_i$ in equation 1 can be regarded as a driving force per unit volume acting on species $i$ and causing it to move with respect to the surrounding fluid. The force per unit volume exerted by species $j$ on species $i$ as a result of their relative motion has been expressed as $K_{ij}(\mathbf{v}_j - \mathbf{v}_i)$, that is, proportional to the difference in velocity of the two species. By Newton's third law of motion (action equals reaction), we find that $K_{ij} = K_{ji}$ or

$$\mathfrak{D}_{ij} = \mathfrak{D}_{ji}. \tag{78-4}$$

Equation 1 thus expresses the balance between the driving force and the total drag exerted by the other species.

The number of independent equations with the form of equation 1 is one less than the number of species. Addition of equation 1 over $i$ gives

$$\sum_i c_i\,\nabla\mu_i = \sum_i \sum_j K_{ij}(\mathbf{v}_j - \mathbf{v}_i). \tag{78-5}$$

The left side is zero by the Gibbs-Duhem relation (at constant temperature and pressure), and the right side is zero since $K_{ij} = K_{ji}$.

Equation 1 avoids the difficulties with the flux equation 69-1 mentioned in section 69. The gradient of the electrochemical potential has been used as the driving force for diffusion and migration, as in section 77. This resolves the question of the electric potential and the activity coefficients of individual ions. The use of the velocity difference $\mathbf{v}_j - \mathbf{v}_i$ in equation 1 avoids or postpones the question of the reference or average velocity on which diffusion and migration fluxes are based. The multicomponent diffusion equation is more general than equation 69-1 because it relates the driving force to a linear combination of resistances instead of just to one resistance, that with the solvent. The number of transport properties $\mathfrak{D}_{ij}$ defined by equation 1 is $\frac{1}{2}n(n-1)$, where $n$ is the number of species present, since $\mathfrak{D}_{ij} = \mathfrak{D}_{ji}$ and $\mathfrak{D}_{ii}$ is not defined. This is different from the number of transport properties $u_i$ and $D_i$ defined by equation 69-1, whether or not the Nernst-Einstein relation 75-1 is used. Thus, for three species (for example, two ions and a solvent), there are three transport properties defined by equation 1; and for four species (for example, three ions and a solvent), there are six transport properties.

Equation 1 is similar to the Stefan-Maxwell equation (see reference

1, p. 570) and is equivalent to one developed by Onsager (equation 14, p. 245, in reference 2). The Stefan-Maxwell equations apply to diffusion in dilute gas mixtures and express the driving force as a mole fraction gradient or a gradient of partial pressure instead of the gradient of the electrochemical potential. Equation 4 is equivalent to the Onsager reciprocal relation. The reciprocals of the $\mathfrak{D}_{ij}$'s can be regarded as friction coefficients similar to those used by Laity[3,4] and Klemm[5,6] to describe transport in ionic solutions and melts. Burgers[7] has also used this concept to treat the conductivity of ionized gases, and Lightfoot *et al.*[8] have applied equation 1 to liquid solutions. Truesdell[9] has discussed the validity of the arguments that $K_{ij} = K_{ji}$. (See also Lamm.[10])

The modification of equation 1 for use in nonisothermal media is indicated in section 85.

## 79. The binary electrolyte

Equation 78-1 expresses the driving forces in terms of the species velocities $v_i$ or the species fluxes $c_i v_i$. For use in the material-balance equation 69-3, it is necessary to invert the set of equations 1 so as to express the species fluxes in terms of the driving forces. Since these are linear, algebraic equations, the inversion is straightforward but lengthy. The general procedure is indicated in section 83.

[1] R. Byron Bird, Warren E. Stewart, and Edwin N. Lightfoot, *Transport Phenomena* (New York: John Wiley & Sons, Inc., 1960).

[2] Lars Onsager, "Theories and Problems of Liquid Diffusion," *Annals of the New York Academy of Sciences, 46* (1945), 241–265.

[3] Richard W. Laity, "General Approach to the Study of Electrical Conductance and its Relation to Mass Transport Phenomena," *Journal of Chemical Physics, 30* (1959), 682–691.

[4] Richard W. Laity, "An Application of Irreversible Thermodynamics to the Study of Diffusion," *Journal of Physical Chemistry, 63* (1959), 80–83.

[5] Alfred Klemm, "Thermodynamik der Transportvorgänge in Ionengemischen und ihre Anwendung auf isotopenhaltige Salze und Metalle," *Zeitschrift für Naturforschung, 8a* (1953), 397–400.

[6] A. Klemm, "Zur Phänomenologie der isothermen Diffusion in Elektrolyten," *Zeitschrift für Naturforschung, 17a* (1962), 805–807.

[7] J. M. Burgers, "Some Problems of Magneto-Gasdynamics." Sydney Goldstein, *Lectures on Fluid Mechanics* (London: Interscience Publishers, Ltd., 1960), pp. 271–299.

[8] E. N. Lightfoot, E. L. Cussler, Jr., and R. L. Rettig, "Applicability of the Stefan-Maxwell Equations to Multicomponent Diffusion in Liquids," *A.I.Ch.E. Journal, 8* (1962), 708–710.

[9] C. Truesdell, "Mechanical Basis of Diffusion," *Journal of Chemical Physics, 37* (1962), 2336–2344.

[10] Ole Lamm, "Studies in the Kinematics of Isothermal Diffusion. A Macro-Dynamical Theory of Multicomponent Fluid Diffusion," I. Prigogine, ed., *Advances in Chemical Physics, 6* (1964), 291–313.

For a binary electrolytic solution composed of anions, cations, and solvent, equation 78-1 yields two independent equations

$$c_+ \nabla \mu_+ = K_{0+}(\mathbf{v}_0 - \mathbf{v}_+) + K_{+-}(\mathbf{v}_- - \mathbf{v}_+). \tag{79-1}$$

$$c_- \nabla \mu_- = K_{0-}(\mathbf{v}_0 - \mathbf{v}_-) + K_{+-}(\mathbf{v}_+ - \mathbf{v}_-). \tag{79-2}$$

These equations can be rearranged, with the introduction of the current density from equation 69-2, to read

$$\mathbf{N}_+ = c_+ \mathbf{v}_+ = -\frac{\nu_+ \mathfrak{D}}{\nu RT}\frac{c_T}{c_0} c \nabla \mu_e + \frac{i t_+^0}{z_+ F} + c_+ \mathbf{v}_0, \tag{79-3}$$

$$\dot{\mathbf{N}}_- = c_- \mathbf{v}_- = -\frac{\nu_- \mathfrak{D}}{\nu RT}\frac{c_T}{c_0} c \nabla \mu_e + \frac{i t_-^0}{z_- F} + c_- \mathbf{v}_0, \tag{79-4}$$

where $\nu = \nu_+ + \nu_-$ (see section 72) and $\mu_e = \nu_+ \mu_+ + \nu_- \mu_- = \nu RT \ln (c f_{+-} a_{+-}^\theta)$. Here $f_{+-}$ is the mean molar activity coefficient of the electrolyte (see equation 14-20). The diffusion coefficient of the electrolyte, based on a thermodynamic driving force, is

$$\mathfrak{D} = \frac{\mathfrak{D}_{0+}\mathfrak{D}_{0-}(z_+ - z_-)}{z_+ \mathfrak{D}_{0+} - z_- \mathfrak{D}_{0-}}. \tag{79-5}$$

The transference numbers (with respect to the solvent velocity) are

$$t_+^0 = 1 - t_-^0 = \frac{z_+ \mathfrak{D}_{0+}}{z_+ \mathfrak{D}_{0+} - z_- \mathfrak{D}_{0-}}. \tag{79-6}$$

The driving force for diffusion used in equations 3 and 4 is the gradient of the chemical potential $\mu_e$ of the electrolyte in the solution. This chemical potential is readily measurable, and no reference to individual ionic activity coefficients is necessary. The diffusion coefficient $D$ of the salt which is usually measured is based on a gradient of the concentration and is related to $\mathfrak{D}$ by[11]

$$D = \mathfrak{D}\frac{c_T}{c_0}\left(1 + \frac{d \ln \gamma_{+-}}{d \ln m}\right), \tag{79-7}$$

where $\gamma_{+-}$ is the mean molal activity coefficient and $m$ is the molality (moles of electrolyte per kilogram of solvent). The gradient of chemical potential can be expressed in terms of the gradient of concentration:

$$\frac{\mathfrak{D}}{\nu RT}\frac{c_T}{c_0} c \nabla \mu_e = D\left(1 - \frac{d \ln c_0}{d \ln c}\right)\nabla c. \tag{79-8}$$

Insertion of equations 3 and 8 into the material-balance equation 69-3 yields

$$\frac{\partial c}{\partial t} + \nabla \cdot (c \mathbf{v}_0) = \nabla \cdot \left[ D\left(1 - \frac{d \ln c_0}{d \ln c}\right)\nabla c\right] - \frac{\mathbf{i} \cdot \nabla t_+^0}{z_+ \nu_+ F}, \tag{79-9}$$

which bears a strong resemblance to equation 72-13. The second term is different because we have not assumed that $\nabla \cdot \mathbf{v}_0 = 0$.

[11]John Newman, Douglas Bennion, and Charles W. Tobias, "Mass Transfer in Concentrated Binary Electrolytes," *Berichte der Bunsengesellschaft für physikalische Chemie*, 69 (1965), 608–612. [For corrections see *ibid.*, 70 (1966), 493.]

## 80. Reference velocities

Diffusion might be defined as a motion of the various components relative to the bulk fluid motion as a result of nonuniform thermodynamic potentials. In order to avoid ambiguity, a velocity characteristic of the bulk motion must be clearly specified, and the diffusion velocities must be referred to this velocity.

In section 79 and, in particular, in equations 79-3 and 79-4, the solvent velocity has been chosen as the reference velocity. Two other possible reference velocities are the mass-average velocity $\mathbf{v}$ and the molar-average velocity $\mathbf{v}^\star$ defined by

$$\mathbf{v} = \frac{1}{\rho} \sum_i \rho_i \mathbf{v}_i \quad \text{and} \quad \mathbf{v}^\star = \frac{1}{c_T} \sum_i c_i \mathbf{v}_i, \tag{80-1}$$

where $\rho_i$ is the mass of species $i$ per unit volume ($\rho_i = M_i c_i$). The choice of which reference velocity to use is arbitrary, and the distinction is less important in sufficiently dilute solutions since the three velocities become the same.

In particular situations, one reference velocity may be more advantageous than another. The solvent velocity becomes less significant in concentrated mixtures and becomes quite inconvenient in a pure fused salt. The mass-average velocity is useful because the fluid mechanical equations (see chapter 15) are invariably written in terms of $\mathbf{v}$. On the other hand, the average velocity is not always determined from momentum considerations, but perhaps from pure stoichiometry (for example, in some porous electrodes). In such a case, the molar-average velocity might be more convenient. Furthermore, chemists more commonly work in molar units than in mass units.

For a binary electrolytic solution, the material-balance equation 79-9 can be written in the equivalent forms

$$c_T\left(\frac{\partial x_e}{\partial t} + \mathbf{v}^\star \cdot \nabla x_e\right) = \nabla \cdot (c_T D \nabla x_e) - \frac{\mathbf{i} \cdot \nabla t_+^\star}{z_+ \nu_+ F} \tag{80-2}$$

and

$$\rho\left(\frac{\partial \omega_e}{\partial t} + \mathbf{v} \cdot \nabla \omega_e\right) = \nabla \cdot (\rho D \nabla \omega_e) - \frac{M_e \mathbf{i} \cdot \nabla t_+}{z_+ \nu_+ F}, \tag{80-3}$$

where $M_e = \nu_+ M_+ + \nu_- M_-$ is the molecular weight of the electrolyte, $x_e = c/c_T$ is the mole fraction of the salt (see the remarks below equation 14-4), $\omega_e = (\rho_+ + \rho_-)/\rho$ is the mass fraction of the salt, $t_+^\star = (c_- + c_0 t_+^0)/c_T$ is the cation transference number with respect to the molar-average velocity, and $t_+ = (\rho_- + \rho_0 t_+^0)/\rho$ is the cation transference number with respect to the mass-average velocity. Equation 2 involves the molar-average velocity, and equation 3 involves the mass-average velocity. These equations can be compared with the corresponding forms for binary solutions of nonelectrolytes (see reference 1, p. 557).

The cation flux referred to the molar-average velocity is

$$\mathbf{N}_+ = -\nu_+ c_T D \nabla x_e + \frac{\mathbf{i} t_+^\star}{z_+ F} + c_+ \mathbf{v}^\star, \tag{80-4}$$

and the cation flux referred to the mass-average velocity is

$$\mathbf{N}_+ = -v_+ \frac{\rho D}{M_e} \nabla \omega_e + \frac{\mathbf{i} t_+}{z_+ F} + c_+ \mathbf{v}. \tag{80-5}$$

Similar equations apply to the anion.

Equations 3 and 5 have been applied to mass transfer to a rotating-disk electrode from a binary electrolytic solution in reference 12.

## 81. The potential

Now we want to introduce a *potential in the solution* for use as a driving force for the current. Various candidates for this rôle were discussed in section 26. We restrict ourselves here to a binary electrolyte.

In order to assure that the potential introduced can be measured, let us first use the potential $\Phi$ of a suitable reference electrode at a point in the solution measured with respect to a similar reference electrode at a fixed point in the solution. By this we mean an actual electrode, not a reference half cell connected to the point in question by a capillary tube filled with an electrolytic solution. The electrode equilibrium must, of course, involve the anions or the cations and possibly the solvent. This electrode reaction can be written, in general, as

$$s_- M_-^{z_-} + s_+ M_+^{z_+} + s_0 M_0 \rightleftharpoons ne^-, \tag{81-1}$$

where $M_i$ is a symbol representing the chemical formula of species $i$ and $s_i$ is the stoichiometric coefficient of species $i$.

In a practical experimental situation, one may want to replace the reference electrode by a reference half cell. The additional diffusion potential thus introduced can be calculated exactly for a reference half cell such as Hg–HgO in a KOH solution if the external electrolyte is also KOH, but not if it is KCl. (See chapters 2 and 6.)

Application of thermodynamic principles to a reference electrode following equation 1 yields

$$s_- \nabla \mu_- + s_+ \nabla \mu_+ + s_0 \nabla \mu_0 = -nF \nabla \Phi. \tag{81-2}$$

This equation can be rearranged so as to replace the electrochemical potentials by the current density and the chemical potential of the electrolyte. Equations 79-3 and 79-4 can be substituted into equation 78-1 to yield

$$\frac{1}{z_-} \nabla \mu_- = -\frac{F}{\kappa} \mathbf{i} - \frac{t_+^0}{z_+ v_+} \nabla \mu_e, \tag{81-3}$$

where $\kappa$ is the conductivity of the solution and is given by

$$\frac{1}{\kappa} = -\frac{RT}{c_T z_+ z_- F^2} \left( \frac{1}{\mathfrak{D}_{+-}} + \frac{c_0 t_-^0}{c_+ \mathfrak{D}_{0-}} \right). \tag{81-4}$$

Equation 3 can be compared with equation 16-3.

12 J. Newman and L. Hsueh, "The Effect of Variable Transport Properties on Mass Transfer to a Rotating Disk," *Electrochimica Acta*, *12* (1967), 417–427.

From equation 2, $\nabla \mu_0$ can be eliminated by means of the Gibbs-Duhem equation, and the terms with the gradients of the electrochemical potentials of the ions can be combined to give

$$s_+ \nabla \mu_+ + s_- \nabla \mu_- = \frac{s_+}{v_+} \nabla \mu_e - \frac{n}{z_-} \nabla \mu_-, \qquad (81\text{-}5)$$

since

$$s_+ z_+ + s_- z_- = -n. \qquad (81\text{-}6)$$

Equation 2 becomes

$$-F \nabla \Phi = \left( \frac{s_+}{nv_+} - \frac{s_0 c}{nc_0} \right) \nabla \mu_e - \frac{1}{z_-} \nabla \mu_-. \qquad (81\text{-}7)$$

Finally, $\nabla \mu_-$ is eliminated by means of equation 3 to yield the desired relation

$$\mathbf{i} = -\kappa \nabla \Phi - \frac{\kappa}{F} \left( \frac{s_+}{nv_+} + \frac{t_+^0}{z_+ v_+} - \frac{s_0 c}{nc_0} \right) \nabla \mu_e. \qquad (81\text{-}8)$$

This result is analogous to equation 72-7, but the potential used here is considerably different from the electrostatic potential used earlier. The new definition avoids the questionable concepts regarding potentials in the solution. When the composition is uniform, the two potentials are similar; but the reference-electrode potential retains a clearly defined physical significance even in the presence of concentration gradients. Equation 8 can be compared directly with equation 17-16.

Another way to avoid the questionable concepts regarding potentials in the solution is to use the quasi-electrostatic potential. In a more general context, this leads eventually to equation 84-3 and will not be pursued further here.

## 82.  Connection with dilute-solution theory

The dilute-solution theory presented in chapter 11 has many useful facets which are only slightly modified by the more complete theory for concentrated solutions. Consequently, it is important to see how the two theories are related. Let us apply equation 78-1 to one of the minor species in a dilute solution. Then $c_i \ll c_0$, and only one of the terms on the right is important:

$$c_i \nabla \mu_i = \frac{RT c_0}{c_T \mathcal{D}_{0i}} (c_i \mathbf{v}_0 - c_i \mathbf{v}_i). \qquad (82\text{-}1)$$

Furthermore, the total concentration $c_T$ is approximately equal to the solvent concentration $c_0$, and equation 1 can be rewritten as

$$\mathbf{N}_i = -\frac{\mathcal{D}_{0i}}{RT} c_i \nabla \mu_i + c_i \mathbf{v}_0. \qquad (82\text{-}2)$$

Equation 2 is only slightly different from equation 69-1. The driving forces for diffusion and migration are both included in the gradient of the electrochemical potential in equation 2, and we see that the applicability of the Nernst-Einstein equation 75-1 is thus implicit in this equation. The further development of equation 2 was carried out in section 77.

The $\mathfrak{D}_{0i}$ correspond to the $D_i$ of the dilute-solution theory, but the interactions of the minor components with each other are not explicitly accounted for in the dilute-solution theory. A different number of transport properties is defined in the two cases.

We have seen that the validity of the Nernst-Einstein relation rests primarily on the fact that the driving force for both migration and diffusion is the gradient of the electrochemical potential, and the decomposition of this into a concentration term and an electrostatic-potential term is without basic physical significance. The Nernst-Einstein relation does not really fail in concentrated solutions; rather, additional composition-dependent transport parameters besides the $\mathfrak{D}_{0i}$ become necessary to describe the processses. It is not sufficient to allow the $D_i$ and $u_i$ to become concentration dependent, even though one might be willing to relax the Nernst-Einstein relation.

We gain additional insight into the validity of the Nernst-Einstein relation from the Debye-Hückel theory of interionic attraction (see section 27) and the theory of the diffuse layer at an interface (see section 52). These both describe equilibrium situations where the ionic fluxes and the convective velocity are zero. Under these conditions, and with the Nernst-Einstein relation, equation 69-1 becomes

$$\mathbf{N}_i = -\frac{z_i D_i F}{RT} c_i \, \nabla\Phi - D_i \, \nabla c_i = 0. \tag{82-3}$$

Integration gives the Boltzmann distribution for the ionic concentrations:

$$c_i = c_{i\infty} \exp\left(-\frac{z_i F \Phi}{RT}\right). \tag{82-4}$$

(See equations 27-1 and 52-1).

Table 1 shows, for binary electrolytes, a comparison of the results of the theories for dilute solutions and concentrated solutions. In order to bring out the similarity, the Nernst-Einstein relation 75-1 has been used in the expression of the transport properties from section 72. The three transport properties $\mathfrak{D}_{0+}$, $\mathfrak{D}_{0-}$, and $\mathfrak{D}_{+-}$ of the theory for concentrated solutions can be

TABLE 82-1. COMPARISON OF RESULTS FOR BINARY ELECTROLYTES

| Dilute-solution theory | Concentrated solutions |
| --- | --- |
| equation 69-1 | equation 78-1 |
| equation 72-13 | equation 79-9 |
| equation 72-7 | equation 81-8 |
| $D = \dfrac{D_+ D_-(z_+ - z_-)}{z_+ D_+ - z_- D_-}$ | $D = \dfrac{\mathfrak{D}_{0+}\mathfrak{D}_{0-}(z_+ - z_-)}{z_+\mathfrak{D}_{0+} - z_-\mathfrak{D}_{0-}}$ |
| $t_+ = \dfrac{z_+ D_+}{z_+ D_+ - z_- D_-}$ | $t_+^0 = \dfrac{z_+\mathfrak{D}_{0+}}{z_+\mathfrak{D}_{0+} - z_-\mathfrak{D}_{0-}}$ |
| $\dfrac{1}{\kappa} = \dfrac{-RT}{c_0 z_+ z_- F^2}\left(\dfrac{c_0 t_-}{c_+ D_-}\right)$ | $\dfrac{1}{\kappa} = \dfrac{-RT}{c_T z_+ z_- F^2}\left(\dfrac{1}{\mathfrak{D}_{+-}} + \dfrac{c_0 t_-^0}{c_+ \mathfrak{D}_{0-}}\right)$ |

calculated as functions of concentration from three independent measurements of $D$, $\kappa$, and $t^0_+$ (see section 90).

## 83. Multicomponent transport

Equation 78-1 expresses the driving forces in terms of the species velocities $\mathbf{v}_i$ or the species fluxes $c_i\mathbf{v}_i$. For use in the material-balance equation 69-3, it is necessary to invert the set of equations 78-1 so as to express the species fluxes in terms of the driving forces. This is carried out in the present section. (See references 13, 14, 15.)

It should first be noted that there are only $n - 1$ independent velocity differences and $n - 1$ independent gradients of electrochemical potentials in a solution with $n$ species (see equation 78-5). Therefore, equation 78-1 can be expressed as

$$c_i \, \nabla \mu_i = \sum_j M_{ij}(\mathbf{v}_j - \mathbf{v}_0) \tag{83-1}$$

where $\mathbf{v}_0$ is the velocity of any one of the species and where

$$
\begin{aligned}
M_{ij} &= K_{ij}, \qquad i \neq j \\
M_{ij} &= K_{ij} - \sum_k K_{ik}, \qquad i = j.
\end{aligned}
\tag{83-2}
$$

It further follows that $M_{ij} = M_{ji}$. Bearing in mind that there are $n - 1$ independent equations of the form of equation 1, one can invert this equation to read

$$\mathbf{v}_j - \mathbf{v}_0 = - \sum_{k \neq 0} L^0_{jk} c_k \, \nabla \mu_k, \qquad j \neq 0, \tag{83-3}$$

where the matrix $\mathbf{L}^0$ is the inverse of the submatrix $\mathbf{M}^0$,

$$\mathbf{L}^0 = -(\mathbf{M}^0)^{-1}, \tag{83-4}$$

and where the submatrix $\mathbf{M}^0$ is obtained from the matrix $\mathbf{M}$ by deleting the row and the column corresponding to the species 0. The inverse matrix $\mathbf{L}^0$ is also symmetric, that is,

$$L^0_{ij} = L^0_{ji}. \tag{83-5}$$

Certain combinations of the $L^0_{ij}$'s are related to measurable transport properties and have particular significance in the treatment of cells with liquid junctions (see section 84). The current density is related to the fluxes

[13] Richard J. Bearman, "The Onsager Thermodynamics of Galvanic Cells with Liquid-Liquid Junctions," *The Journal of Chemical Physics*, 22 (1954), 585–587.

[14] Richard J. Bearman and John G. Kirkwood, "Statistical Mechanics of Transport Processes. XI. Equations of Transport in Multicomponent Systems," *The Journal of Chemical Physics*, 28 (1958), 136–145.

[15] William H. Smyrl and John Newman, "Potentials of Cells with Liquid Junctions," *The Journal of Physical Chemistry*, 72 (1968), 4660–4671.

of ionic species by equation 69-2, which can be rewritten as

$$\mathbf{i} = F \sum_i z_i c_i \mathbf{v}_i = F \sum_i z_i c_i (\mathbf{v}_i - \mathbf{v}_0), \tag{83-6}$$

the equivalence of the last two expressions being assured by the electroneutrality of the solution. Substitution of equation 3 yields

$$\mathbf{i} = -F \sum_{i \neq 0} z_i c_i \sum_{k \neq 0} L_{ik}^0 c_k \, \nabla \mu_k. \tag{83-7}$$

In a solution of uniform composition,

$$\nabla \mu_k = z_k F \, \nabla \Phi, \tag{83-8}$$

where $\nabla \Phi$ is the gradient of the electric potential. Equation 7 becomes in this case

$$\mathbf{i} = -F^2 \, \nabla \Phi \sum_{i \neq 0} z_i c_i \sum_{k \neq 0} L_{ik}^0 z_k c_k. \tag{83-9}$$

Comparison with Ohm's law (see equation 70-2), also applicable to a solution of uniform composition,

$$\mathbf{i} = -\kappa \, \nabla \Phi, \tag{83-10}$$

allows us to identify the conductivity

$$\kappa = F^2 \sum_{i \neq 0} \sum_{k \neq 0} L_{ik}^0 z_i c_i z_k c_k. \tag{83-11}$$

Although the $L_{ik}^0$'s depend upon the reference velocity chosen, the conductivity $\kappa$ is invariant with respect to this choice.

Next we can identify the transference numbers. Again, for a solution of uniform composition, equation 8 is valid, and equation 3 becomes

$$\mathbf{v}_j - \mathbf{v}_0 = -F \, \nabla \Phi \sum_{k \neq 0} L_{jk}^0 z_k c_k. \tag{83-12}$$

For this case of uniform composition, the species flux is related to the current density and the transference number by the expression

$$t_j^0 \mathbf{i} = z_j F c_j (\mathbf{v}_j - \mathbf{v}_0) = -t_j^0 \kappa \, \nabla \Phi. \tag{83-13}$$

Comparison of equations 12 and 13 shows that the transference number $t_j^0$ of species $j$ with respect to the velocity of species 0 is given by

$$t_j^0 = \frac{z_j c_j F^2}{\kappa} \sum_{k \neq 0} L_{jk}^0 z_k c_k. \tag{83-14}$$

It is to be noted that the transference number has been defined as the fraction of the current carried by an ion in a solution of uniform composition. In a solution in which there are concentration gradients, the transference number is still a transport property related to the $L_{ij}^0$'s by equation 14, but it no longer represents the fraction of current carried by an ion (compare with the remarks at the end of section 70). A different choice of the reference species will change the $L_{ij}$'s, and hence the transference numbers with respect to the velocities of different reference species will be different (see problem 2).

Comparison of equation 14 with equation 11 or of equation 13 with equation 6 shows that the transference numbers sum to unity:

$$\sum_i t_i^0 = 1. \tag{83-15}$$

One could go on to describe diffusion of electrolytes in terms of the inverted transport equations. However, this becomes cumbersome, and the symmetry of the coefficients becomes obscured if one tries to eliminate the special place occupied by the species 0 in the inversion process. One of the primary purposes of the present section is to lead to the development of equation 84-2 in the next section. This equation was used as the basis to treat irreversible diffusion effects in electrochemical cells in chapters 2 and 6.

In general, the $\frac{1}{2}n(n-1)$ coefficients $\mathfrak{D}_{ij}$ yield one conductivity and $n-1$ transference numbers or ratios $t_i^0/z_i$ in the inverted formulation. The remainder of the coefficients generate diffusion coefficients for neutral combinations of species. For example, for a solution containing a solvent and $K^+$, $Na^+$, and $Cl^-$ ions, there is one conductivity, two independent transference numbers, and three diffusion coefficients required to describe diffusion of NaCl and KCl in the solvent. These six transport properties correspond to, and are derivable from, the six coefficients $\mathfrak{D}_{ij}$ for the system.

## 84. Liquid-junction potentials

It was shown in chapter 2 that many electrochemical cells involve junction regions where the composition is nonuniform and diffusion therefore occurs. The evaluation of the open-circuit potentials of these cells, and, in particular, the evaluation of the variation of the electrochemical potentials of ions in such junctions, requires consideration of these transport processes.

Equation 83-7 is applicable even in a nonuniform solution, and it can now be rewritten in terms of the conductivity and the transference numbers. Inversion of the order of summation in equation 83-7 gives

$$\mathbf{i} = -F\sum_{i\neq 0} c_i \nabla\mu_i \sum_{k\neq 0} L_{ki}^0 z_k c_k, \tag{84-1}$$

where we have also relabeled the subscripts. Since $L_{ik}^0 = L_{ki}^0$, substitution of equation 83-14 into equation 1 yields

$$\frac{F}{\kappa}\mathbf{i} = -\sum_i \frac{t_i^0}{z_i}\nabla\mu_i. \tag{84-2}$$

As already noted in section 83, a different choice of the reference species will change the transference numbers, but it is apparent from the derivation that equation 2 still applies. However, equation 83-14 shows that the ratio $t_j^0/z_j$ is not zero even for a neutral species. While the reference velocity can be chosen arbitrarily to be that of any one of the species, charged or uncharged, it is usually taken to be the velocity of the solvent. In this case there is no problem if there are no other neutral components, since the ratio $t_i^0/z_i$ is always zero for the reference species.

It is shown in problem 7 that equation 2 also has the same form if other reference velocities, such as the mass-average velocity or the molar-average velocity, are used. Again, care should be exercised since the ratio $t_i/z_i$ is then not zero for neutral species.

Equation 2 is quite useful in the calculation of the potential of cells with liquid junctions. It was presented and discussed in section 16, and it was applied to the problem of liquid junctions in chapters 2 and 6. In the cases of interest, the current density is supposed to be zero, but equation 2 also allows one to estimate the effect of the passage of small amounts of current. Equation 2 is generally useful only if the concentration profiles in the liquid junction are known. These are determined not from equation 2 but from the laws of diffusion (equation 78-1 or equation 83-3) and the method of forming the junction.

Substitution of equation 26-4 into equation 2 gives

$$F \, \nabla \Phi = -\frac{F}{\kappa} \mathbf{i} - RT \sum_i \frac{t_i^0}{z_i} \nabla \ln c_i$$
$$- RT \sum_i \frac{t_i^0}{z_i} \nabla \left( \ln f_i - \frac{z_i}{z_n} \ln f_n \right), \qquad (84\text{-}3)$$

where $\Phi$ is the quasi-electrostatic potential referred to species $n$. This equation, which was used in chapter 6, can be compared with equation 75-4 or equation 70-7 and provides an additional connection with the dilute-solution theory. It also suggests the validity of the Nernst-Einstein relation (see section 82), since this relation was necessary in the derivation of equation 75-4.

## PROBLEMS

**1.** Derive equations 79-3 and 79-4 from equations 79-1, 79-2, 69-2, and 78-3.

**2.** Let the transference number $t_i$ of a species with respect to the velocity $\mathbf{v}$ be defined by the equation

$$t_i \mathbf{i} = z_i F c_i (\mathbf{v}_i - \mathbf{v})$$

for a solution of uniform composition. This equation says that the flux of species $\mathbf{i}$ relative to the velocity $\mathbf{v}$ accounts for the fraction $t_i$ of the current density.

(a) Let $t_i'$ be the transference number of species $i$ relative to the velocity $\mathbf{v}'$. Show that the transference numbers of two species $i$ and $j$ relative to the velocities $\mathbf{v}$ and $\mathbf{v}'$ are related by

$$\frac{t_i' - t_i}{z_i c_i} = \frac{t_j' - t_j}{z_j c_j}.$$

(b) For a binary electrolyte, show that

$$\frac{t_0^+}{z_0} = -\frac{c_0 t_+^0}{z_+ c_+},$$

thus demonstrating that the ratio $t_i/z_i$ is not always zero for a neutral species.

Here $t_0^+$ is the transference number of the solvent relative to the cation velocity.

(c) Show for a binary electrolytic solution that

$$\mathbf{v} - \mathbf{v}_0 = \frac{1}{\rho}\left[-\frac{M_e \mathfrak{D}}{\nu RT}\frac{c_T}{c_0}c\nabla\mu_e + \frac{\mathbf{i}}{F}\left(\frac{M_+ t_+^0}{z_+} + \frac{M_- t_-^0}{z_-}\right)\right],$$

where $\mathbf{v}$ is the mass-average velocity, and derive the relation between $t_+$ and $t_+^0$, given below equation 80-3.

(d) In a similar manner, derive the relation between $t_+^*$ and $t_+^0$ given below equation 80-3.

**3.** Derive equation 17-16 from equation 81-8.

**4.** Derive equation 81-8 from equation 84-3, bearing in mind that $\Phi$ represents different quantities in the two equations.

**5.** For a binary electrolytic solution,

(a) State the form of the matrices $\mathbf{M}$ and $\mathbf{M}^0$.

(b) Invert $\mathbf{M}^0$ to obtain $\mathbf{L}^0$.

(c) By substitution of the result from part (b) into equations 83-11 and 83-14, verify equation 79-6 for the transference number and equation 81-4 for the conductivity.

(d) By substitution of the result from part (b) into equation 83-3 and elimination of the electrochemical potential of individual ions by means of equation 16-3, derive the expressions 79-3 and 79-4 for the fluxes of the ions.

**6.** Apply the development of section 83 to a four-component system, $0, +, -$, and 3. Take species 3 to be charged. In subsequent applications, one can set $z_3 = 0$ in order to treat mixed solvents or membranes. One can set $c_0$ equal to zero to treat fused salts.

**7.** Use the result of problem 2a to show that $t_i^0$ in equation 84-2 can be replaced by the transference numbers relative to any reference velocity; that is, show that

$$\sum_i \frac{t_i}{z_i}\nabla\mu_i = \sum_i \frac{t_i'}{z_i}\nabla\mu_i.$$

**8.** Calculate the magnitude of a diffusion velocity $D_i \nabla \ln c_i$ and a migration velocity $z_i u_i F \nabla \Phi$ and compare with the magnitude of a typical convective velocity.

**9.** Derive equation 80-3 from equation 79-9 using also the continuity equation

$$\frac{\partial \rho}{\partial t} + \nabla\cdot(\rho\mathbf{v}) = 0$$

(see equation 93-2) and the expression for $\mathbf{v} - \mathbf{v}_0$ in problem 2c.

**10.** Show that $1 - d\ln c_0/d\ln c$, appearing in equation 79-8, can also be written as

$$\frac{\rho}{c_0 M_0}\left(1 - \frac{d\ln\rho}{d\ln c}\right) = \frac{1}{c_0 \bar{V}_0}.$$

Appendix A may be helpful here. (See also problem 2-1.)

## NOTATION

| | |
|---|---|
| $a^\theta_{+-}$ | property expressing secondary reference state, liter/mole |
| $c$ | molar concentration of a single electrolyte, mole/cm$^3$ |
| $c_i$ | concentration of species $i$, mole/cm$^3$ |
| $c_T$ | total solution concentration, mole/cm$^3$ |
| $D$ | measured diffusion coefficient of electrolyte, cm$^2$/sec |
| $D_i$ | diffusion coefficient of species $i$, cm$^2$/sec |
| $\mathfrak{D}$ | diffusion coefficient of electrolyte, based on a thermodynamic driving force, cm$^2$/sec |
| $\mathfrak{D}_{ij}$ | diffusion coefficient for interaction of species $i$ and $j$, cm$^2$/sec |
| $f_{+-}$ | mean molar activity coefficient of an electrolyte |
| $F$ | Faraday's constant, 96,487 C/equiv |
| $\mathbf{i}$ | current density, A/cm$^2$ |
| $K_{ij}$ | friction coefficient for interaction of species $i$ and $j$, J-sec/cm$^5$ |
| $L^0_{ij}$ | inverted transport coefficient, cm$^5$/J-sec |
| $m$ | molality of a single electrolyte, mole/kg |
| $M_i$ | symbol for the chemical formula of species $i$ |
| $M_i$ | molecular weight of species $i$, g/mole |
| $M_e$ | molecular weight of electrolyte, g/mole |
| $M_{ij}$ | modified friction coefficient, J-sec/cm$^5$ |
| $n$ | number of electrons involved in electrode reaction |
| $n$ | number of species present in solution |
| $\mathbf{N}_i$ | flux of species $i$, mole/cm$^2$-sec |
| $R$ | universal gas constant, 8.3143 J/mole-deg |
| $s_i$ | stoichiometric coefficient of species $i$ in electrode reaction |
| $t$ | time, sec |
| $t_i$ | transference number of species $i$ with respect to the mass-average velocity |
| $t^0_i$ | transference number of species $i$ with respect to the velocity of species 0 |
| $t^\star_i$ | transference number of species $i$ with respect to the molar-average velocity |
| $T$ | absolute temperature, deg K |
| $u_i$ | mobility of species $i$, cm$^2$-mole/J-sec |
| $\mathbf{v}$ | mass-average velocity, cm/sec |
| $\mathbf{v}_i$ | velocity of species $i$, cm/sec |
| $\mathbf{v}^\star$ | molar-average velocity, cm/sec |
| $x_e$ | mole fraction of electrolyte $= c/c_T$ |
| $z_i$ | charge number of species $i$ |
| $\gamma_{+-}$ | mean molal activity coefficient of an electrolyte |
| $\kappa$ | conductivity, mho/cm |
| $\mu_e$ | chemical potential of an electrolyte, J/mole |

$\mu_i$      electrochemical potential of species $i$, J/mole

$\nu$      number of moles of ions into which a mole of electrolyte dissociates

$\nu_+, \nu_-$      numbers of cations and anions into which a molecule of electrolyte dissociates

$\rho$      density, g/cm³

$\rho_i$      mass of species $i$ per unit volume, g/cm³

$\Phi$      electric potential, V

$\omega_e$      mass fraction of electrolyte

subscript

0      species 0, generally the solvent

# Thermal Effects 13

## 85. Thermal diffusion

Equation 78-1 applies at constant temperature and pressure. The generalization would be[1]

$$c_i \left( \nabla \mu_i + \bar{S}_i \nabla T - \frac{M_i}{\rho} \nabla p \right)$$

$$= RT \sum_j \frac{c_i c_j}{c_T \mathfrak{D}_{ij}} \left[ \mathbf{v}_j - \mathbf{v}_i + \left( \frac{D_j^T}{\rho_j} - \frac{D_i^T}{\rho_i} \right) \nabla \ln T \right], \qquad (85\text{-}1)$$

where $\bar{S}_i$ is the partial molar entropy of species $i$ and $D_i^T$ is the thermal diffusion coefficient of species $i$. The first modification is that the driving force for diffusion and migration is now written according to the left side of this equation. These driving forces now sum to zero even when the temperature and pressure vary. Furthermore, they describe properly equilibrium in a gravitational or centrifugal field (see problem 1).

The second modification is the inclusion of thermal diffusion, represented by the two terms in the bracket in equation 1. The gradient of the temperature is a new driving force in the system, in addition to the $n - 1$ driving forces $c_i[\nabla \mu_i + \bar{S}_i \nabla T - (M_i/\rho) \nabla p]$. The gradient of pressure is not really an independent driving force for heat and mass transfer; rather, it is a driving force for fluid flow, as discussed in chapter 15. The temperature gradient can also contribute to mass transport, as shown in equation 1.

[1] Joseph O. Hirschfelder, Charles F. Curtiss, and R. Byron Bird, *Molecular Theory of Gases and Liquids* (New York: John Wiley & Sons, Inc., 1954), p. 718.

This process is called *thermal diffusion*, whereby a temperature gradient maintained across a solution can lead to a variation in composition. Thermal diffusion is not, however, usually important in industrial systems. The converse process, called the Dufour effect, is mentioned in the next section. The thermal diffusion coefficients $D_i^T$ are additional transport properties, of which only $n - 1$ are independent since they always appear as differences as in equation 1. It would make more sense to call $D_i^T/\rho_i$ the thermal diffusion coefficient, since this quantity has the units of $cm^2/sec$ and $D_A^T/\rho_A - D_B^T/\rho_B$ is approximately constant for a binary solution.

For a binary electrolyte, we can combine equations 1 to yield

$$\mathbf{N}_+ = -\frac{v_+\mathfrak{D}}{vRT}\frac{c_T}{c_0}c\left(\nabla\mu_e + \bar{S}_e\,\nabla T - \frac{M_e}{\rho}\nabla p\right) + \frac{it_+^0}{z_+F} + c_+\mathbf{v}_0$$

$$+ c_+\left(\frac{D_0^T}{\rho_0} - t_-^0\frac{D_+^T}{\rho_+} - t_+^0\frac{D_-^T}{\rho_-}\right)\nabla\ln T. \qquad (85\text{-}2)$$

This is a generalization of equation 79-3, $\mathfrak{D}$ and $t_+^0$ being defined by equations 79-5 and 79-6. Equation 80-5 now becomes

$$\mathbf{N}_+ = -v_+\frac{\rho D}{M_e}\nabla\omega_e + \frac{it_+}{z_+F} + c_+\mathbf{v}$$

$$- \frac{v_+\mathfrak{D}c_Tc}{v}\left[\frac{M_0}{RT\rho}\left(\bar{V}_e - \frac{M_e}{\rho}\right)\nabla p - \frac{\omega_0\sigma}{c_0+c}\nabla T\right], \qquad (85\text{-}3)$$

where $D$ is given by equation 79-7, $t_+$ is given below equation 80-3, and $\sigma$, called the Soret coefficient, is

$$\sigma = \frac{v(c_0+c)}{c_T\mathfrak{D}T}\left[\frac{D_0^T}{\rho_0} - t_-^0\frac{D_+^T}{\rho_+} - t_+^0\frac{D_-^T}{\rho_-}\right]. \qquad (85\text{-}4)$$

Consequently, equation 80-3 is replaced by

$$\rho\left(\frac{\partial\omega_e}{\partial t} + \mathbf{v}\cdot\nabla\omega_e\right) = \nabla\cdot(\rho D\,\nabla\omega_e) - \frac{M_e\mathbf{i}\cdot\nabla t_+}{z_+v_+F}$$

$$+ \frac{M_0}{v}\nabla\cdot\mathfrak{D}c_T\omega_e\left[\left(\bar{V}_e - \frac{M_e}{\rho}\right)\frac{\nabla p}{RT} - \frac{c_0\sigma}{c_0+c}\nabla T\right]. \qquad (85\text{-}5)$$

Consideration of entropy production and the second law of thermodynamics show that $\sigma$ is governed by the inequality

$$\sigma^2 \le \frac{(c_0+c)^2}{c_0c}\frac{vk'}{c_T\mathfrak{D}RT^2}, \qquad (85\text{-}6)$$

where $k'$ is the thermal conductivity (see section 86). Thus, $\sigma$ can be either positive or negative, depending on whether the solvent or the electrolyte migrates toward the hot wall under the influence of thermal diffusion.

By following the development in sections 83 and 84, we can invert equation 1 and then derive the result

$$\mathbf{i} = -\frac{\kappa}{F}\sum_i\frac{t_i^0}{z_i}\left(\nabla\mu_i + \bar{S}_i\,\nabla T - \frac{M_i}{\rho}\nabla p\right)$$

$$- F\sum_i z_ic_i\left(\frac{D_i^T}{\rho_i} - \frac{D_0^T}{\rho_0}\right)\nabla\ln T, \qquad (85\text{-}7)$$

where the transference numbers $t_i^0$ and the conductivity $\kappa$ are given by equations 83-14 and 83-11, respectively. This result shows how the temperature gradient can affect the flow of electric current.

For a binary electrolyte, this equation becomes

$$\mathbf{i} = -\frac{\kappa t_+^0}{z_+ v_+ F}\left(\nabla\mu_e + \bar{S}_e\,\nabla T - \frac{M_e}{\rho}\,\nabla p\right) - z_+ v_+ Fc\left(\frac{D_+^T}{\rho_+} - \frac{D_-^T}{\rho_-}\right)\nabla\ln T$$
$$- \frac{\kappa}{z_- F}\left(\nabla\mu_- + \bar{S}_-\,\nabla T - \frac{M_-}{\rho}\,\nabla p\right). \qquad (85\text{-}8)$$

In this equation, the gradient of the chemical potential of the electrolyte could also be written as

$$\nabla\mu_e + \bar{S}_e\,\nabla T - \frac{M_e}{\rho}\,\nabla p = \frac{vRT}{\omega_e\omega_0}\left(1 + \frac{d\ln\gamma_{+-}}{d\ln m}\right)_{T,\,p}\nabla\omega_e + \left(\bar{V}_e - \frac{M_e}{\rho}\right)\nabla p.$$
$$(85\text{-}9)$$

Note that a different combination of thermal diffusion coefficients occurs in equation 8 from that which occurs in equations 2, 3, and 5.

It is even more difficult to define an electric potential in a solution of varying temperature than in one of varying composition. Even with reference electrodes, thermal electric effects between the electrode leads and the potential-measuring device must be taken into account. Equation 8 is the analogue of equation 72-7 or 81-8 representing conduction effects, or the generalization of Ohm's law. For the present, we shall let $\nabla\mu_- + \bar{S}_-\,\nabla T - (M_-/\rho)\,\nabla p$ represent the effect of the gradient of the electrical state of the solution.

Tyrrell[2] has summarized some measurements of Soret coefficients in electrolytic solutions. The value of $\sigma$ is about 2 to $5 \times 10^{-3}$ deg$^{-1}$.

## 86. Heat generation, conservation, and transfer

Electrolytic solutions are described by the same basic equations as nonelectrolytic solutions, caution being used to regard the chemical potential $\mu_i$ as the electrochemical potential if it applies to an ionic species. The first law of thermodynamics is used to deduce a differential energy balance, which includes the kinetic energy of the flowing fluid. The momentum equation 93-4 is used to subtract this mechanical energy, yielding a thermal energy balance. By means of the appropriate thermodynamic relationships for mixtures, this can be put into the form

$$\rho\hat{C}_p\left(\frac{\partial T}{\partial t} + \mathbf{v}\cdot\nabla T\right) + \left(\frac{\partial\ln\rho}{\partial\ln T}\right)_{p,\,\omega_i}\left(\frac{\partial p}{\partial t} + \mathbf{v}\cdot\nabla p\right)$$
$$= -\nabla\cdot\mathbf{q} - \boldsymbol{\tau}:\nabla\mathbf{v} + \sum_i \bar{H}_i\,(\nabla\cdot\mathbf{J}_i - R_i). \qquad (86\text{-}1)$$

[2]H. J. V. Tyrrell, "Thermal-Diffusion Phenomena in Electrolytes and the Constants Involved," *Electrochemical Constants*, pp. 119–129. National Bureau of Standards (U.S.) Circular 524 (1953).

The last term on the right represents thermal effects due to diffusion, migration, and chemical reaction. Here, $\mathbf{J}_i$ is the flux of species $i$ relative to the mass-average velocity:

$$\mathbf{J}_i = \mathbf{N}_i - c_i \mathbf{v}. \tag{86-2}$$

The second term on the right and the last term on the left represent irreversible and reversible conversions of mechanical energy into thermal energy, $\boldsymbol{\tau}$ being the stress (see section 94). The term $-\boldsymbol{\tau}:\nabla\mathbf{v}$ is referred to as the *viscous dissipation*.

The heat flux $\mathbf{q}$ can be expressed as

$$\mathbf{q} = \sum_i \bar{H}_i \mathbf{J}_i - k \, \nabla T + \mathbf{q}^{(x)}, \tag{86-3}$$

the first term representing heat carried by the interdiffusion of the species, the second term being heat transfer by conduction with the thermal conductivity $k$, and the last term being the Dufour energy flux given by

$$\mathbf{q}^{(x)} = -\sum_i \frac{D_i^T}{\rho_i} c_i \left( \nabla \mu_i + \bar{S}_i \, \nabla T - \frac{M_i}{\rho} \nabla p \right). \tag{86-4}$$

The Dufour effect is the converse of thermal diffusion, treated in section 85, and accounts for the balance of the heat induced by interdiffusion. The thermal diffusion coefficients $D_i^T$ are the same as those introduced in the preceding section. Again, the Dufour effect is not usually important in industrial systems.

Substitution of equation 3 into equation 1 yields

$$\rho \hat{C}_p \left( \frac{\partial T}{\partial t} + \mathbf{v}\cdot\nabla T \right) + \left( \frac{\partial \ln \rho}{\partial \ln T} \right)_{p,\omega_i} \left( \frac{\partial p}{\partial t} + \mathbf{v}\cdot\nabla p \right)$$
$$= \nabla\cdot(k \, \nabla T) - \nabla\cdot\mathbf{q}^{(x)} - \sum_i \mathbf{J}_i\cdot\nabla\bar{H}_i - \boldsymbol{\tau}:\nabla\mathbf{v} - \sum_i \bar{H}_i R_i. \tag{86-5}$$

The thermal effect due to diffusion and migration now appears in a modified form.

For water at 20°C, $\partial \ln \rho/\partial T = -0.207 \times 10^{-3} \, \text{deg}^{-1}$. Since $\hat{C}_p = 4.1819$ J/g-deg, a change in pressure of one atmosphere in the last term on the left in equation 5 corresponds to a temperature change of only 0.00146° C in the first term. Consequently, the pressure term in equation 5 is usually negligible for condensed phases. More generally, the pressure changes in a system would be calculated by solving the fluid mechanics (chapter 15).

For a binary electrolyte, the heat flux can be expressed as

$$-k \, \nabla T + \mathbf{q}^{(x)} = -k' \, \nabla T + \frac{c_T \mathfrak{D} T \sigma}{v(c_0 + c)} c \left( \nabla \mu_e + \bar{S}_e \, \nabla T - \frac{M_e}{\rho} \nabla p \right)$$
$$+ \frac{z_+ c_+ F}{\kappa} \left( \frac{D_+^T}{\rho_+} - \frac{D_-^T}{\rho_-} \right) \mathbf{i}. \tag{86-6}$$

Here, the current density has been introduced from equation 85-8 in order to avoid using the potential in the solution. This also means that a somewhat different thermal conductivity is measured if no current is ever allowed to pass through the solution:

$$k' = k - \frac{(z_+ v_+ Fc)^2}{\kappa T}\left(\frac{D_+^T}{\rho_+} - \frac{D_-^T}{\rho_-}\right)^2.$$  (86-7)

An important difference between electrical and nonelectrical systems is the conversion of electrical energy to thermal energy, called *Joule heating*, due to the passage of electric current. This arises from the first term in the heat flux equation 3. For example, in a solution of uniform temperature, pressure, and composition, we have

$$\mathbf{J}_i = \frac{t_i \mathbf{i}}{z_i F} \quad \text{and} \quad \nabla \bar{H}_i = \nabla \mu_i,$$  (86-8)

and, consequently,

$$-\nabla \cdot \sum_i \bar{H}_i \mathbf{J}_i = -\sum_i \mathbf{J}_i \cdot \nabla \bar{H}_i = -\mathbf{i} \cdot \nabla \Phi = \frac{\mathbf{i} \cdot \mathbf{i}}{\kappa}.$$  (86-9)

Thus we conclude that the first term in equation 3 is by no means negligible and that the third term on the right in equation 5 can be associated in part with Joule heating.

For a constant value of $k'$ or for constant values of $\rho$ and $\hat{C}_p$, it is appropriate to define a thermal diffusivity as $k'/\rho\hat{C}_p$. Equation 5 then resembles the equation of convective diffusion 72-5 but with source terms for generation of heat by Joule heating, viscous dissipation, and chemical reaction. For water at 20°C, the value of the thermal diffusivity is $1.43 \times 10^{-3}$ cm²/sec, about one hundred times larger than diffusion coefficients encountered in aqueous solutions.

## 87.  Heat generation at an interface

Let us make an energy balance on an interface where a single electrode reaction

$$\sum_i s_i M_i^{z_i} \longrightarrow ne^-$$  (87-1)

is occurring at a steady state (thereby excluding the storage of energy in a charged electric double layer, see chapter 7). The total energy flux must then be continuous:

$$\Delta[(\tfrac{1}{2}\rho v^2 + \rho \hat{U} + p)v_y + q_y] = 0,$$  (87-2)

where $y$ is the distance from the electrode into the solution and $v_y$ is measured relative to the interface.

With neglect of the kinetic energy, we have

$$q_y' - q_{1y}' = -\frac{i_y}{F}\bar{H}_{e^-} + \sum_i \bar{H}_i \frac{s_i i_y}{nF},$$  (87-3)

where

$$\mathbf{q}' = \mathbf{q} - \sum_i \bar{H}_i \mathbf{J}_i$$  (87-4)

and I denotes the electrode phase. Let us neglect the Dufour energy flux and use the relationship

$$\bar{H}_i = \mu_i - T\bar{S}_i. \tag{87-5}$$

Equation 3 becomes

$$-k\frac{\partial T}{\partial y} + k_1\frac{\partial T_1}{\partial y} = -\frac{i_y}{F}\left(\mu_{e^-} - \sum_i \frac{s_i\mu_i}{n}\right) - \frac{i_y T}{nF}\left(n\bar{S}_{e^-} - \sum_i s_i\bar{S}_i\right)$$

$$= i_y\eta_s - \frac{i_y T}{nF}\left(n\bar{S}_{e^-} - \sum_i s_i\bar{S}_i\right), \tag{87-6}$$

where we have used the definition 55-2 of the surface overpotential.

The terms on the right represent the generation of heat at an interface. The first term is clearly the irreversible generation of heat; the second term is the reversible heat generation for a single electrode reaction. For the copper reaction, equation 6 is

$$-k\frac{\partial T}{\partial y} + k_1\frac{\partial T_1}{\partial y} = i_y\eta_s - \frac{i_y T}{F}\left(\bar{S}_{e^-} + \tfrac{1}{2}\bar{S}_{Cu^{++}} - \tfrac{1}{2}\bar{S}_{Cu}\right). \tag{87-7}$$

One could speculate as to how well defined or measurable the partial molar entropy of a charged species is. This quantity represents the reversible heat transferred when a mole of the species is added to a large volume of the phase in question. On this basis one might conclude that the partial molar entropy is independent of the electrical state of the phase since the concept of the partial molar entropy involves heat release rather than reversible work. On the other hand, we have the relationship

$$\bar{S}_i = -\left(\frac{\partial\mu_i}{\partial T}\right)_{p,\omega_i}. \tag{87-8}$$

Should this differentiation be carried out at constant electrical state also, and if so how should the electrical state be defined? Equations 6 and 7 involve the partial molar entropies of charged species in different phases. At equilibrium the electrical states of the two phases are related by

$$\mu_{e^-} = \sum_i \frac{s_i\mu_i}{n}, \tag{87-9}$$

but the electrical states are shifted relative to equilibrium when the electrode is polarized.

For a whole electrochemical cell, the reversible heat effects at the two electrodes add to yield a quantity that is frequently measured. For example, for the cell 18-12, involving a silver-silver chloride electrode and a hydrogen electrode, the reversible heat is

$$Q_{rev} = \frac{IT}{F}\left(\bar{S}^\beta_{HCl} + \bar{S}^\lambda_{Ag} - \frac{1}{2}\bar{S}^\alpha_{H_2} - \bar{S}^\epsilon_{AgCl}\right), \tag{87-10}$$

the quantity in parentheses being the entropy change for the overall cell reaction:

$$\tfrac{1}{2}H_2 + AgCl \rightleftharpoons Ag + HCl. \tag{87-11}$$

Here we have ignored the difference between phases $\beta$ and $\delta$. The entropy change for the overall cell recation can be obtained from the temperature coefficient of the reversible cell potential (see equation 18-13):

$$\bar{S}^{\beta}_{HCl} + \bar{S}^{\lambda}_{Ag} - \frac{1}{2}\bar{S}^{\alpha}_{H_2} - \bar{S}^{\epsilon}_{AgCl} = F\left(\frac{\partial U}{\partial T}\right)_p, \qquad (87\text{-}12)$$

so that equation 10 becomes

$$Q_{rev} = IT\left(\frac{\partial U}{\partial T}\right)_p. \qquad (87\text{-}13)$$

It should be noted that the reversible heat for an electrochemical cell is not related to the enthalpy change for the reaction, but to the entropy change. The entropy change is appropriate for pressure-volume systems but not to electrical systems.

## 88. Thermogalvanic cells

Figure 88-1 shows a simple thermogalvanic cell where an aqueous solution of copper sulfate is confined between two horizontal copper electrodes,

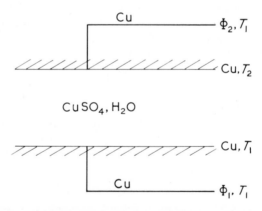

**Figure 88-1.** Thermogalvanic cell.

the upper of which is maintained at a higher temperature than the lower. What will be the potential difference measured between the two copper wires, which are brought to the same temperature at the potentiometer? In this experiment we consider the steady state in the absence of current, so that all the species fluxes are zero. In addition, we shall neglect the pressure variation induced by the gravitational field.

In this situation, equation 85-2 or 85-3 reduces to

$$\nabla \mu_e + \bar{S}_e \nabla T = \frac{c_0 RT\sigma}{c_0 + c} \nabla T \qquad (88\text{-}1)$$

or
$$\nabla \omega_e = \frac{\mathfrak{D} c_T}{Dv} \frac{\omega_e \omega_0 \sigma}{c_0 + c} \nabla T \qquad (88\text{-}2)$$

or
$$\left(1 + \frac{d \ln \gamma_{+-}}{d \ln m}\right)_{T,p} \nabla \omega_e = \frac{\omega_e \omega_0 c_0 \sigma}{v(c_0 + c)} \nabla T. \qquad (88\text{-}3)$$

These equations thus describe how the variation in the solution composition is related to the Soret coefficient $\sigma$ or how the Soret coefficient can be determined by measuring this variation in composition.

We might note in passing that equation 86-3 becomes

$$\mathbf{q} = -k'' \nabla T, \qquad (88\text{-}4)$$

where
$$k'' = k' - \frac{c_T \mathfrak{D} R c_0 c}{v(c_0 + c)^2} T^2 \sigma^2. \qquad (88\text{-}5)$$

Thus, a measurement of the heat flux yields the thermal conductivity $k''$ rather than $k'$ if no special attention is given to the fact that the composition varies within the system. A separate measurement of the Soret coefficient is necessary in order to yield $k'$. Thus we see that caution needs to be used in reporting or using thermal conductivities in the literature. The inequality 85-6 shows that

$$k'' \geq 0. \qquad (88\text{-}6)$$

Equation 85-8 becomes

$$\frac{\nabla \mu_+ + \bar{S}_+ \nabla T}{z_+ F} = \frac{t^0_-}{z_+ v_+ F} \frac{c_0 R T \sigma}{c_0 + c} \nabla T - \frac{z_- v_- Fc}{\kappa T} \left(\frac{D^T_-}{\rho_-} - \frac{D^T_+}{\rho_+}\right) \nabla T \qquad (88\text{-}7)$$

or
$$\frac{\nabla \mu_+}{z_+ F} = -\left(\frac{t^0_-}{z_- v_- F} \frac{c_0 R T \sigma}{c_0 + c} + \xi - \frac{\bar{S}_e}{z_+ v_+ F}\right) \nabla T, \qquad (88\text{-}8)$$

where
$$\xi = \left(\frac{D^T_-}{\rho_-} - \frac{D^T_+}{\rho_+}\right) \frac{z_- c_- F}{\kappa T} - \frac{\bar{S}_-}{z_- F} \qquad (88\text{-}9)$$

and might be called a thermoelectric coefficient (see problem 2). Equation 8 is used to assess the variation of the electrical state within the solution.

The procedure now is this. The copper ions are equilibrated across the two electrode interfaces. The variation in the electrical state within the solution is given by equation 8, and a similar equation describes the variation in the electrical state within the copper wire attached to the electrode at $T_2$. When we put these facts together, we obtain for the measured potential

$$\Phi_2 - \Phi_1 = \frac{\mu_{Cu}(T_1) - \mu_{Cu}(T_2)}{2F}$$
$$+ \int_{T_1}^{T_2} \left[\xi_{Cu} - \xi + \frac{\bar{S}_e}{z_+ v_+ F} + \frac{t^0_-}{z_+ v_+ F} \frac{c_0 R T \sigma}{c_0 + c}\right] dT, \qquad (88\text{-}10)$$

where $\xi_{Cu}$ is the thermoelectric coefficient for the copper wire. Strictly speaking, the variation of solution properties with composition as well as temperature should be considered in evaluating the integral. Since

$$\bar{S}_{Cu} = -\left(\frac{\partial \mu_{Cu}}{\partial T}\right)_p, \qquad (88\text{-}11)$$

equation 10 can be written as

$$\Phi_2 - \Phi_1 = \int_{T_1}^{T_2} \left[ \xi_{Cu} - \xi - \frac{\bar{S}_{Cu}}{2F} + \frac{\bar{S}_e}{z_+ v_+ F} + \frac{t_-^0}{z_+ v_+ F} \frac{c_0 R T \sigma}{c_0 + c} \right] dT.$$

(88-12)

Since the Soret coefficient $\sigma$ can be determined by an independent measurement, the measurement of the potential of such thermogalvanic cells allows $\xi - \xi_{Cu}$ to be determined. We can observe that such measurements do not aid in any sensible way in the establishment of electrode potentials at one temperature relative to those at a different temperature.

## PROBLEMS

**1.** (a) Show that, for gravitational equilibrium in a region of uniform temperature, the variation of the chemical potential of a species is given by

$$\nabla \mu_i = \frac{M_i}{\rho} \nabla p.$$

Do this by consideration of a reversible process of removing a mole of the species at one point in the field, moving it against the force of gravity, and reintroducing it at another point. If possible, avoid assuming that the gravitational field is uniform.

Consideration of the process gives some justification for the expression for the driving force for diffusion, $c_i[\nabla \mu_i + \bar{S}_i \nabla T - (M_i/\rho) \nabla p]$, since this driving force should reduce to zero in such an equilibrium situation.

(b) Consider whether this relation correctly describes the equilibrium distributions of concentration in a centrifuge.

(c) The molecular weight of NaCl is 58.44 and that of $H_2O$ is 18.015. If sea water is 0.5 $M$ in NaCl and 55 $M$ in $H_2O$ at the surface, would the equilibrium concentrations of both $H_2O$ and NaCl be higher at a depth of one mile?

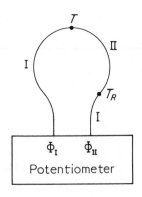

**Figure P13-1.** Thermocouple.

**2.** (a) A thermocouple consists of metal I and metal II, one junction being at temperature $T$ and the other junction being at temperature $T_R$, at which temperature the measurement is made. With no current being passed, show that the potential difference measured (see sketch) is

$$\Phi_{\text{II}} - \Phi_{\text{I}} = \int_{T_R}^{T} (\xi_{\text{II}} - \xi_{\text{I}})\, dT,$$

where, for each metal,

$$\xi = \left(\frac{D_-^T}{\rho_-} - \frac{D_+^T}{\rho_+}\right)\frac{z_- c_- F}{\kappa T} - \frac{\bar{S}_-}{z_- F},$$

the metals being regarded as composed of electrons and metal ions. The electrons are equilibrated between the two metals at the temperature $T$. This problem gives some insight into what quantity $\xi$ can be determined from the properties of a thermocouple.

(b) For given sizes of the wires, what current would be passed through the thermocouple when the ends are shorted together?

(c) It is desired to study the potential of a concentration cell in liquid ammonia at $223°K$:

$$\text{Cu} \mid \text{W} \mid \text{Pb} \mid \begin{array}{c} \text{NaSCN,} \\ \text{NH}_3, \text{Pb}^{++} \\ 223 \end{array} \Big\} \begin{array}{c} \text{transition} \\ \text{region} \end{array} \Big\{ \begin{array}{c} \text{NaSCN,} \\ \text{NH}_3, \text{Pb}^{++} \\ 223 \end{array} \Big| \text{Pb} \mid \text{W} \mid \text{Cu}$$

with temperatures 283 and 293 at the ends.

The tungsten rods coming from the cell make contact with copper wires, but through inadvertence these contacts are made at slightly different temperatures, 283 and $293°K$. Show that the error introduced into the measurement of the cell potential at $223°K$ is

$$\Delta\Phi = \int_{283}^{293} (\xi_{\text{W}} - \xi_{\text{Cu}})\, dT.$$

## NOTATION

| | |
|---|---|
| $c$ | molar concentration of a single electrolyte, mole/cm³ |
| $c_i$ | concentration of species $i$, mole/cm³ |
| $c_T$ | total solution concentration, mole/cm³ |
| $\hat{C}_p$ | heat capacity at constant pressure, J/g-deg |
| $D$ | measured diffusion coefficient of electrolyte, cm²/sec |
| $D_i^T$ | thermal diffusion coefficient of species $i$, g/cm-sec |
| $\mathfrak{D}$ | diffusion coefficient of electrolyte, based on a thermodynamic driving force, cm²/sec |
| $\mathfrak{D}_{ij}$ | diffusion coefficient for interaction of species $i$ and $j$, cm²/sec |
| $e^-$ | symbol for the electron |
| $F$ | Faraday's constant, 96,487 C/equiv |
| $\bar{H}_i$ | partial molar enthalpy of species $i$, J/mole |
| $\mathbf{i}$ | current density, A/cm² |

| $I$ | cell current, A |
| $\mathbf{J}_i$ | molar flux of species $i$ relative to the mass-average velocity, mole/cm$^2$-sec |
| $k$ | thermal conductivity, J/cm-sec-deg |
| $k'$ | thermal conductivity, J/cm-sec-deg |
| $k''$ | thermal conductivity, J/cm-sec-deg |
| $m$ | molality of a single electrolyte, mole/kg |
| $M_i$ | symbol for the chemical formula of species $i$ |
| $M_i$ | molecular weight of species $i$, g/mole |
| $n$ | number of electrons involved in electrode reaction |
| $n$ | number of species present in the solution |
| $\mathbf{N}_i$ | flux of species $i$, mole/cm$^2$-sec |
| $p$ | pressure, dyne/cm$^2$ |
| $\mathbf{q}$ | heat flux, J/cm$^2$-sec |
| $\mathbf{q}^{(x)}$ | Dufour energy flux, J/cm$^2$-sec |
| $\mathbf{q}'$ | conduction and Dufour energy flux, J/cm$^2$-sec |
| $Q_{\mathrm{rev}}$ | reversible heat transfer rate, J/sec |
| $R$ | universal gas constant, 8.3143 J/mole-deg |
| $R_i$ | rate of homogeneous production of species $i$, mole/cm$^3$-sec |
| $s_i$ | stoichiometric coefficient of species $i$ in electrode reaction |
| $\bar{S}_i$ | partial molar entropy of species $i$, J/mole-deg |
| $t$ | time, sec |
| $t_i$ | transference number of species $i$ with respect to the mass-average velocity |
| $t_i^0$ | transference number of species $i$ with respect to the velocity of species 0 |
| $T$ | absolute temperature, deg K |
| $U$ | reversible cell potential, V |
| $\hat{U}$ | internal energy per unit mass, J/g |
| $\mathbf{v}$ | mass-average velocity, cm/sec |
| $\mathbf{v}_i$ | velocity of species $i$, cm/sec |
| $\bar{V}_i$ | partial molar volume of species $i$, cm$^3$/mole |
| $y$ | distance from electrode, cm |
| $z_i$ | charge number of species $i$ |
| $\gamma_{+-}$ | mean molal activity coefficient of an electrolyte |
| $\eta_s$ | surface overpotential, V |
| $\kappa$ | conductivity, mho/cm |
| $\mu_e$ | chemical potential of an electrolyte, J/mole |
| $\mu_i$ | electrochemical potential of species $i$, J/mole |
| $\nu$ | number of moles of ions into which a mole of electrolyte dissociates |
| $\nu_+, \nu_-$ | numbers of cations and anions into which a molecule of electrolyte dissociates |

| $\zeta$ | thermoelectric coefficient, V/deg |
|---|---|
| $\rho$ | density, g/cm$^3$ |
| $\rho_i$ | mass of species $i$ per unit volume, g/cm$^3$ |
| $\sigma$ | Soret coefficient, deg$^{-1}$ |
| $\tau$ | stress, dyne/cm$^2$ |
| $\Phi$ | electric potential, V |
| $\omega_i$ | mass fraction of species $i$ |

subscripts

| $e$ | electrolyte |
|---|---|
| 0 | solvent |

# Transport Properties 14

## 89. Infinitely dilute solutions

In infinitely dilute solutions, there is one diffusion coefficient $D_i$ for each solute species. This transport property describes interaction between this species and the solvent. The properties of aqueous solutions were reviewed in section 75, where it was indicated that the mobility $u_i$ is related to the diffusion coefficient by the Nernst-Einstein relation 75-1.

## 90. Solutions of a single salt

Sections 79 and 81 indicate that the solutions of a single salt are characterized by three transport properties: the conductivity $\kappa$, the diffusion coefficient $D$, and the transference number $t_+^0$. These can be measured as functions of the concentration as well as the temperature, as reviewed by Robinson and Stokes.[1] The conductivity is commonly measured in terms of the alternating current resistance between two electrodes placed in the solution. The Hittorf method of measuring transference numbers involves the determination of the concentration changes near the anode and the cathode when a current is passed. The moving-boundary method, generally regarded as being more accurate than the Hittorf method, measures the rate of movement of the boundary between, say, solutions of $NH_4NO_3$ and $AgNO_3$ when

[1] R. A. Robinson and R. H. Stokes, *Electrolyte Solutions* (London: Butterworths, 1965).

266

a current is passed through that boundary.[2] Diffusion coefficients can be measured by following the concentration changes across a porous glass diaphragm. Also, accurate results can be obtained by measuring optically the concentration changes which take place when two solutions of different concentration are placed in contact with each other. This can be done either at very short times, in which case the initial boundary should be sharp, or at very long times[3] with the diffusion taking place in a restricted space about 7 cm high.

Over the years, a surprisingly large amount of data has been taken on the transport properties of solutions of single salts. Landolt-Börnstein[4] is a good source of conductivity data, and Kaimakov and Varshavskaya[5] have searched the literature for transference numbers. Robinson and Stokes[1] have compiled data on diffusion coefficients and activity coefficients, and Chapman and Newman[6] have collected data for a number of systems.

The conductivity, diffusion coefficient, and transference number represent three quite different transport properties. We might hope to find a more unified treatment by dealing with the equivalent transport coefficients $\mathfrak{D}_{0+}$, $\mathfrak{D}_{0-}$, and $\mathfrak{D}_{+-}$ defined by equation 78-1. These can be obtained from the measured values of $\kappa$, $D$, and $t_+^0$ by solving equations 79-5, 79-6, and 81-4:

$$\mathfrak{D}_{0-} = \frac{z_+}{z_+ - z_-} \frac{\mathfrak{D}}{t_+^0}, \tag{90-1}$$

$$\mathfrak{D}_{0+} = \frac{-z_-}{z_+ - z_-} \frac{\mathfrak{D}}{1 - t_+^0}, \tag{90-2}$$

$$\frac{1}{\mathfrak{D}_{+-}} = -\frac{z_+ z_- c_T F^2}{RT\kappa} - \frac{z_+ - z_-}{z_+ v_+} \frac{c_0 t_+^0 t_-^0}{c\mathfrak{D}}. \tag{90-3}$$

We see that first we need to determine $\mathfrak{D}$ from $D$ according to equation 79-7:

$$D = \mathfrak{D}\frac{c_T}{c_0}\left(1 + \frac{d \ln \gamma_{+-}}{d \ln m}\right), \tag{90-4}$$

which requires a knowledge of the activity coefficient (see problem 2-1).

There are more than 30 binary systems[3] with sufficient data to justify calculating values of $\mathfrak{D}_{ij}$. Figure 90-1 shows the multicomponent diffusion coefficients of KCl in water at 25°C. We may note that the coefficients for interactions of the ions with the solvent are reasonably constant, while that

[2] Paul Milios and John Newman, "Moving Boundary Measurement of Transference Numbers," *Journal of Physical Chemistry*, *73* (1969), 298–303.

[3] Thomas W. Chapman, *The Transport Properties of Concentrated Electrolytic Solutions.* Ph.D. thesis, University of California, Berkeley, November, 1967 (UCRL-17768).

[4] A. Eucken, ed., *Landolt-Börnstein, Zahlenwerte und Funktionen aus Physik, Chemie, Astronomie, Geophysik und Technik*, 6th ed., vol. 2, part 7 (Berlin: Springer-Verlag, 1960).

[5] E. A. Kaimakov and N. L. Varshavskaya, "Measurement of Transport Numbers in Aqueous Solutions of Electrolytes," *Uspekhi Khimii*, *35* (1966), 201–288.

[6] Thomas W. Chapman and John Newman, *A Compilation of Selected Thermodynamic and Transport Properties of Binary Electrolytes in Aqueous Solution*, Lawrence Radiation Laboratory, University of California, Berkeley, May, 1968 (UCRL-17767).

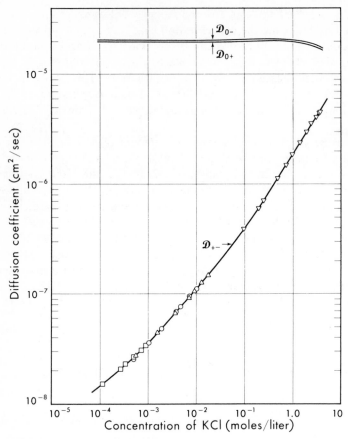

**Figure 90-1.** Multicomponent diffusion coefficients of KCl-$H_2O$ at 25°C.

for ion-ion interactions shows roughly a square-root-of-concentration dependence characteristic of the Debye-Hückel-Onsager theory of ionic interactions in dilute solutions.

On this basis, we define a function $G$:

$$G = \frac{z_+\mathscr{D}_{0+} - z_-\mathscr{D}_{0-}}{\mathscr{D}_{+-}} \frac{\sqrt{c}}{c_0} \frac{1 + \sqrt{q}}{z_+^2 z_-^2 q} \left(\frac{z_+\nu_+}{z_+ - z_-}\right)^{1/2} T^{3/2}, \qquad (90\text{-}5)$$

where
$$q = \frac{-z_+ z_-}{z_+ - z_-} \frac{\lambda_+^0 + \lambda_-^0}{z_+\lambda_+^0 - z_-\lambda_-^0}, \qquad (90\text{-}6)$$

and $\lambda_i^0$ is the ionic equivalent conductance at infinite dilution. $G$ is essentially $\sqrt{c}/\mathscr{D}_{+-}$, the other factors being based on the theory for dilute solutions. Figure 90-2 shows some calculated $G$ values for several chloride systems. For electrostatic ionic interactions, with neglect of electrophoresis, the limiting value for dilute solutions would be about 2860. We see that this value appears to be more characteristic of concentrated solutions.

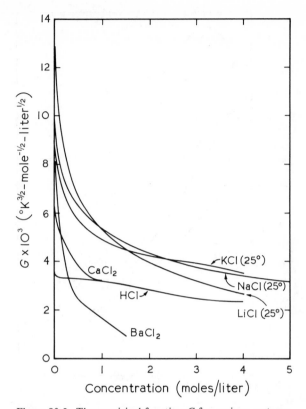

**Figure 90-2.** The empirical function $G$ for various systems.

Figure 90-3 shows $\mathfrak{D}_{0-}$ values for several chloride solutions and accentuates the concentration dependence relative to the logarithmic scale used in figure 1. The behavior definitely depends on the nature of the counter ion. We might think that multiplication by a viscosity factor would help (see equation 75-10). The reader can judge for himself from figure 90-4.

Figure 90-5 and 90-6 show the dependence on temperature. To a first approximation, the temperature dependence is given by the temperature dependence of the limiting value $\mathfrak{D}_{0i}^0$ as $c$ approaches zero.

## 91. Multicomponent solutions

The multicomponent diffusion equation 78-1 provides us with the proper macroscopic equation for defining transport properties in multicomponent solutions, just as thermodynamics provides the proper macroscopic framework for studying the equilibrium properties of solutions. The transport coefficients $\mathfrak{D}_{ij}$ represent, at least grossly, interactions between the two species

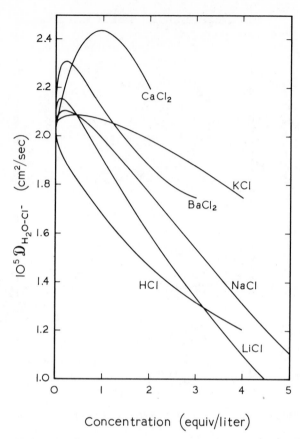

**Figure 90-3.** The diffusion coefficient of chloride ion in various aqueous
solutions at 25°C.

*i* and *j*. They thus give us some hope of discovering some systematic behavior
of transport properties and perhaps of extending results for binary solutions
to multicomponent solutions, since the multicomponent solutions still involve interactions between species *i* and *j*. In contrast, conductivities, transference numbers, and conventional diffusion coefficients represent averages
over more complex interactions. Furthermore, these interaction coefficients
are more directly related to ionic diffusion coefficients applicable in dilute
solutions.

However, in most cases, data for all the necessary transport properties
are incomplete.[7] For multicomponent solutions, the moving-boundary
method cannot be used, and transference numbers must be obtained by the
less accurate Hittorf method. Similarly, optical methods are not directly

[7]Donald G. Miller, "Application of Irreversible Thermodynamics to Electrolyte
Solutions. II. Ionic Coefficients $\ell_{ij}$ for Isothermal Vector Transport Processes in Ternary
System," *The Journal of Physical Chemistry, 71* (1967), 616–632.

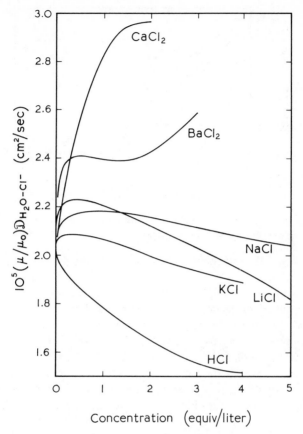

**Figure 90-4.** The diffusion coefficient of chloride ion with a viscosity factor.

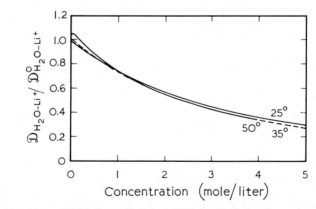

**Figure 90-5.** Lithium ion diffusion coefficients in lithium chloride solutions at various temperatures.

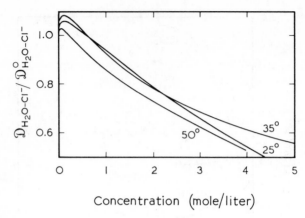

Concentration (mole/liter)

**Figure 90-6.** Chloride ion diffusion coefficients in lithium chloride solutions at various temperatures.

applicable, and diffusion coefficients must be obtained with the less accurate diaphragm cell.

If the data were available, multicomponent transport theory could be rigorously applied to certain simple geometries that would involve numerical solution of ordinary differential equations for the concentration profiles, with the computer being used to invert the transport matrix, as shown in section 83. In the absence of such data, we are frequently forced to go back to dilute-solution theory by way of sections 82 and 77. However, in certain mass-transfer situations, it is possible to make reliable estimates of the transport properties to be used, as indicated in the next section.

## 92. Integral diffusion coefficients for mass transfer

In practice, we are often interested in mass transfer to an electrode from a multicomponent solution, such as deposition of copper from a solution of copper sulfate and sulfuric acid. However, all the transport properties of such a solution may not be known. What solution property can we measure that will allow us to predict accurately the behavior of the system?

Usually the process will obey approximately the equation of convective diffusion 73-2 with an effective diffusion coefficient that we desire to predict. This will be called an *integral diffusion coefficient* because it represents an average over the behavior and properties encountered in the diffusion layer near an electrode. The thesis is suggested that this diffusion coefficient should be measured in a system with similar hydrodynamic conditions. For example, an integral diffusion coefficient measured with a rotating disk electrode (see section 103) should be applicable to mass transfer in an annulus or pipe (see section 105). This integral diffusion coefficient is different from the integral diffusion coefficient measured with a diaphragm cell.[8] Similarly, the integral diffusion coefficient measured in transient mass transfer to an electrode at the end of a stagnant diffusion cell should be applicable to a growing mercury drop (see section 110). These latter diffusion coefficients are called *polarographic diffusion coefficients*.

The validity of the use of an integral diffusion coefficient is influenced by the following effects: the value of the Schmidt number $v/D_i$, the nonzero interfacial velocity, the effect of ionic migration, and the variations of transport properties with composition. Each effect has been treated individually by several workers, mostly in nonelectrolytic systems.

The treatment of mass transfer is simplified at the high Schmidt numbers which prevail in electrolytic solutions (see sections 106 and 107). The correction for the fact that the Schmidt number is not infinite can differ among several hydrodynamic situations; the correction is usually no more than a few percent. The high value of the Schmidt number allows justification for assuming that the other effects are properly accounted for.

The effect of nonzero interfacial velocity due to the high mass-transfer rate can also be expressed as a correction factor to the mass-transfer coefficient in the absence of an interfacial velocity. This correction factor depends on the mass-flux ratio and, in the limit of large Schmidt numbers, has been shown to be the same for arbitrary, two-dimensional boundary layers[9] and for the rotating disk.[8]

Similarly, the effect of ionic migration in the diffusion layer can also be expressed as a correction factor for the mass-transfer rate in the absence of migration[10] (see also chapter 19). For large Schmidt numbers, one correction factor has been shown to apply to arbitrary two-dimensional and axisymmetric diffusion layers,[11] including the rotating disk, and another to the transient processes of a growing mercury drop and an electrode at the end of a stagnant diffusion cell.[10]

Acrivos[12] has shown that in the limit of high Schmidt numbers one effective diffusion coefficient should apply to mass transfer at the limiting rate from a given solution for arbitrary boundary layer flows, even though the physical properties vary with composition in the diffusion layer.

[8]J. Newman and L. Hsueh, "The Effect of Variable Transport Properties on Mass Transfer to a Rotating Disk," *Electrochimica Acta, 12* (1967), 417–427.

[9]Andreas Acrivos, "The asymptotic form of the laminar boundary-layer mass-transfer rate for large interfacial velocities," *Journal of Fluid Mechanics, 12* (1962), 337–357.

[10]John Newman, "Effect of Ionic Migration on Limiting Currents," *Industrial and Engineering Chemistry Fundamentals, 5* (1966), 525–529.

[11]John Newman, "The Effect of Migration in Laminar Diffusion Layers," *International Journal of Heat and Mass Transfer, 10* (1967), 983–997.

[12]Andreas Acrivos, "Solution of the Laminar Boundary Layer Energy Equation at High Prandtl Numbers," *The Physics of Fluids, 3* (1960), 657–658.

These considerations lead us to conclude that one effective or integral diffusion coefficient should describe mass transfer at the limiting current from a given solution for arbitrary two-dimensional and axisymmetric diffusion layers in laminar forced convection. This integral diffusion coefficient will depend upon the bulk composition of the solution. Somewhat different integral diffusion coefficients may apply to free convection, turbulent flow, or the transient processes cited above. However, these diffusion coefficients should be closer to each other than to the value obtained with a diaphragm cell, which is a completely different situation from mass transfer to an electrode with the flow of current.

Two electrode reactions have proved to be particularly popular for experimental mass-transfer studies. These are deposition of copper,

$$Cu^{++} + 2e^- \longrightarrow Cu, \tag{92-1}$$

from solutions of copper sulfate and sulfuric acid and the reduction of ferricyanide ions,

$$Fe(CN)_6^{3-} + e^- \longrightarrow Fe(CN)_6^{4-}, \tag{92-2}$$

from solutions using NaOH, KOH, or $KNO_3$ as a supporting electrolyte. Selman[13] has analyzed the available literature on intergral diffusion coefficients for these solutions.

## PROBLEM

1. The simplest ternary ionic solution is one containing a single electrolyte, one of whose ions is present in two isotopic forms. An example would be a sodium chloride solution containing stable sodium ions and radioactive sodium ions. We assume that these ions are identical except that the radioactive ions are tagged. Let the solvent be denoted by 0, the cations by 1 and 2, and the anion by 3. There are six transport properties for this system, $\mathfrak{D}_{01}, \mathfrak{D}_{02}, \mathfrak{D}_{03}, \mathfrak{D}_{12}, \mathfrak{D}_{13}$, and $\mathfrak{D}_{23}$. On the assumption that there is no isotope effect, five of these can be predicted from the values of $\mathfrak{D}_{0+}, \mathfrak{D}_{0-}$, and $\mathfrak{D}_{+-}$ of the binary untagged solutions, while the last can be obtained from the value of the *self diffusion coefficient* $D_*$ describing the diffusion of tagged electrolyte in a solution whose total electrolyte concentration is uniform. This gives a way of getting at the concentration dependence of $\mathfrak{D}_{12}$ related to interactions of ions of the same charge, something that cannot be ascertained from binary solutions of a single salt. In the following, assume that

$$f_1^{\nu+} f_3^{\nu-} = f_2^{\nu+} f_3^{\nu-} = f_{+-}^{\nu}.$$

(a) Show that $\mathfrak{D}_{01}, \mathfrak{D}_{02}, \mathfrak{D}_{03}, \mathfrak{D}_{12}$, and $\mathfrak{D}_{13}$ are given by

$$\mathfrak{D}_{13} = \mathfrak{D}_{23} = \mathfrak{D}_{+-}, \quad \mathfrak{D}_{01} = \mathfrak{D}_{02} = \mathfrak{D}_{0+}, \quad \text{and} \quad \mathfrak{D}_{03} = \mathfrak{D}_{0-},$$

[13]Jan Robert Selman, *Measurement and Interpretation of Limiting Currents*. Ph.D. thesis, University of California, Berkeley, June, 1971 (UCRL-20557).

where $\mathfrak{D}_{+-}$, $\mathfrak{D}_{0+}$, and $\mathfrak{D}_{0-}$ are to be evaluated at the total electrolyte concentration, $c = (c_1 + c_2)/\nu_+$.

(b) Show that $\mathfrak{D}_{12}$ can be obtained from measured values of $D_*$ according to the relation

$$\mathfrak{D}_{12} = \frac{c_+}{\dfrac{c_T}{D_*} - \dfrac{c_0}{\mathfrak{D}_{0+}} - \dfrac{c_-}{\mathfrak{D}_{+-}}}.$$

(c) Show that as the total electrolyte concentration approaches zero, $D_*$ approaches $\mathfrak{D}_{0+}$.

## NOTATION

| | |
|---|---|
| $c$ | concentration of a single electrolyte, mole/cm³ |
| $c_0$ | concentration of solvent, mole/cm³ |
| $c_T$ | total solution concentration, mole/cm³ |
| $D$ | measured diffusion coefficient of electrolyte, cm²/sec |
| $D_i$ | diffusion coefficient of species $i$, cm²/sec |
| $\mathfrak{D}$ | diffusion coefficient of electrolyte, based on a thermodynamic driving force, cm²/sec |
| $\mathfrak{D}_{ij}$ | diffusion coefficient for interaction of species $i$ and $j$, cm²/sec |
| $F$ | Faraday's constant, 96,487 C/equiv |
| $G$ | function related to $\mathfrak{D}_{+-}$, deg K³ᐟ²-(liter/mole)¹ᐟ² |
| $m$ | molality of a single electrolyte, mole-kg |
| $R$ | universal gas constant, 8.3143 J/mole-deg |
| $q$ | see equation 90-6 |
| $t_i^0$ | transference number of species $i$ relative to the solvent velocity |
| $T$ | absolute temperature, deg K |
| $u_i$ | mobility of species $i$, cm²-mole/J-sec |
| $z_i$ | charge number of species $i$ |
| $\gamma_{+-}$ | mean molal activity coefficient of an electrolyte |
| $\kappa$ | conductivity, mho/cm |
| $\lambda_i^0$ | equivalent ionic conductance of species $i$ at infinite dilution, mho-cm²/equiv |
| $\mu$ | viscosity, g/cm-sec |
| $\nu$ | kinematic viscosity, cm²/sec |
| $\nu_+$ | number of cations into which a molecule of electrolye dissociates |

# Fluid Mechanics

# 15

Since diffusion and migration fluxes are expressed relative to an average velocity of the fluid, mass-transfer calculations require a previous or a simultaneous determination of the velocity. In many systems, the velocity distribution is governed by momentum considerations. The mechanical behavior of fluids is briefly described in this chapter. For more details, one should consult the literature.[1,2] The velocity profiles for various specific systems will be taken as a basis for determining mass-transfer rates in part D.

## 93. Mass and momentum balances

The mass-average velocity is defined as

$$\mathbf{v} = \frac{1}{\rho} \sum_i c_i M_i \mathbf{v}_i, \qquad (93\text{-}1)$$

where $c_i \mathbf{v}_i$ is the molar flux of species $i$, $M_i$ is the molecular weight, and $\rho$ is the density of the medium. The mass-average velocity is useful in fluid mechanics because $\rho \mathbf{v}$ is both the mass flux and the momentum density in the fluid.

The law of conservation of mass can be expressed in a differential

[1] Hermann Schlichting, *Boundary-Layer Theory* (New York: McGraw-Hill Book Company, 1968).

[2] R. Byron Bird, Warren E. Stewart, and Edwin N. Lightfoot, *Transport Phenomena* (New York: John Wiley & Sons, Inc., 1960).

form as

$$\frac{\partial \rho}{\partial t} = -\nabla \cdot (\rho \mathbf{v}).$$ (93-2)

This equation can be obtained from the species material balance, equation 69-3, by multiplying that equation by the molecular weight $M_i$ and summing over species. When the density is constant in space and time, equation 2 reduces to

$$\nabla \cdot \mathbf{v} = 0.$$ (93-3)

This is frequently an adequate approximation for dilute liquid solutions.

The law of conservation of momentum can be expressed in a differential form as

$$\frac{\partial \rho \mathbf{v}}{\partial t} + \nabla \cdot (\rho \mathbf{v} \mathbf{v}) = \rho \left( \frac{\partial \mathbf{v}}{\partial t} + \mathbf{v} \cdot \nabla \mathbf{v} \right) = -\nabla p - \nabla \cdot \boldsymbol{\tau} + \rho \mathbf{g},$$ (93-4)

where $p$ is the thermodynamic pressure, $\boldsymbol{\tau}$ is the stress tensor, and $\mathbf{g}$ is the acceleration due to gravity. This equation is an expression of Newton's second law of motion; the rate of change of momentum of a fluid element is equal to the force applied. Here the forces are the pressure gradient, the stress in the fluid, and the force of gravity. The divergence of the stress appears because one needs the net force—the difference between the forces on opposite sides of the fluid element (compare figure 69-1). The stress tensor will be considered in the following section.

Other forces could be included in the momentum balance. If the fluid is not electrically neutral, we should add to the right side of equation 4 the electrical force

$$\rho_e \mathbf{E} = \epsilon (\nabla \cdot \mathbf{E}) \mathbf{E} = \epsilon (\nabla^2 \Phi) \nabla \Phi.$$ (93-5)

This term is usually omitted because electrolytic solutions are electrically neutral to a very good approximation. However, this conclusion was arrived at on the basis of the large magnitude of electrical forces, and it is not immediately obvious that the electrical force can be omitted from the momentum balance. This question will be reconsidered in section 97. In some electrochemical systems the magnetic force

$$\mathbf{i} \times \mathbf{B}$$

should also be included on the right side of equation 4. Here, $\mathbf{i}$ is the current density within the solution, and $\mathbf{B}$ is the magnetic induction (weber/m$^2$). The magnetic field may itself be due to the flow of current in the system.

For a fluid of constant density, it may be advantageous to define the dynamic pressure $\mathcal{P}$ by

$$\nabla \mathcal{P} = \nabla p - \rho \mathbf{g}.$$ (93-6)

Essentially, this equation subtracts the hydrostatic pressure from the thermodynamic pressure to yield the dynamic pressure, changes in which are directly related to the fluid motion.

## 94. Stress in a Newtonian fluid

The stress $\tau$ is related to velocity gradients within the fluid. For Newtonian fluids, which include most electrolytic solutions, the appropriate expression is

$$\tau = -\mu[\nabla v + (\nabla v)^*] + \tfrac{2}{3}\mu\, I\nabla\cdot v, \qquad (94\text{-}1)$$

where $I$ is the unit tensor and $\mu$ is the viscosity, a transport property which depends on temperature, pressure, and composition. A basic physical law requires the stress to be symmetric; this is assured by the presence in equation 1 of the transpose $(\nabla v)^*$ of the velocity gradient.

To be specific, a diagonal element of the stress looks like

$$\tau_{xx} = -2\mu\frac{\partial v_x}{\partial x} + \frac{2}{3}\mu\,\nabla\cdot v, \qquad (94\text{-}2)$$

while an off-diagonal element looks like

$$\tau_{xy} = \tau_{yx} = -\mu\left(\frac{\partial v_x}{\partial y} + \frac{\partial v_y}{\partial x}\right). \qquad (94\text{-}3)$$

For a fluid of constant density and viscosity, substitution of equation 1 into equation 93-4 yields

$$\frac{\partial v}{\partial t} + v\cdot\nabla v = -\frac{1}{\rho}\nabla p + \nu\,\nabla^2 v + g, \qquad (94\text{-}4)$$

where $\nu = \mu/\rho$ is the *kinematic viscosity* of the fluid. This equation, known as the Navier-Stokes equation, is written out in rectangular coordinates in appendix B and in cylindrical coordinates in section 96.

## 95. Boundary conditions

On solid surfaces the velocity $v$ is zero, or, more generally, the velocity is continuous at an interface. An exception to this was encountered in electrokinetic phenomena (see section 62), where the discontinuity in velocity was related to the tangential electric field. However, this was after taking account of the behavior of the diffuse double layer, and in the details of the analysis the velocity was continuous. Electrokinetic phenomena do not need to be considered in all processes.

For a fluid-fluid interface, the velocity at the surface may not be known in advance, and then the relationship between the shear stress in the two phases must be considered. If the interface is of negligible mass (see remark below equation 50-4), the forces at the interface must balance. In the simplest case, this means that the tangential (shear) stress is continuous.

Let the two phases be denoted by superscripts $\alpha$ and $\beta$, and let the force per unit area exerted by these phases on the interface be $f^\alpha$ and $f^\beta$, respectively.

These forces **f** are the product of the stress **τ** with the unit normal vector of the interface:

$$\mathbf{f}^{\alpha} = \mathbf{n} \cdot \mathbf{\tau}^{\alpha} + \mathbf{n}p^{\alpha} \tag{95-1}$$

and

$$\mathbf{f}^{\beta} = -\mathbf{n} \cdot \mathbf{\tau}^{\beta} - \mathbf{n}p^{\beta}, \tag{95-2}$$

where **n** points into phase $\beta$. (The stress **τ** is expressed in various coordinate systems in reference 2.)

Figure 95-1 shows a surface element lying in the plane of the paper. A force balance in the $x$-direction yields

$$(f_x^{\alpha} + f_x^{\beta}) \Delta x \, \Delta z + \Delta z (\sigma|_{x+\Delta x} - \sigma|_{x}) = 0, \tag{95-3}$$

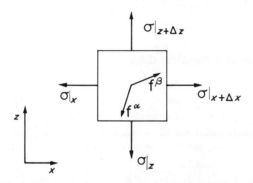

**Figure 95-1.** Tangential forces on an interfacial element lying in the $x,z$-plane. (From J. Newman, "Transport Processes in Electrolytic Solutions," in C. W. Tobias, *Advances in Electrochemistry and Electrochemical Engineering*, Vol. 5. Copyright © 1967 by John Wiley & Sons, Inc. Reprinted by permission.)

where $\sigma$ is the interfacial tension (dyne/cm). If we divide by $\Delta x \, \Delta z$ and let $\Delta x$ approach zero, we obtain

$$f_x^{\alpha} + f_x^{\beta} + \frac{\partial \sigma}{\partial x} = 0. \tag{95-4}$$

A similar equation applies in the $z$-direction. Together, we have

$$\mathbf{f}_s^{\alpha} + \mathbf{f}_s^{\beta} + \nabla_s \sigma = 0, \tag{95-5}$$

where $\mathbf{f}_s^{\alpha}$ and $\mathbf{f}_s^{\beta}$ denote the components of $\mathbf{f}^{\alpha}$ and $\mathbf{f}^{\beta}$ lying in the surface and $\nabla_s$ denotes the surface gradient.

The tangential parts of the forces $\mathbf{f}^{\alpha}$ and $\mathbf{f}^{\beta}$ are viscous in nature, and, if the interfacial tension is independent of position in the surface, equation 5 says that the tangential viscous stress is continuous.

For the normal component of the force balance we have

$$f_n^{\alpha} + f_n^{\beta} = \sigma\left(\frac{1}{r_1} + \frac{1}{r_2}\right), \tag{95-6}$$

where $r_1$ and $r_2$ are the principal radii of curvature of the surface. The normal components $f_n^\alpha$ and $f_n^\beta$ include the thermodynamic pressure $p$ as well as the normal viscous stress. The appropriate signs on the radii of curvature in equation 6 would be such that the pressure inside a drop or bubble is greater than the pressure outside.

These elements of surface dynamics enter into the treatment of electrocapillary phenomena (see chapter 10).

Only the gradient of the pressure appears in equation 93-4 or equation 94-4. Consequently, it is sometimes possible to solve a problem in terms of this gradient without ever requiring the pressure itself. Then it is sufficient to specify the pressure at only one point.

## 96. Fluid flow to a rotating disk

The rotating-disk electrode is very popular in electrochemical studies, partly because the hydrodynamic conditions are well known and partly because the experimental setup is small and simple. The rotating disk is also one of the few systems for which a nontrivial solution of the equations of fluid mechanics is possible.

We consider the steady flow of an incompressible fluid caused by the rotation of a large disk about an axis through its center. For this purpose we use cylindrical coordinates $r$, $\theta$, and $z$, where $z$ is the perpendicular distance from the disk and $r$ is the radial distance from the axis of rotation. The velocity on the surface of the disk is

$$v_r = 0, \qquad v_z = 0, \qquad v_\theta = r\Omega . \tag{96-1}$$

The last condition expresses the fact that the rotating disk drags the adjacent fluid with it at an angular velocity $\Omega$ (radian/sec).

Because of the rotation, there is a centrifugal effect which tends to throw the fluid out in a radial direction. This will result in a radial component of the velocity which is zero at the surface, has a maximum value near the surface, and then goes to zero again at greater distances from the disk. In order to replace the liquid flowing out in the radial direction, it is necessary to have a $z$-component of the velocity which brings fluid toward the disk from far away. This gives us a qualitative picture of the flow field in which none of the velocity components is zero.

In cylindrical coordinates, the equation of continuity 93-3 is[2]

$$\frac{1}{r}\frac{\partial}{\partial r}(rv_r) + \frac{1}{r}\frac{\partial v_\theta}{\partial \theta} + \frac{\partial v_z}{\partial z} = 0, \tag{96-2}$$

and the components of the equation of motion 94-4 are[2]

($r$-component)

$$\frac{\partial v_r}{\partial t} + v_r\frac{\partial v_r}{\partial r} + \frac{v_\theta}{r}\frac{\partial v_r}{\partial \theta} - \frac{v_\theta^2}{r} + v_z\frac{\partial v_r}{\partial z} = -\frac{1}{\rho}\frac{\partial \mathcal{P}}{\partial r}$$

$$+ v\left[\frac{\partial}{\partial r}\left(\frac{1}{r}\frac{\partial}{\partial r}(rv_r)\right) + \frac{1}{r^2}\frac{\partial^2 v_r}{\partial \theta^2} - \frac{2}{r^2}\frac{\partial v_\theta}{\partial \theta} + \frac{\partial^2 v_r}{\partial z^2}\right], \qquad (96\text{-}3)$$

($\theta$-component)

$$\frac{\partial v_\theta}{\partial t} + v_r\frac{\partial v_\theta}{\partial r} + \frac{v_\theta}{r}\frac{\partial v_\theta}{\partial \theta} + \frac{v_r v_\theta}{r} + v_z\frac{\partial v_\theta}{\partial z} = -\frac{1}{\rho r}\frac{\partial \mathcal{P}}{\partial \theta}$$

$$+ v\left[\frac{\partial}{\partial r}\left(\frac{1}{r}\frac{\partial}{\partial r}(rv_\theta)\right) + \frac{1}{r^2}\frac{\partial^2 v_\theta}{\partial \theta^2} + \frac{2}{r^2}\frac{\partial v_r}{\partial \theta} + \frac{\partial^2 v_\theta}{\partial z^2}\right], \qquad (96\text{-}4)$$

($z$-component)

$$\frac{\partial v_z}{\partial t} + v_r\frac{\partial v_z}{\partial r} + \frac{v_\theta}{r}\frac{\partial v_z}{\partial \theta} + v_z\frac{\partial v_z}{\partial z} = -\frac{1}{\rho}\frac{\partial \mathcal{P}}{\partial z}$$

$$+ v\left[\frac{1}{r}\frac{\partial}{\partial r}\left(r\frac{\partial v_z}{\partial r}\right) + \frac{1}{r^2}\frac{\partial^2 v_z}{\partial \theta^2} + \frac{\partial^2 v_z}{\partial z^2}\right], \qquad (96\text{-}5)$$

where we have used the dynamic pressure $\mathcal{P}$ introduced in equation 93-6. In our problem with axial symmetry and steady flow, the derivatives with respect to $t$ and $\theta$ are zero in these equations.

In 1921 von Kármán[3] suggested that these partial differential equations could be reduced to ordinary differential equations by seeking a solution of the form

$$v_\theta = rg(z), \qquad v_r = rf(z), \qquad v_z = h(z), \qquad \mathcal{P} = \mathcal{P}(z), \qquad (96\text{-}6)$$

which is a separation of variables. If these expressions are substituted into equations 2 to 5, one obtains

$$\left.\begin{aligned} 2f + h' &= 0, \\ f^2 - g^2 + hf' &= vf'', \\ 2fg + hg' &= vg'', \\ phh' + \mathcal{P}' &= \mu h'', \end{aligned}\right\} \qquad (96\text{-}7)$$

where the primes denote differentiation with respect to $z$.

The boundary conditions are

$$\left.\begin{aligned} h = f &= 0, \qquad g = \Omega \quad \text{at} \quad z = 0. \\ f = g &= 0 \quad \text{at} \quad z = \infty. \end{aligned}\right\} \qquad (96\text{-}8)$$

In addition, the value of $\mathcal{P}$ needs to be specified at one point.

The von Kármán transformation is successful in reducing the problem to ordinary differential equations. It should be noted, however, that this solution does not take into account the fact that the radius of the disk might be

[3] Th. v. Kármán, "Über laminare und turbulente Reibung," *Zeitschrift für angewandte Mathematik und Mechanik, 1* (1921), 233–252.

finite. In practice, these edge effects can frequently be neglected, and the resulting solution is quite useful.[4]

The remaining parameters, $v$, $\rho$, $\Omega$, can be eliminated by introducing a dimensionless distance, dimensionless velocities, and a dimensionless pressure as follows:

$$\zeta = z\sqrt{\frac{\Omega}{v}}, \qquad \mathcal{P} = \mu\Omega P, \qquad v_\theta = r\Omega G, \\ v_r = r\Omega F, \qquad v_z = \sqrt{v\Omega}H. \tag{96-9}$$

The differential equations 7 become

$$2F + H' = 0, \\ F^2 - G^2 + HF' = F'', \\ 2FG + HG' = G'', \\ HH' + P' = H'', \tag{96-10}$$

where the primes now denote differentiation with respect to $\zeta$. The boundary conditions are

$$H = F = 0, \qquad G = 1 \quad \text{at} \quad \zeta = 0. \\ F = G = 0 \quad \text{at} \quad \zeta = \infty. \tag{96-11}$$

Since these equations are nonlinear, it seems necessary to obtain the solution numerically. Cochran[5] originally solved these equations by forming series expansions for small values of $\zeta$ and for large values of $\zeta$ and then adjusting the unknown coefficients in the series until agreement between the two sets of series was obtained at an intermediate value of $\zeta$. However, it is fairly simple to solve coupled, nonlinear, ordinary differential equations by direct numerical techniques (see appendix C). The solution to equations 10 subject to conditions 11 is shown in figure 96-1. After the velocity profiles have been determined, the pressure can be obtained by integrating the last of equations 10:

$$P = P(0) + H' - \tfrac{1}{2}H^2. \tag{96-12}$$

The normal component $v_z$ of the velocity will be important for the calculation of rates of mass transfer to the rotating disk (see section 103). For small distances from the disk, the dimensionless velocity can be expressed as a power series:

$$H = -a\zeta^2 + \frac{1}{3}\zeta^3 + \frac{b}{6}\zeta^4 + \cdots, \tag{96-13}$$

with the coefficients[5,6]

$$a = 0.51023 \quad \text{and} \quad b = -0.616. \tag{96-14}$$

[4]A. C. Riddiford, "The Rotating Disk System," *Advances in Electrochemistry and Electrochemical Engineering, 4* (1966), 47–116.

[5]W. G. Cochran, "The flow due to a rotating disc," *Proceedings of the Cambridge Philosophical Society, 30* (1934), 365–375.

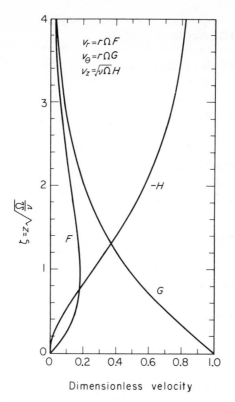

**Figure 96-1.** Velocity profiles for a rotating disk.

On the other hand, for large distances from the disk, the dimensionless velocity can be expressed as

$$H = -\alpha + \frac{2A}{\alpha}e^{-\alpha\zeta} + \cdots,\qquad(96\text{-}15)$$

where[5,6]            $\alpha = 0.88447$   and   $A = 0.934.$            (96-16)

The fact that the normal component of the velocity $v_z$ depends only on the normal distance $z$ and not on the radial distance $r$ is another reason for the popularity of the rotating-disk electrode among electrochemists.

The flow in the boundary layer remains laminar for a Reynolds number, $Re = r^2\Omega/v$, up to about $2 \times 10^5$. For larger radial distances, the flow becomes turbulent.

[6]M. H. Rogers and G. N. Lance, "The rotationally symmetric flow of a viscous fluid in the presence of an infinite rotating disk," *Journal of Fluid Mechanics*, 7 (1960), 617–631.

## 97.  Magnitude of electrical forces

Let us now include the electrical force, equation 93-5, in the momentum balance, equation 93-4. For a rotating disk we might at first imagine that **E** lies in the $z$-direction because of the uniform accessibility of the surface (see section 103). Then

$$\epsilon(\nabla \cdot \mathbf{E})E_z = \frac{1}{2}\epsilon\frac{dE_z^2}{dz}. \tag{97-1}$$

This enters only into the $z$-component of the equation of motion; that is, the last of equations 96-10 becomes

$$\frac{dP}{d\zeta} = \frac{d^2H}{d\zeta^2} - H\frac{dH}{d\zeta} + \frac{1}{2}\frac{\epsilon}{\mu\Omega}\frac{dE_z^2}{d\zeta} = \frac{d}{d\zeta}\left(\frac{dH}{d\zeta} - \frac{1}{2}H^2 + \frac{1}{2}\frac{\epsilon}{\mu\Omega}E_z^2\right). \tag{97-2}$$

For $\Omega = 300$ rpm, $\epsilon = 78.3\,\epsilon_0$, and $\mu = 0.8903$ cp, $\sqrt{\mu\Omega/\epsilon} = 63.5$ V/cm. This gives us a basis for evaluating the relative importance of electrical forces since, for example, $H^2$ is of order unity. With $i = 0.1$ A/cm$^2$ and $\kappa = 0.01$ mho/cm, we can expect electric fields of the order of $i/\kappa = 10$ V/cm. If the variation in the field were small compared to $\sqrt{\mu\Omega/\epsilon}$, we could neglect the electrical force altogether.

From equation 70-1, we can express the electric field as

$$\mathbf{E} = \frac{\mathbf{i}}{\kappa} + \frac{F}{\kappa}\sum_i z_i D_i \nabla c_i, \tag{97-3}$$

and hence the electric charge density is

$$\frac{\rho_e}{\epsilon} = -\left(\mathbf{i} + F\sum_i z_i D_i \nabla c_i\right)\cdot\frac{\nabla\kappa}{\kappa^2} + \frac{F}{\kappa}\sum_i z_i D_i \nabla^2 c_i. \tag{97-4}$$

From these equations, we can make several observations. The electric charge density is different from zero only in the thin diffusion layers near electrodes, since it is only in these regions that the concentrations and conductivity vary. The charge density here is still small (see section 76) since $\epsilon$ is small, but outside these regions it is identically zero. The electric effect will also be largest with a binary electrolyte, since, with a supporting electrolyte, $\kappa$ will be large compared to the variations in $\kappa$.

Furthermore, equation 3 shows that in the diffusion layers, where $\rho_e$ is different from zero, **E** lies mainly in the direction perpendicular to the electrode. This means that the electric force enters most dominantly into the normal component of the equation of motion, where it affects the pressure distribution without altering the velocity distribution. This effect is relatively unimportant since it is the velocity profiles which determine the mass-transfer rates. This we can see clearly in the case of the rotating disk where, as formulated above, the entire electrical effect can be absorbed into the variation of the dynamic pressure $\mathcal{P}$, and the velocity profiles are not affected at all.

Any part of $\rho_e\mathbf{E}$ which can be expressed as the gradient of some quantity can, in general, be absorbed into $\mathcal{P}$. This is the part of $\rho_e \mathbf{E}$ whose curl is zero.

$$\nabla \times \rho_e \mathbf{E} = \rho_e \nabla \times \mathbf{E} + (\nabla \rho_e) \times \mathbf{E} = (\nabla \rho_e) \times \mathbf{E}. \qquad (97\text{-}5)$$

The curl of $\mathbf{E}$ is zero since $\mathbf{E}$ is minus the gradient of the potential (see entry 5b of table B-1). In this manner, we arrive at the conclusion that it is the quantity in equation 5 which affects the velocity profiles. As observed above, $\rho_e$ is nonzero only in the diffusion layers, and here we can expect that $\mathbf{E}$ and $\nabla \rho_e$ are nearly parallel to each other so that their cross product is small. This reinforces the conclusion that the electrical force will have most of its effect on the pressure and less effect on the velocity.

Furthermore, the diffusion layer is much thinner than the hydrodynamic boundary layer at high Schmidt numbers. Here the viscous forces are important, and the electric force might be expected to have less effect on the velocity profile if exerted here than if exerted farther from the wall. On the other hand, the velocities are much smaller here and are important in determining the mass-transfer rate. Consequently, the effect could still be important.

The way to ascertain the effect of the electrical force on the velocity profiles, while excluding the effect on the dynamic pressure, is to take the curl of the equation of motion 94-4. This eliminates the pressure. We shall assume that the von Kármán transformation 96-9 is still valid and examine the magnitude of the neglected electrical force. Taking the curl of the equation of motion now yields

$$2FG + H\frac{\partial G}{\partial \zeta} = \frac{\partial^2 G}{\partial \zeta^2}, \qquad (97\text{-}6)$$

$$\frac{d}{d\zeta}\left(F^2 - G^2 + H\frac{dF}{d\zeta} - \frac{d^2F}{d\zeta^2}\right)r\Omega^2 = \frac{E_r}{\rho}\frac{\partial \rho_e}{\partial \zeta} - \frac{E_z}{\rho}\sqrt{\frac{\nu}{\Omega}}\frac{\partial \rho_e}{\partial r}, \qquad (97\text{-}7)$$

and the continuity equation 93-3 becomes

$$2F + \frac{dH}{d\zeta} = 0. \qquad (97\text{-}8)$$

The term on the right in equation 7 comes from the cross product of $\nabla \rho_e$ and $\mathbf{E}$ (see equation 5). As observed before, $E_r$ should be much less than $E_z$, and $\partial \rho_e/\partial r$ should be much less than $\partial \rho_e/\partial z$. This makes it difficult to assess the magnitude of these terms. In order to continue the analysis, let us take $\rho_e$ to be independent of $r$ and take $E_r$ to be independent of $\zeta$ in the diffusion layer and given by

$$E_r = A\frac{ir}{\kappa_\infty r_0}, \qquad (97\text{-}9)$$

where $A$ is approximately equal to 0.73. This is an approximation to the radial dependence of the tangential electric field just outside the diffusion layer when a uniform current density $i$ prevails over the surface of a disk electrode embedded in an insulating plane (see figure 117-4). The importance of the tangential electric field here finds analogy in the electrokinetic phenomena treated in chapter 9.

Equation 7 can now be integrated to read

$$F^2 - G^2 + H\frac{dF}{d\zeta} = \frac{d^2 F}{d\zeta^2} + \frac{Ai\rho_e}{\rho\kappa_\infty r_0 \Omega^2},$$  (97-10)

the integration constant being zero since $F$, $G$, and $\rho_e$ approach zero as $\zeta$ approaches infinity.

For a binary electrolyte, equation 4 becomes

$$\frac{\rho_e}{\epsilon} = -\frac{\mathbf{i}\cdot\nabla c}{z_+ \nu_+ \Lambda c^2} + \frac{RT}{F}\left(\frac{t_+}{z_+} + \frac{t_-}{z_-}\right)\nabla^2 \ln c,$$  (97-11)

where $\Lambda$ is the equivalent conductance (see equation 75-8) and is taken to be constant. We can take the solution of the equation of convective diffusion 72-5 to be

$$c = c_0 + \frac{c_\infty - c_0}{\Gamma(4/3)}\int_0^\zeta e^{-x^3}\,dx,$$  (97-12)

where

$$\xi = \left(\frac{a\nu}{3D}\right)^{1/3}\zeta$$  (97-13)

and $a$ is given by equation 96-14. This is the appropriate form of equation 103-8 for high Schmidt numbers $\nu/D$, $c_0$ and $c_\infty$ being the concentrations at the electrode surface and in the bulk solution, respectively. We shall restrict ourselves to a metal deposition reaction where the current density is given by equation 72-11.

With these assumptions, the last term in equation 10 becomes

$$\frac{Ai\rho_e}{\rho\kappa_\infty r_0 \Omega^2} = \frac{Ai\epsilon}{r_0\kappa_\infty\mu\Omega}\frac{RT}{-z_+z_-F}\left(\frac{a\nu}{3D}\right)^{2/3}\left[\frac{c_\infty - c_0}{c\Gamma(4/3)}\right]^2 e^{-\xi^3}$$
$$\times \left\{z_+ - (z_+t_- + z_-t_+)\left[1 + e^{-\xi^3} + 3\xi^2\frac{c\Gamma(4/3)}{c_\infty - c_0}\right]\right\}.$$  (97-14)

The coefficient in this expression can be estimated to be

$$\frac{Ai\epsilon}{r_0\kappa_\infty\mu\Omega}\frac{RT}{F}\left(\frac{a\nu}{3D}\right)^{2/3} = 0.0057,$$  (97-15)

where we have used, in addition to the values below equation 2, $r_0 = 0.25$ cm for the electrode radius and $Sc = \nu/D = 1000$ for the Schmidt number.

The factor $(a\nu/3D)^{2/3}$ in equation 14 or 15 accounts for the fact that the electrical force is applied only within the diffusion layer, which is much thinner than the hydrodynamic boundary layer. The coefficient in equation 15 suggests that the neglected electrical force is only 0.6 percent as large as the terms which were retained in equation 10 when it was solved in section 96 (the retained term $G^2$ being equal to 1 at $\zeta = 0$). The factor involving $(c_\infty - c_0)/c$ in equation 14 indicates that the electrical effect becomes relatively more important near the limiting current, since $c$ then becomes zero at the electrode. (This remark does not, of course, apply when a supporting electrolyte is present.)

## 98. Turbulent flow

Turbulent flow is characterized by rapid and random fluctuations of velocity, pressure, and concentration about their average values. One usually is interested in these fluctuations only in a statistical sense. Consequently, a first step in the study of turbulent flow usually involves an average of the equations presumed to describe the flow. This yields differential equations for certain average quantities, but with the involvement of higher order averages. This procedure thus does not lead to any straightforward means of calculating any average quantities. The problem has a strong analogue in the kinetic theory of gases, where one is not interested in the details of the random motion of the molecules, but only in certain average, measurable quantities.

There are many situations for which a simple, laminar solution of the equation of motion 94-4 can be found, but the actual flow is observed to be turbulent. This has led people to investigate the stability of the laminar flow; if the flow is disturbed by an infinitesimal amount, will the disturbance grow in time or distance or will the disturbance die away and leave the laminar flow? This analysis usually proceeds by linearizing the problem about the basic laminar solution. Sometimes the results agree with experimentally observed conditions of transition to turbulence or a more complex laminar flow, as in the case of Taylor vortices in the flow between rotating cylinders (see section 4); but sometimes there is a considerable discrepancy, as in the case of Poiseuille flow in a pipe.

Mean values in turbulent flow can be defined by a time average, for example,

$$\bar{v}_z = \frac{1}{t_0} \int_t^{t+t_0} v_z \, dt. \tag{98-1}$$

The time $t_0$ over which the average is taken should be long compared to the period of the fluctuations, which might be estimated as 0.01 sec.

In laminar flow, the stress is given by Newton's law of viscosity, equation 94-1. However, in turbulent flow there is an additional mechanism of momentum transfer. The random fluctuations of velocity tend to carry momentum toward regions of lower momentum. Thus, the total mean stress or momentum flux is the sum of a viscous stress and a turbulent momentum flux:

$$\bar{\tau} = \bar{\tau}^{(l)} + \bar{\tau}^{(t)}, \tag{98-2}$$

where the viscous momentum flux $\bar{\tau}^{(l)}$ is given by the time average of equation 94-1 and the turbulent momentum flux $\bar{\tau}^{(t)}$ will be derived later in this section.

Far from a solid wall, momentum transfer by the turbulent mechanism predominates. However, near a solid wall the turbulent fluctuations are damped, and viscous momentum transfer predominates, so that the shear stress at the wall is still given by

$$\tau_0 = -\mu \frac{\partial \bar{v}_z}{\partial r}\bigg|_{r=R} \tag{98-3}$$

for flow in a pipe of radius $R$. It seems reasonable that the turbulent fluctuations should be damped near the wall since the fluid cannot penetrate the wall.

The origin of the turbulent momentum flux is revealed by taking the time average of the equation of motion 93-4

$$\frac{\partial \rho \mathbf{v}}{\partial t} = -\nabla \cdot (\rho \mathbf{v} \mathbf{v}) - \nabla p - \nabla \cdot \boldsymbol{\tau}^{(l)} + \rho \mathbf{g}. \tag{98-4}$$

Here $\boldsymbol{\tau}^{(l)}$ denotes the same stress tensor which had previously been called $\boldsymbol{\tau}$ and is given by equation 94-1 for Newtonian fluids.

The deviation of a flow quantity from its time average is defined as follows for the velocity and pressure:

$$\left.\begin{aligned} \mathbf{v} &= \bar{\mathbf{v}} + \mathbf{v}'. \\ p &= \bar{p} + p'. \end{aligned}\right\} \tag{98-5}$$

We call $\mathbf{v}'$ the velocity fluctuation or the fluctuating part of the velocity. Several rules of time averaging follow simply from the definition 1. The time average of a sum is equal to the sum of the time averages:

$$\overline{A + B} = \bar{A} + \bar{B}.$$

The time average of a derivative is equal to the derivative of the time average: $\overline{dA/dx} = d\bar{A}/dx$. In general, the time average of a nonlinear term will give more than one term. For example, $\overline{AB} = \bar{A}\bar{B} + \overline{A'B'}$. Of course, the time average of a fluctuation is zero, $\bar{A}' = 0$.

In this discussion, the fluid is assumed to have constant properties, $\rho$, $\mu$, etc., since, even with this assumption, the turbulent-flow problem remains intractable and since incompressible fluids do exhibit turbulent flow. In fact, a compressible, laminar boundary layer may be more stable than an incompressible one. With this assumption, the time average of the equation of motion 4 yields

$$\frac{\partial \rho \bar{\mathbf{v}}}{\partial t} = -\nabla \cdot (\rho \bar{\mathbf{v}} \bar{\mathbf{v}}) - \nabla \bar{p} - \nabla \cdot (\bar{\boldsymbol{\tau}}^{(l)} + \rho \overline{\mathbf{v}' \mathbf{v}'}) + \rho \mathbf{g}. \tag{98-6}$$

The time-averaged continuity equation 93-3 is

$$\nabla \cdot \bar{\mathbf{v}} = 0. \tag{98-7}$$

The mean viscous stress is given by the time average of equation 94-1:

$$\bar{\boldsymbol{\tau}}^{(l)} = -\mu [\nabla \bar{\mathbf{v}} + (\nabla \bar{\mathbf{v}})^*]. \tag{98-8}$$

These equations are the same as the equations before averaging, except for the appearance of the term $-\nabla \cdot (\rho \overline{\mathbf{v}' \mathbf{v}'})$ in the equation of motion 6. If we identify the turbulent momentum flux as

$$\bar{\boldsymbol{\tau}}^{(t)} = \rho \overline{\mathbf{v}' \mathbf{v}'} \tag{98-9}$$

and write the total mean stress according to equation 2, then the equation of motion becomes

$$\frac{\partial \rho \bar{\mathbf{v}}}{\partial t} = -\nabla \cdot (\rho \bar{\mathbf{v}} \bar{\mathbf{v}}) - \nabla \bar{p} - \nabla \cdot \bar{\boldsymbol{\tau}} + \rho \mathbf{g} \qquad (98\text{-}10)$$

and bears a strong resemblance to the equation before averaging.

These maneuvers illustrate the origin of the turbulent momentum flux or so-called *Reynolds stress*, given by equation 9. The turbulent mechanism of momentum transfer is somewhat similar to the molecular mechanism in gases; one is due to random motion of molecules, and the other is due to random motion of larger, coherent aggregations of molecules.

We observe that the averaging process provides no reliable route to the prediction of the Reynolds stress. In the absence of a fundamental theory, many people have written empirical expressions for $\bar{\boldsymbol{\tau}}^{(t)}$ with various degrees of success. It should, perhaps, be emphasized that there is no simple relationship between turbulent stress and velocity derivatives, as there is for the viscous stress in a Newtonian fluid, where $\mu$ is a state property depending only on temperature, pressure, and composition.

Many practical problems of turbulence involve the region near a solid wall, since this is, in a sense, the origin of the turbulence and because it is in this region that we want to calculate shear stresses and rates of mass transfer. Experimental data have been studied extensively in order to draw some generalization about the behavior near the wall of the turbulent transport terms, these being the higher-order averages, such as the Reynolds stress, resulting from the averaging of the equations of motion and convective diffusion. This generalization takes the form of a universal law of velocity distribution near the wall, and the results can also be expressed in terms of the eddy viscosity and the eddy kinematic viscosity—coefficients relating the turbulent transport terms to gradients of velocity. These coefficients are strong functions of the distance from the wall and, thus, are not fundamental fluid properties. This type of information is frequently deduced from studies of fully developed pipe flow or certain simple boundary layers.

In studying turbulent flow near the wall, it is found that a correlation called the *universal velocity profile* results if the mean tangential velocity is plotted against the distance from the wall as shown in figure 98-1. This describes fully developed turbulent flow near a smooth wall and applies both to pipe flow and to turbulent boundary layers. The information is correlated by means of the shear stress $\tau_0$ at the wall:

$$v^+ = \frac{\bar{v}_x}{v_*}, \qquad y^+ = \frac{y v_*}{\nu}, \qquad v_* = \sqrt{\frac{\tau_0}{\rho}}. \qquad (98\text{-}11)$$

Note that away from the wall the mean velocity depends linearly on the logarithm of the distance, while near the wall it increases linearly with the distance. The essential features of the curve are represented by the rough

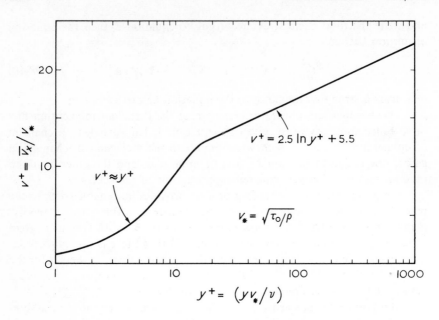

**Figure 98-1.** Universal velocity profile for fully developed turbulent flow.

approximations

$$v^+ \approx y^+ \quad \text{for} \quad y^+ < 20 \tag{98-12}$$

and $\qquad v^+ \approx 2.5 \ln y^+ + 5.5 \quad \text{for} \quad y^+ > 20.$ (98-13)

In the logarithmic region,

$$\bar{v}_x = 2.5\sqrt{\frac{\tau_0}{\rho}} \ln y + \left(2.5 \ln \frac{\sqrt{\tau_0/\rho}}{\nu} + 5.5\right)\sqrt{\frac{\tau_0}{\rho}}. \tag{98-14}$$

Here, the term with the $y$-dependence of the velocity profile is independent of the viscosity; the viscosity of the fluid enters only into the additive constant.

From the data summarized in figure 1, it should be apparent that the Reynolds stress depends strongly on the distance from the wall. A common way to express this is to introduce an *eddy viscosity* $\mu^{(t)}$ by the relation

$$\bar{\tau}_{xy}^{(t)} = -\mu^{(t)} \frac{\partial \bar{v}_x}{\partial y}. \tag{98-15}$$

The empirical results for $\bar{\tau}_{xy}^{(t)}$ are then expressed in terms of the eddy viscosity. Since the turbulent shear flow near a wall should not be expected to be isotropic, other components of the Reynolds stress probably require different values of the eddy viscosity, even at the same distance from the wall.

The universal velocity profile of figure 1 probably applies only to a region near the wall where the shear stress is essentially constant, but not to the region near the center of a pipe, say, where the stress goes to zero. If we assume that the shear stress is constant over the region where the universal velocity profile is applicable, then we can obtain an idea of the variation of $\mu^{(t)}$ with distance from the wall.

$$\bar{\tau}_{xy} \approx -\tau_0 = -(\mu + \mu^{(t)})\frac{\partial \bar{v}_x}{\partial y} = -\frac{\mu + \mu^{(t)}}{\mu}\tau_0\frac{dv^+}{dy^+}$$

or
$$1 = \left[1 + \frac{\mu^{(t)}}{\mu}\right]\frac{dv^+}{dy^+}. \tag{98-16}$$

This result shows that the ratio $\mu^{(t)}/\mu$ should also be a universal function of the wall variable $y^+$. Figure 98-2 is obtained by differentiation of the universal

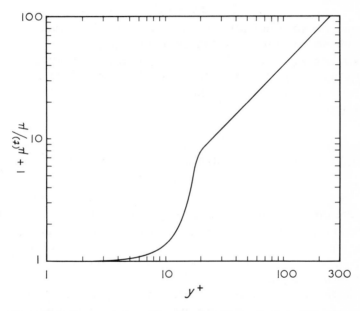

**Figure 98-2.** Representation of the eddy viscosity as a "universal" function of the distance from the wall.

velocity profile of figure 1. It is not possible to obtain accurate values for $\mu^{(t)}$ near the wall by this method because in this region $\mu^{(t)} \ll \mu$. However, this problem should not be of immediate concern, since it is the sum $\mu + \mu^{(t)}$ which enters into problems of fluid mechanics.

The universal velocity profile is one of the few generalizations possible in turbulent shear flow, and it is widely applied in the analysis of problems for which experimental observations are not available. Thus it is the basis of a semi-empirical theory of turbulent flow which can be applied to the hydrodynamics of turbulent boundary layers, mass transfer in turbulent boundary layers, and the beginning of a mass-transfer section in fully developed pipe flow.

## 99. Mass transfer in turbulent flow

For the consideration of mass transfer in turbulent flow, one can average the equation of convective diffusion 73-2. The nonlinear term $\mathbf{v} \cdot \nabla c_i$ again yields a new term in the averaged equation:

$$\frac{\partial \bar{c}_i}{\partial t} + \bar{\mathbf{v}} \cdot \nabla \bar{c}_i = D_i \, \nabla^2 \bar{c}_i - \nabla \cdot \mathbf{J}_i^{(t)}, \qquad (99\text{-}1)$$

where

$$\mathbf{J}_i^{(t)} = \overline{\mathbf{v}' c_i'} \qquad (99\text{-}2)$$

represents the mean turbulent mass flux due to the fluctuations of the concentration and the velocity about their mean values.

Next one might address himself to the problem of estimating mass-transfer rates in turbulent flow, but with the use of as little additional information as possible. One usually starts with the assumption that the eddy diffusivity $D^{(t)}$, defined by the relation

$$J_{iy}^{(t)} = \overline{c_i' v_y'} = -D^{(t)} \frac{\partial \bar{c}_i}{\partial y}, \qquad (99\text{-}3)$$

is related to or equal to the eddy kinematic viscosity

$$D^{(t)} = \nu^{(t)} = \frac{\mu^{(t)}}{\rho}. \qquad (99\text{-}4)$$

This is based on the idea that transfer of momentum and mass is similar, whether it is by a molecular or by a turbulent mechanism. Thus, figure 98-2 can be used to get information about the variation of the eddy diffusivity in fully developed, turbulent flow near a wall. As stated earlier, $\nu^{(t)}$ is much smaller than $\nu$ very close to the wall, and figure 98-2 gives no information about it. But for mass transfer at large Schmidt numbers, it is necessary to know $D^{(t)}$ closer to the wall. Thus, even if $\nu^{(t)} = D^{(t)}$, we can have at some distance $D^{(t)} \gg D_i$ even where $\nu^{(t)} \ll \nu$ if $Sc = \nu/D_i$ is large.

Thus, much of our information about $D^{(t)}$ and possibly $\nu^{(t)}$ near the wall comes from mass-transfer experiments. Actually it probably results more from fitting average mass-transfer rates at the wall than from examination in detail of actual concentration profiles. If, for mass transfer in a pipe, we measure the rate by means of the Stanton number

$$St = \frac{D_i}{\langle \bar{v}_z \rangle \Delta c_i} \left. \frac{\partial \bar{c}_i}{\partial r} \right|_{r=R}, \qquad (99\text{-}5)$$

then the dependence of the Stanton number on the Schmidt number at large Schmidt numbers is determined by the variation of the eddy diffusivity close to the wall as follows:

| $D^{(t)}$ near the wall | $St$ for large $Sc$ |
|---|---|
| $D^{(t)} \propto y^2$ | $St \propto Sc^{-1/2}$ |
| $y^3$ | $Sc^{-2/3}$ |
| $y^4$ | $Sc^{-3/4}$ |

It is easy to show that $v^{(t)}$ must go to zero as $y^3$ or a higher power of $y$ but cannot vary as $y^2$. If $v'_x$ and $v'_y$ are expanded in power series near the wall, then $v'_x$ is proportional to $y$ and, by the continuity equation, $v'_y$ is proportional to $y^2$. Hence $\bar{\tau}^{(t)}_{xy} = \overline{\rho v'_x v'_y}$ is proportional to $y^3$, and the same must be true of $v^{(t)}$. A controversy between $y^3$ and $y^4$ remains. Levich[7] had $D^{(t)}$ proportional to $y^3$ but later[8] changed this to $y^4$.

Sherwood[9] has reviewed the attempts at describing $D^{(t)}$. The goal of such work is to determine how the eddy diffusivity depends on distance from the wall, that is, to relate $D^{(t)}/v$ to $y^+$, the dimensionless distance from the wall. This is based on the universal velocity profile and on information gleaned from mass-transfer experiments. Following Wasan et al.,[10] we write

$$v^+ = y^+ - A_1(y^+)^4 + A_2(y^+)^5 \quad \text{for} \quad y^+ \leq 20,$$
$$v^+ = 2.5 \ln y^+ + 5.5 \quad \text{for} \quad y^+ \geq 20. \tag{99-6}$$

The constants $A_1$ and $A_2$ are selected so that $v^+$ and its derivative are continuous at $y^+ = 20$:

$$A_1 = 1.0972 \times 10^{-4} \quad \text{and} \quad A_2 = 3.295 \times 10^{-6}. \tag{99-7}$$

The corresponding expressions for the eddy diffusivity are

$$\left.\begin{aligned}\frac{D^{(t)}}{v} &= \frac{4A_1(y^+)^3 - 5A_2(y^+)^4}{1 - 4A_1(y^+)^3 + 5A_2(y^+)^4} \quad \text{for} \quad y^+ \leq 20, \\ \frac{D^{(t)}}{v} &= \frac{y^+}{2.5} - 1 \quad \text{for} \quad y^+ \geq 20 \end{aligned}\right\} \tag{99-8}$$

The concept of the universal velocity profile and the variation of eddy diffusivity with distance from the wall as given in figure 99-1 form the basis of a semi-empirical theory widely used to calculate mass-transfer rates in turbulent boundary layers, near the beginning of a mass-transfer section in a pipe, and for similar problems (see, for example, reference 11).

[7] B. Levich, "The Theory of Concentration Polarization," Acta Physicochimica U.R.S.S., 17 (1942), 257–307.

[8] Ibid., 19, 117–132 (1944).

[9] T. K. Sherwood, "Mass, Heat, and Momentum Transfer between Phases," Chemical Engineering Progress Symposium Series, no 25, vol. 55 (1959), pp. 71–85.

[10] D. T. Wasan, C. L. Tien, and C. R. Wilke, "Theoretical Correlation of Velocity and Eddy Viscosity for Flow Close to a Pipe Wall," A.I.Ch.E. Journal, 9 (1963), 567–568.

[11] B. T. Ellison and I. Cornet, "Mass Transfer to a Rotating Disk," Journal of the Electrochemical Society, 118 (1971), 68–72.

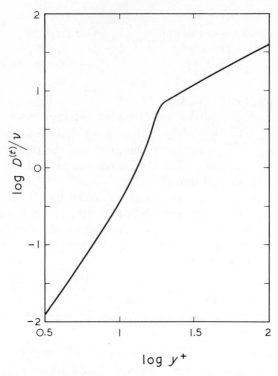

**Figure 99-1.** Variation of eddy diffusivity near a wall for fully-developed turbulent flow.

## PROBLEM

**1.** Show that equation 93-6 implies that

$$\nabla \times \mathbf{g} = 0.$$

This condition is satisfied by most gravitational fields.

## NOTATION

| | |
|---|---|
| $a$ | constant $= 0.51023$ |
| $A$ | constant |
| $\mathbf{B}$ | magnetic induction, weber/m$^2$ |
| $c$ | molar concentration of a single electrolyte, mole/cm$^3$ |
| $c_i$ | concentration of species $i$, mole/cm$^3$ |
| $D$ | diffusion coefficient of electrolyte, cm$^2$/sec |
| $D_i$ | diffusion coefficient of species $i$, cm$^2$/sec |
| $D^{(t)}$ | eddy diffusivity, cm$^2$/sec |

| | |
|---|---|
| $\mathbf{E}$ | electric field, V/cm |
| $f$ | function for radial velocity, $\sec^{-1}$ |
| $\mathbf{f}$ | force per unit area, dyne/cm$^2$ |
| $F$ | Faraday's constant, 96,487 C/equiv |
| $F$ | dimensionless radial velocity |
| $g$ | function for velocity in $\theta$-direction, $\sec^{-1}$ |
| $\mathbf{g}$ | acceleration of gravity, cm/sec$^2$ |
| $G$ | dimensionless velocity in $\theta$-direction |
| $h$ | axial velocity, cm/sec |
| $H$ | dimensionless axial velocity |
| $i$ | uniform current density on disk electrode, A/cm$^2$ |
| $\mathbf{i}$ | current density, A/cm$^2$ |
| $\mathbf{I}$ | unit tensor |
| $\mathbf{J}_i^{(t)}$ | mean turbulent mass flux of species $i$, mole/cm$^2$-sec |
| $M_i$ | molecular weight of species $i$, g/mole |
| $\mathbf{n}$ | unit vector normal to surface |
| $p$ | pressure, dyne/cm$^2$ |
| $P$ | dimensionless dynamic pressure |
| $\mathcal{P}$ | dynamic pressure, dyne/cm$^2$ |
| $r$ | radial distance, cm |
| $r_0$ | radius of disk electrode, cm |
| $r_1, r_2$ | principal radii of curvature of surface, cm |
| $R$ | universal gas constant, 8.3143 J/mole-deg |
| $Re$ | Reynolds number |
| $Sc$ | Schmidt number |
| $St$ | Stanton number |
| $t$ | time, sec |
| $t_0$ | time over which quantities are averaged, sec |
| $t_i$ | transference number of species $i$ |
| $T$ | absolute temperature, deg K |
| $\mathbf{v}$ | mass-average velocity, cm/sec |
| $\mathbf{v}_i$ | velocity of species $i$, cm/sec |
| $\langle v_z \rangle$ | average velocity in a pipe, cm/sec |
| $x$ | tangential distance, cm |
| $y$ | distance from surface, cm |
| $z$ | axial distance, cm |
| $z_i$ | charge number of species $i$ |
| $\Gamma(4/3)$ | $= 0.89298$, the gamma function of 4/3 |
| $\epsilon$ | permittivity, farad/cm |
| $\epsilon_0$ | permittivity of free space, $8.8542 \times 10^{-14}$ farad/cm |
| $\zeta$ | dimensionless axial distance from disk |
| $\theta$ | angle in cylindrical coordinates, radian |
| $\kappa$ | conductivity, mho/cm |

| | |
|---|---|
| $\Lambda$ | equivalent conductance of binary electrolyte, mho-cm²/equiv |
| $\mu$ | viscosity, g/cm-sec |
| $\mu^{(t)}$ | eddy viscosity, g/cm-sec |
| $\nu$ | kinematic viscosity, cm²/sec |
| $\nu^{(t)}$ | eddy kinematic viscosity, cm²/sec |
| $\nu_+, \nu_-$ | numbers of cations and anions into which a molecule of electrolyte dissociates |
| $\zeta$ | dimensionless axial distance from disk |
| $\rho$ | density, g/cm³ |
| $\rho_e$ | electric charge density, C/cm³ |
| $\sigma$ | interfacial tension, dyne/cm |
| $\tau$ | stress, dyne/cm² |
| $\Phi$ | electric potential, V |
| $\Omega$ | rotation speed of disk, radian/sec |

subscripts

| | |
|---|---|
| 0 | at electrode surface |
| $\infty$ | in bulk solution |
| $s$ | surface |

superscripts

| | |
|---|---|
| ' | fluctuation |
| $(l)$ | viscous |
| $(t)$ | turbulent |
| $+$ | related to universal velocity profile |
| overbar | time average |

*Part* **D**

# Current Distribution and Mass Transfer in Electrochemical Systems

Electrochemical systems find widespread technical application. Industrial electrolytic processes include electroplating and refining, electropolishing and machining, and the electrochemical production of chlorine, caustic soda, aluminum, and other products. Energy conversion in fuel cells and in primary and secondary batteries has received increasing attention. Electrochemical corrosion should not be neglected, and some systems for desalting water involve electrochemical processes. Electrochemical methods are used for qualitative and quantitative analysis. Idealized electrochemical systems are also of interest for studies of mass transfer processes and the mechanisms of electrode reactions and for the determination of basic data on transport properties.

Engineering design procedures for electrochemical systems have not been developed as thoroughly as for mass-transfer operations such as distillation. Nevertheless, the fundamental laws governing electrochemical systems are known and have been developed in parts A, B, and C of this book. The purpose of part D is to review the analysis of certain electrochemical systems in relation to these fundamental laws. To a greater or lesser extent, one is concerned with fluid flow patterns, ohmic potential drop in solutions, restricted rates of mass transfer, and the kinetics of electrode reactions. The situation is complicated as well by the variety of specific chemical systems. These examples provide some of the main tools of the electrochemical engineer.

Application of the fundamental laws has followed two main courses. There are systems where the ohmic potential drop can be neglected. The current distribution is then determined by the same principles that apply to heat transfer and nonelectrolytic mass transfer. This usually involves systems operated at the limiting current with an excess of supporting electrolyte, because, below the limiting current, neglecting the ohmic potential drop is usually not justified and because the presence of the supporting electrolyte allows the effect of ionic migration in the diffusion layer to be ignored. Furthermore, the concentration of the limiting reactant is zero at the electrode surface, and the treatment becomes simplified. Let us call these *convective-transport problems*, treated in chapter 17.

At currents much below the limiting current, it is possible to neglect concentration variations near the electrodes. The current distribution is then determined by the ohmic potential drop in the solution and by electrode overpotentials. Mathematically, this means that the potential satisfies Laplace's equation; and many results of potential theory, developed in electrostatics, the flow of inviscid fluids, and steady heat conduction in solids, are directly applicable. Let us call these *potential-theory problems*, treated in chapter 18. The electrode kinetics provide boundary conditions that are usually different from those encountered in other applications of potential theory.

Existing work on current distribution and mass transfer in electrochemical systems is reviewed here, with emphasis being placed on how each contribution is related to these limiting cases of convective-transport problems and applications of potential theory. This framework can be compared with Wagner's discussion[1] of the scope of electrochemical engineering. Much work either fits into the extreme cases or takes into account phenomena neglected in the extreme cases.

We also discuss problems which do not fall into either of these two classes. Some problems can be regarded as an extension of the convective-transport problems. At the limiting current, the ohmic potential drop in the bulk of the solution may still be negligible, but the electric field in the diffusion layer near electrodes may lead to an enhancement of the limiting current. The current density is then distributed along the electrode in the same manner as when migration is neglected, but the magnitude of the current density at all points is increased or diminished by a constant factor which depends upon the bulk composition of the solution. (See chapter 19.) Free convection in a supported electrolytic solution also involves this migration effect. In addition, the nonuniform concentration of the added electrolyte affects the

---

[1] Carl Wagner, "The Scope of Electrochemical Engineering," Charles W. Tobias, ed., *Advances in Electrochemistry and Electrochemical Engineering*, 2 (1962), 1–14.

density distribution and hence the velocity profiles in the system. This effect, which does not disappear with a large excess of supporting electrolyte, is treated in the last section of chapter 19.

At currents below, but at an appreciable fraction of, the limiting current, diffusion and convective transport are essential, but neither concentration variations near the electrode nor the ohmic potential drop in the bulk solution can generally be neglected. These problems are complex because all the factors are involved at once. They span the limiting cases of convective-transport problems and applications of potential theory and are treated in chapter 21. Prior to this, the concept of the concentration overpotential is further developed in chapter 20.

In technical electrochemical systems, the ohmic potential drop is of great importance, and potential-theory problems find applications here. Nevertheless, concentration variations near electrodes frequently provide limitations on reaction rates and current efficiencies in industrial operations. In view of the complexity of simultaneously treating concentration variations and ohmic potential drop, qualitative or semiquantitative application of these concepts may have to suffice for some time. Thus, only a limited number of systems can be discussed in chapter 21.

Many electrodes found in fuel cells and primary and secondary batteries are porous in order to provide an extensive surface area for electrochemical reactions. In such electrodes, convection may not be present, but it is usually necessary to consider the ohmic potential drop, concentration variations, and electrode kinetics. Most treatments adopt a macroscopic model that does not take account of the detailed, random geometry of the porous structure. Results of potential theory are then not applicable since Laplace's equation does not hold. Porous-electrode problems thus do not fall within the framework of convective-transport problems and applications of potential theory and are not treated here.

Earlier reviews of current distribution and mass transfer in electrochemical systems are given in references 2, 3, 4, and 5. Reference 6 contains more

[2]John Newman, "Engineering Design of Electrochemical Systems," *Industrial and Engineering Chemistry*, *60* (no. 4) (April, 1968), 12–27.

[3]C. W. Tobias, M. Eisenberg, and C. R. Wilke, "Diffusion and Convection in Electrolysis—A Theoretical Review," *Journal of the Electrochemical Society*, *99* (1952), 359C–365C.

[4]Wolf Vielstich, "Der Zusammenhang zwischen Nernstscher Diffusionsschicht und Prandtlscher Strömungsgrenzschicht," *Zeitschrift für Elektrochemie*, *57* (1953), 646–655.

[5]N. Ibl, "Probleme des Stofftransportes in der angewandten Elektrochemie," *Chemie- Ingenieur-Technik*, *35* (1963), 353–361.

[6]John Newman, "The Fundamental Principles of Current Distribution and Mass Transport in Electrochemical Cells," Allen J. Bard, ed., *Electroanalytical Chemistry* (New York: Marcel Dekker, Inc., to be published).

mathematical development than will be included here. Some of the work on porous electrodes, which will not be reviewed here, can be found in references 7, 8, 9, 10, and 11.

[7]John S. Newman and Charles W. Tobias, "Theoretical Analysis of Current Distribution in Porous Electrodes," *Journal of the Electrochemical Society, 109* (1962), 1183–1191.

[8]Edward A. Grens II, "Analysis of Operation of Porous Gas Electrodes with Two Superimposed Scales of Pore Structure," *Industrial and Engineering Chemistry Fundamentals, 5* (1966), 542–547.

[9]Robert de Levie, "Electrochemical Response of Porous and Rough Electrodes," *Advances in Electrochemistry and Electrochemical Engineering, 6* (1967), 329–397.

[10]E. A. Grens, II, "On the Assumptions Underlying Theoretical Models for Flooded Porous Electrodes," *Electrochimica Acta, 15* (1970), 1047–1057.

[11]A. M. Johnson and John Newman, "Desalting by Means of Porous Carbon Electrodes," *Journal of the Electrochemical Society, 118* (1971), 510–517.

# Fundamental Equations  **16**

## 100. Transport in dilute solutions

The laws of transport in dilute solutions have been known for many years and have been developed in chapter 11. The four principal equations in dilute-solution theory are presented in section 69. The flux of a solute species is due to migration in an electric field, diffusion in a concentration gradient, and convection with the fluid velocity.

$$\mathbf{N}_i = -z_i u_i F c_i \, \nabla \Phi - D_i \, \nabla c_i + \mathbf{v} c_i. \tag{100-1}$$

A material balance for a small volume element leads to the differential conservation law:

$$\frac{\partial c_i}{\partial t} = -\nabla \cdot \mathbf{N}_i + R_i. \tag{100-2}$$

Since reactions are frequently restricted to the surfaces of electrodes, the bulk reaction term $R_i$ is often zero in electrochemical systems. To a very good approximation, the solution is electrically neutral,

$$\sum_i z_i c_i = 0, \tag{100-3}$$

except in the diffuse part of the double layer very close to an interface. The current density in an electrolytic solution is due to the motion of charged species:

$$\mathbf{i} = F \sum_i z_i \mathbf{N}_i. \tag{100-4}$$

These laws provide the basis for the analysis of electrochemical systems.

301

The flux relation, equation 1, defines transport coefficients—the mobility $u_i$ and the diffusion coefficient $D_i$ of an ion in a dilute solution. The dilute-solution theory has been applied fruitfully to many electrochemical systems. We shall furthermore assume that the physical properties are constant. The dominant factors are then revealed in the simplest manner, and the results have the widest range of applicability.

Many electrochemical systems involve flow of the electrolytic solution. For a fluid of constant density $\rho$ and viscosity $\mu$, the fluid velocity is to be determined from the Navier-Stokes equation (see chapter 15)

$$\rho \left( \frac{\partial \mathbf{v}}{\partial t} + \mathbf{v} \cdot \nabla \mathbf{v} \right) = - \nabla p + \mu \nabla^2 \mathbf{v} + \rho \mathbf{g} \tag{100-5}$$

and the continuity equation

$$\nabla \cdot \mathbf{v} = 0. \tag{100-6}$$

## 101. Electrode kinetics

The differential equations describing the electrolytic solution require boundary conditions for the behavior of an electrochemical system to be predicted. The most complex of these concerns the kinetics of electrode reactions, treated in chapter 8. A single electrode reaction can be written in symbolic form as

$$\sum_i s_i M_i^{z_i} \longrightarrow ne^-. \tag{101-1}$$

Then the normal component of the flux of a species is related to the normal component of the current density, that which contributes to the external current to the electrode.

$$N_{in} = - \frac{s_i}{nF} i_n. \tag{101-2}$$

This equation is restricted not only to a single electrode reaction but also to the absence of an appreciable charging of the double layer, a process which does not follow Faraday's law.

Next, one needs an equation describing the kinetics of the electrode reaction, that is, an equation which relates the normal component of the current density to the surface overpotential at that point and the composition of the solution just outside the diffuse part of the double layer. The motivation of the electrochemical engineer in this regard is basically different from that of an electrochemist. The object is to predict the behavior of a complex electrochemical system rather than to elucidate the mechanism of an electrode reaction. To accomplish this objective, one needs an equation which describes accurately how the interface behaves during the passage of current, and, for this purpose, the interface includes the diffuse part of the double layer.

The surface overpotential $\eta_s$ can be defined as the potential of the working electrode relative to a reference electrode of the same kind located just outside the double layer. Then one seeks a kinetic expression of the form

$$i_n = f(\eta_s, c_i),\qquad (101\text{-}3)$$

where charging of the double layer is again ignored. The concentrations $c_i$ here refer to the point just outside the double layer. Such an expression thus describes the interface since $i_n$, $\eta_s$, and $c_i$ are all local quantities. In particular, the concentration variation between the interface and the bulk solution, and the ohmic potential drop in the solution have only an incidental bearing on events at the interface. At the same time, no attempt is made to give a separate account of the diffuse part of the double layer.

The function $f$ in equation 3 is, in general, complicated. However, there is ample evidence that there is a large class of electrode reactions for which the current density depends exponentially on the surface overpotential in the following form:

$$i_n = i_0\left[\exp\left(\frac{\alpha_a F}{RT}\eta_s\right) - \exp\left(-\frac{\alpha_c F}{RT}\eta_s\right)\right],\qquad (101\text{-}4)$$

where $i_0$ is the exchange current density and depends on the concentrations $c_i$. This latter dependence can frequently be expressed as a product of powers of the concentrations (see section 57). In this equation, $i_n$ and $\eta_s$ are positive for anodic processes, negative for cathodic processes. Both $\alpha_a$ and $\alpha_c$ are kinetic parameters and must be determined to agree with experimental data. A rereading of section 8 may be useful at this point.

## NOTATION

| | |
|---|---|
| $c_i$ | concentration of species $i$, mole/cm³ |
| $D_i$ | diffusion coefficient of species $i$, cm²/sec |
| $e^-$ | symbol for the electron |
| $f$ | function in expression of electrode kinetics |
| $F$ | Faraday's constant, 96,487 C/equiv |
| $g$ | acceleration of gravity, cm/sec² |
| $i$ | current density, A/cm² |
| $i_0$ | exchange current density, A/cm² |
| $M_i$ | symbol for the chemical formula of species $i$ |
| $n$ | number of electrons transferred in electrode reaction |
| $\mathbf{N}_i$ | flux of species $i$, mole/cm²-sec |
| $p$ | pressure, dyne/cm² |
| $R$ | universal gas constant, 8.3143 J/mole-deg |
| $R_i$ | rate of homogeneous production of species $i$, mole/cm³-sec |
| $s_i$ | stoichiometric coefficient of species $i$ in electrode reaction |
| $t$ | time, sec |
| $T$ | absolute temperature, deg K |

$u_i$        mobility of species $i$, cm²-mole/J-sec
$\mathbf{v}$         fluid velocity, cm/sec
$z_i$         charge number of species $i$
$\alpha_a, \alpha_c$      transfer coefficients
$\eta_s$         surface overpotential, V
$\mu$         viscosity, g/cm-sec
$\rho$         density, g/cm³
$\Phi$         electric potential, V

# Convective-transport Problems

**17**

## 102. Simplifications for convective transport

For the reaction of minor ionic species in a solution containing excess supporting electrolyte, it should be permissible to neglect the contribution of ionic migration to the flux of the reacting ions (see section 73), so that equation 100-1 becomes

$$\mathbf{N}_i = -D_i \nabla c_i + \mathbf{v} c_i,$$

(102-1)

and substitution into equation 100-2 yields

$$\frac{\partial c_i}{\partial t} + \mathbf{v} \cdot \nabla c_i = D_i \nabla^2 c_i.$$

(102-2)

This may be called the equation of *convective diffusion*. A similar equation applies to convective heat transfer and convective mass transfer in nonelectrolytic solutions. Since these fields have been studied in detail, it is possible to apply many results to electrochemical systems which obey equation 2. At the same time, electrochemical systems sometimes provide the most convenient experimental means of testing these results or of arriving at new results for systems too complex to analyze.

Essential to the understanding of convective-transport problems is the concept of the diffusion layer. Frequently, due to the small value of the diffusion coefficient, the concentrations differ significantly from their bulk values only in a thin region near the surface of an electrode. In this region,

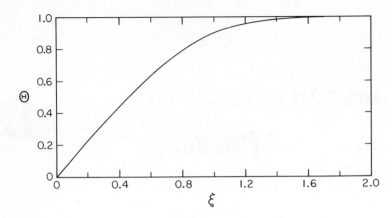

**Figure 102-1.** Concentration profile in the diffusion layer.

the velocity is small, and diffusion is of primary importance to the transport process. The thinness of this region permits a simplification in the analysis, but it is erroneous to treat the diffusion layer as a stagnant region. Figure 102-1 shows the concentration profile in the diffusion layer, with the electrode surface at the left. Far from the suface, convective transport dominates, while at the surface itself, there is only diffusion.

The systems typically studied in heat and mass transfer involve laminar and turbulent flow with various geometric arrangements. The flow may be due to some more or less well-characterized stirring (forced convection) or may be the result of density differences created in the solution as part of the transfer process (free convection). We shall discuss a few examples, although there is no need to be exhaustive since convective heat and mass transfer are thoroughly treated in many texts and monographs.[1,2,3,4] There are also several reviews of mass transfer in electrochemical systems.[5,6,7] The examples selected are primarily those which have been studied with electrochemical systems. In addition, certain theoretical results of general

[1]R. Byron Bird, Warren E. Stewart, and Edwin N. Lightfoot, *Transport Phenomena* (New York: John Wiley & Sons, Inc., 1960).

[2]W. M. Kays, *Convective Heat and Mass Transfer* (New York: McGraw-Hill Book Company, 1966).

[3]Veniamin G. Levich, *Physicochemical Hydrodynamics* (Englewood Cliffs, N.J.: Prentice-Hall, Inc., 1962).

[4]Hermann Schlichting, *Boundary-Layer Theory* (New York: McGraw-Hill Book Company, 1968).

[5]C. W. Tobias, M. Eisenberg, and C. R. Wilke, "Diffusion and Convection in Electrolysis—A Theoretical Review," *Journal of the Electrochemical Society, 99* (1952), 359C–365C.

[6]Wolf Vielstich, "Der Zusammenhang zwischen Nernstscher Diffusionsschicht und Prandtlscher Strömungsgrenzschicht," *Zeitschrift für Elektrochemie, 57* (1953), 646–655.

[7]N. Ibl, "Probleme des Stofftransportes in der angewandten Elektrochemie," *Chemie-Ingenieur-Technik, 35* (1963), 353–361.

validity are included because they are particularly applicable to electrolytic solutions, where the Schmidt numbers are invariably large.

## 103. The rotating disk

Our first example of a convective-transport problem, the rotating-disk electrode, is well-known to electrochemists. Imagine a large, or infinite, disk rotating about its axis in an infinite fluid medium, so that wall and end effects may be ignored. Actually, these edge effects can be neglected for a suitable design of the disk. Thus, we consider the electrode shown in figure 103-1, a disk electrode embedded in a larger insulating plane which also rotates. This system has been reviewed by Riddiford.[8]

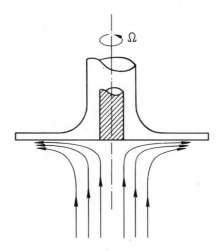

**Figure 103-1.** Rotating disk electrode.

The rotation of the disk provides the stirring of the fluid. The hydrodynamic aspects of the problem were presented in section 96. The pertinent feature of those results is that the velocity normal to the disk, which brings fresh reactant to the surface, depends on $z$ but not on $r$:

$$v_z = \sqrt{\nu\Omega}\, H\!\left(z\sqrt{\frac{\Omega}{\nu}}\right). \qquad (103\text{-}1)$$

Consequently, there is no reason for the concentration to depend on anything besides the normal distance from the disk, and the equation of convective diffusion 102-2 reduces to

$$v_z \frac{dc_i}{dz} = D_i \frac{d^2 c_i}{dz^2}, \qquad (103\text{-}2)$$

[8]A. C. Riddiford, "The Rotating Disk System," *Advances in Electrochemistry and Electrochemical Engineering*, **4** (1966), 47–116.

with boundary conditions

$$c_i = c_0 \quad \text{at} \quad z = 0 \quad \text{and} \quad c_i = c_\infty \quad \text{at} \quad z = \infty. \qquad (103\text{-}3)$$

At the limiting current, $c_0 = 0$. Thus, the fact that the convective velocity bringing fresh reactant to the electrode is the same over the entire surface of the disk has the mathematical advantage of reducing the equation of convective diffusion to an ordinary differential equation and the practical advantage that the reaction rate at the electrode will be everywhere the same, independent of the distance from the axis of rotation.

Levich[9] has analyzed the mass transfer to a rotating disk with the fluid motion described above. The analogous heat-transfer problem was not treated by Wagner[10] until 1948.

Equation 2 is a first-order differential equation for $dc_i/dz$ and can be integrated to give

$$\ln \frac{dc_i}{dz} = \frac{1}{D_i} \int_0^z v_z \, dz + \ln K \qquad (103\text{-}4)$$

or

$$\frac{dc_i}{dz} = K \exp \left( \frac{1}{D_i} \int_0^z v_z \, dz \right). \qquad (103\text{-}5)$$

A second integration gives

$$c_i = c_0 + K \int_0^z \exp \left( \int_0^z \frac{v_z}{D_i} \, dz \right) dz. \qquad (103\text{-}6)$$

The constant $K$ is determined now from boundary condition 3:

$$\frac{c_\infty - c_0}{K} = \int_0^\infty \exp \left( \int_0^z \frac{v_z}{D_i} \, dz \right) dz$$

$$= \sqrt{\frac{\nu}{\Omega}} \int_0^\infty \exp \left[ Sc \int_0^\eta H(\zeta) \, d\zeta \right] d\eta, \qquad (103\text{-}7)$$

where the last expression is obtained with the use of equation 1 and the definition of the Schmidt number $Sc = \nu/D_i$. The solution thus can be expressed as

$$\Theta = \frac{\displaystyle\int_0^z \exp \left( \int_0^z \frac{v_z}{D_i} \, dz \right) dz}{\displaystyle\int_0^\infty \exp \left( \int_0^z \frac{v_z}{D_i} \, dz \right) dz}, \qquad (103\text{-}8)$$

where

$$\Theta = \frac{c_i - c_0}{c_\infty - c_0} \qquad (103\text{-}9)$$

is a dimensionless concentration.

[9]B. Levich, "The Theory of Concentration Polarization," *Acta Physicochimica U.R.S.S.*, *17* (1942), 257–307.

[10]Carl Wagner, "Heat Transfer from a Rotating Disk to Ambient Air," *Journal of Applied Physics*, *19* (1948), 837–839.

The flux from the disk surface is

$$N_{in} = - D_i \frac{dc_i}{dz}\bigg|_{z=0} = - D_i K, \qquad (103\text{-}10)$$

and by means of equation 101-2 the current density can be expressed as

$$\frac{i_n s_i}{nF(c_\infty - c_0)\sqrt{\nu\Omega}} = \frac{1}{Sc} \Theta'(0), \qquad (103\text{-}11)$$

where the prime denotes the derivative with respect to $\zeta$ and

$$\frac{1}{Sc}\Theta'(0) = \frac{1}{Sc \int_0^\infty \exp\left[ Sc \int_0^\eta H(\zeta)\, d\zeta \right] d\eta}. \qquad (103\text{-}12)$$

The dimensionless mass-transfer rate in equation 12 is seen to depend only on the Schmidt number $Sc = \nu/D_i$ and is plotted in figure 103-2 (see

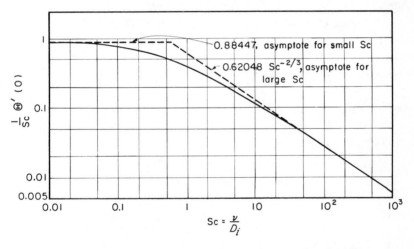

**Figure 103-2.** Dimensionless mass-transfer rates for a rotating disk. (From J. Newman, "Transport Processes in Electrolytic Solutions," in C. W. Tobias, *Advances in Electrochemistry and Electrochemical Engineering*, Vol. 5. Copyright © 1967 by John Wiley & Sons, Inc. Reprinted by permission.)

Sparrow and Gregg[11]). If the mass flux or current density is known, then the ordinate is independent of the diffusion coefficient (see equation 11). Hence, this method of plotting is advantageous for the determination of diffusion coefficients by the rotating-disk method. From the limiting current density, the ordinate can be calculated directly. The Schmidt number can

[11]E. M. Sparrow and J. L. Gregg, "Heat Transfer From a Rotating Disk to Fluids of Any Prandtl Number," *Journal of Heat Transfer, 81C* (1959), 249–251.

then be obtained from the graph without a trial-and-error calculation, and the diffusion coefficient is then given by $D_i = v/Sc$.

The asymptote for large Schmidt numbers was first derived by Levich in 1942. In this case, the diffusion coefficient $D_i$ is very small and the concentration variation occurs very close to the surface of the disk (at small values of $\zeta$ in figure 96-1). Therefore, it is appropriate to use in equation 12 the first term of the velocity profile for small values of $\zeta$ as given in equation 96-13. The behavior for large Schmidt numbers is particularly important for diffusion in liquids, since here the Schmidt number is on the order of 1000. Corrections to this asymptote can be obtained by expansion of the mass-transfer rate for large Schmidt numbers, with the result[12]

$$\frac{1}{Sc}\Theta'(0) = \frac{0.62048Sc^{-2/3}}{1 + 0.2980Sc^{-1/3} + 0.14514Sc^{-2/3} + 0(Sc^{-1})}. \quad (103\text{-}13)$$

This expression adequately represents the curve in figure 2 for $Sc > 100$ (in this region, the maximum error is about 0.1 percent). See also reference 13.

On the other hand, for very low Schmidt numbers, the diffusion layer extends a large distance from the disk, and it is appropriate to use the velocity profile of equation 96-15. At very low Schmidt numbers, equation 12 becomes

$$\frac{1}{Sc}\Theta'(0) = 0.88447e^{-1.611Sc}[1 + 1.961Sc^2 + 0(Sc^3)]. \quad (103\text{-}14)$$

The first term of equation 14 tells us that the maximum flux to the disk for very large diffusion coefficients is completely determined by the rate of convection of material from infinity:

$$N_{i\,\text{max}} = -(c_\infty - c_0)\sqrt{v\Omega}\,H(\infty) = 0.88447(c_\infty - c_0)\sqrt{v\Omega}. \quad (103\text{-}15)$$

Because of the well-defined fluid motion, the rotating disk electrode has been used extensively for the determination of diffusion coefficients and the parameters of electrode kinetics. It can also be used for quantitative analysis (polarography) in electrolytic solutions. The edge effect for the rotating disk is treated in reference 14.

The rotating ring-disk system is popular because active intermediates produced on the disk electrode can be detected with the ring electrode. The amount so detected can be compared with the theoretical collection efficiency[15,16] for the system.

[12]John Newman, "Schmidt Number Correction for the Rotating Disk," *The Journal of Physical Chemistry, 70* (1966), 1327–1328.

[13]D. P. Gregory and A. C. Riddiford, "Transport to the Surface of a Rotating Disc," *Journal of the Chemical Society* (1956), pp. 3756–3764.

[14]William H. Smyrl and John Newman, "Limiting Current on a Rotating Disk with Radial Diffusion," *Journal of the Electrochemical Society, 118* (1971), 1079–1081.

[15]W. J. Albery and S. Bruckenstein, "Ring-Disc Electrodes. Part 2.—Theoretical and Experimental Collection Efficiencies," *Transactions of the Faraday Society, 62* (1966), 1920–1931.

[16]William H. Smyrl and John Newman. "Ring-Disk and Sectioned Disk Electrodes." *Journal of the Electrochemical Society, 119* (1972), 212–219.

## 104.  The Graetz problem

An important problem which received early analytic treatment is that of mass transfer to the wall of a tube in which Poiseuille flow is presumed to prevail:

$$v_z = 2\langle v_z \rangle \left( 1 - \frac{r^2}{R^2} \right),$$                                            (104-1)

$$v_r = v_\theta = 0.$$                                            (104-2)

Here $r$, $\theta$, and $z$ refer to cylindrical coordinates, with $z$ measured along the tube and $r$ being the radial distance from the center of the tube. Although the Reynolds number can attain values of 2000 before the flow becomes turbulent, this is not a boundary-layer flow.

The equation of convective diffusion is

$$v_z \frac{\partial c_i}{\partial z} = D_i \left[ \frac{1}{r} \frac{\partial}{\partial r} \left( r \frac{\partial c_i}{\partial r} \right) + \frac{\partial^2 c_i}{\partial z^2} \right].$$                                            (104-3)

We treat this for mass transfer to a section with a constant wall concentration,

$$c_i = c_0 \quad \text{at} \quad r = R,$$                                            (104-4)

beginning at $z = 0$ after the Poiseuille flow is fully developed. At the limiting current, $c_0 = 0$. For other boundary conditions, we may state

$$c_i = c_b \quad \text{at} \quad z = 0 \quad \text{and} \quad \frac{\partial c_i}{\partial r} = 0 \quad \text{at} \quad r = 0.$$                                            (104-5)

The inlet concentration is $c_b$, and symmetry dictates that the derivative should be zero at the center of the tube.

Let us introduce dimensionless variables

$$\xi = \frac{r}{R}, \qquad \Theta = \frac{c_i - c_0}{c_b - c_0}, \qquad \zeta = \frac{z D_i}{2 \langle v_z \rangle R^2}.$$                                            (104-6)

The equation of convective diffusion becomes

$$(1 - \xi^2) \frac{\partial \Theta}{\partial \zeta} = \frac{1}{\xi} \frac{\partial}{\partial \xi} \left( \xi \frac{\partial \Theta}{\partial \xi} \right) + \frac{1}{Pe^2} \frac{\partial^2 \Theta}{\partial \zeta^2},$$                                            (104-7)

where

$$Pe = Re \cdot Sc = \frac{2R \langle v_z \rangle}{D_i} = \frac{2R \langle v_z \rangle}{\nu} \frac{\nu}{D_i}$$                                            (104-8)

is the Péclet number. On the assumption that the Péclet number is large, we discard the second derivative with respect to $\zeta$:

$$(1 - \xi^2) \frac{\partial \Theta}{\partial \zeta} = \frac{\partial^2 \Theta}{\partial \xi^2} + \frac{1}{\xi} \frac{\partial \Theta}{\partial \xi},$$                                            (104-9)

$$\Theta = 1 \quad \text{at} \quad \zeta = 0.$$

$$\Theta = 0 \quad \text{at} \quad \xi = 1.$$

$$\partial \Theta / \partial \xi = 0 \quad \text{at} \quad \xi = 0.$$

The total amount of material transferred to the wall in a length $z$ is

$$J = - \int_0^z D_i \frac{\partial c_i}{\partial r} \bigg|_{r=R} 2\pi R \, dz$$                                            (104-10)

or
$$\frac{J}{\pi R^2(c_b - c_0)\langle v_z \rangle} = 2\frac{Nu_{avg}}{Pe}\frac{z}{R} = -4\int_0^\zeta \left.\frac{\partial\Theta}{\partial\xi}\right|_{\xi=1} d\zeta, \qquad (104\text{-}11)$$

where $Nu_{avg}$ is the average Nusselt number based on the concentration difference at the inlet.

The Nusselt number is a dimensionless mass-transfer rate:

$$Nu = -\frac{N_{in}2R}{D_i \Delta c_i}, \qquad (104\text{-}12)$$

where the flux $N_i$ is made dimensionless with a characteristic length, here $2R$, the diffusion coefficient $D_i$, and a concentration difference $\Delta c_i$, here equal to the value at the inlet, $c_b - c_0$. For the local Nusselt number, the local flux is used; for the average Nusselt number, the average flux is used. For convective-transport problems where the contribution of ionic migration is negligible, the flux is related to the concentration derivative at the wall. For a single electrode reaction following equation 101-1, equation 101-2 allows the local Nusselt number to be related to the current density:

$$Nu = \frac{s_i 2R}{D_i \Delta c_i}\frac{i_n}{nF}. \qquad (104\text{-}13)$$

### Solution by separation of variables

Graetz,[17] followed by Nusselt,[18] treated this problem by the method of separation of variables:

$$\Theta = \sum_{k=1}^{\infty} C_k e^{-\lambda_k^2 \zeta} R_k(\xi), \qquad (104\text{-}14)$$

in which the $R_k$ satisfy the equation

$$\frac{1}{\xi}\frac{d}{d\xi}\left(\xi\frac{dR_k}{d\xi}\right) + \lambda_k^2(1 - \xi^2)R_k = 0, \qquad (104\text{-}15)$$

with the boundary conditions

$$R_k = 0 \quad \text{at} \quad \xi = 1.$$

$$R_k = 1, \qquad \frac{dR_k}{d\xi} = 0 \quad \text{at} \quad \xi = 0. \qquad (104\text{-}16)$$

The solution of this Sturm-Liouville system has been calculated, and $R_1$, $R_2$, and $R_3$ are reproduced in figure 104-1.

The total amount of material transferred to the wall can be calculated from the expression:

[17]L. Graetz, "Ueber die Wärmeleitungsfähigkeit von Flüssigkeiten," *Annalen der Physik und Chemie, 18* (1883), 79–94, *25* (1885), 337–357.

[18]Wilhelm Nusselt, "Die Abhängigkeit der Wärmeübergangszahl von der Rohrlänge," *Zeitschrift des Vereines deutscher Ingenieure, 54* (1910), 1154–1158.

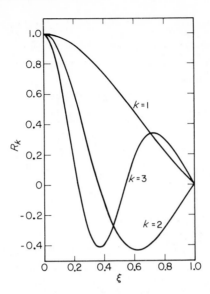

**Figure 104-1.** Graetz functions.

$$1 - \frac{J}{\pi R^2 (c_b - c_0)\langle v_z \rangle} = \sum_{k=1}^{\infty} 4C_k e^{-\lambda_k^2 \zeta} \int_0^1 \xi(1 - \xi^2) R_k(\xi)\, d\xi$$

$$= \sum_{k=1}^{\infty} M_k e^{-\lambda_k^2 \zeta}, \tag{104-17}$$

where the values of $M_k$ and $\lambda_k$ are given for ten terms in table 1 (see Brown[19]).

### Solution for very short distances

For small values of $\zeta$, Lévêque[20] recognized that there is a diffusion layer near the wall, and derivatives with respect to $\xi$ become large. Within the diffusion layer, the following approximations apply

$$1 - \xi^2 = (1 - \xi)(1 + \xi) \approx 2(1 - \xi) \tag{104-18}$$

and

$$\frac{1}{\xi} \frac{\partial \Theta}{\partial \xi} \ll \frac{\partial^2 \Theta}{\partial \xi^2}, \tag{104-19}$$

and the diffusion equation becomes

$$2(1 - \xi) \frac{\partial \Theta}{\partial \zeta} = \frac{\partial^2 \Theta}{\partial \xi^2} \tag{104-20}$$

[19]George Martin Brown, "Heat or Mass Transfer in a Fluid in Laminar Flow in a Circular or Flat Conduit," *A.I.Ch.E. Journal*, 6 (1960), 179–183.

[20]M. A. Lévêque, "Les Lois de la Transmission de Chaleur par Convection," *Annales des Mines, Memoires*, ser. 12, *13* (1928), 201–299, 305–362, 381–415.

with boundary conditions
$$\Theta = 0 \quad \text{at} \quad \xi = 1 \quad \text{and} \quad \Theta = 1 \quad \text{at} \quad \zeta = 0. \qquad (104\text{-}21)$$
In addition, $\Theta$ approaches 1 outside the diffusion layer.

The similarity transformation
$$\eta = (1 - \xi)(2/9\zeta)^{1/3} \qquad (104\text{-}22)$$
reduces the diffusion equation to an ordinary differential equation
$$\Theta'' + 3\eta^2\Theta' = 0, \qquad (104\text{-}23)$$
with the solution
$$\Theta = \frac{1}{\Gamma(4/3)} \int_0^\eta e^{-x^3} \, dx. \qquad (104\text{-}24)$$

TABLE 104-1. EIGENVALUES AND COEFFICIENTS FOR THE GRAETZ SERIES.

| $k$ | $\lambda_k$ | $M_k$ |
|---|---|---|
| 1 | 2.7043644 | 0.8190504 |
| 2 | 6.6790315 | 0.0975269 |
| 3 | 10.6733795 | 0.0325040 |
| 4 | 14.6710785 | 0.0154402 |
| 5 | 18.6698719 | 0.0087885 |
| 6 | 22.6691434 | 0.0055838 |
| 7 | 26.6686620 | 0.0038202 |
| 8 | 30.6683233 | 0.0027564 |
| 9 | 34.6680738 | 0.0020702 |
| 10 | 38.6678834 | 0.0016043 |

In terms of the physical variables, the similarity variable $\eta$ is
$$\eta = y\left(\frac{4\langle v_z\rangle}{9zD_iR}\right)^{1/3}, \qquad (104\text{-}25)$$
where
$$y = R - r \qquad (104\text{-}26)$$
is the distance from the wall. The function given by equation 24 is plotted in figure 102-1, where $\eta$ is rendered as $\xi$ on the abscissa.

The total amount of material transferred to the wall is given by
$$\frac{J}{\pi R^2(c_b - c_0)\langle v_z\rangle} = \frac{(48)^{1/3}}{\Gamma(4/3)}\zeta^{2/3} = 4.070\zeta^{2/3}. \qquad (104\text{-}27)$$
This result shows more clearly than the Graetz series how the mass-transfer rate becomes infinite near the beginning of the mass-transfer section.

### Extension of the Lévêque solution

With an approximate solution for short distances, it should be possible[21] to obtain correction terms which account for the approximations 18 and 19

---

[21]John Newman, "Extension of the Lévêque Solution," *Journal of Heat Transfer, 91C* (1969), 177–178.

and justify their validity. On this basis, the average Nusselt number referred to the concentration difference at the inlet can be expressed as

$$Nu_{avg} = 1.6151\left(\frac{Sc\,Re}{z/2R}\right)^{1/3} - 1.2 - 0.28057\left(\frac{z/2R}{Sc\,Re}\right)^{1/3} + 0\left(\frac{z/2R}{Sc\,Re}\right)^{2/3},$$

(104-28)

and the local Nusselt number is

$$Nu(\zeta) = -2\frac{\partial\Theta}{\partial\xi}\bigg|_{\xi=1} = 1.3566\zeta^{-1/3} - 1.2 - 0.296919\zeta^{1/3} + 0(\zeta^{2/3}).$$

(104-29)

In contrast to the Graetz series, the Lévêque series cannot be expected to converge for all values of $z$. It is useful for small values of $z$.

Figure 104-2 shows the local Nusselt number, divided by the first term

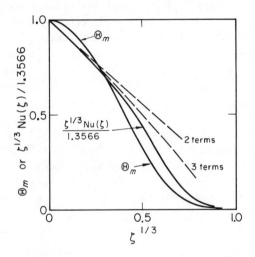

**Figure 104-2.** Dimensionless cup-mixing concentration difference $\Theta_m$ and the local Nusselt number (divided by Lévêque's solution). For comparison with the latter, the corresponding form of the Lévêque series is shown for 2 and 3 terms.

of the Lévêque series so that this ratio approaches one as $\zeta$ approaches zero. The dashed lines indicate how well the Lévêque series approximates the exact solution. The dimensionless cup-mixing concentration difference $\Theta_m$, related to the average Nusselt number by

$$\Theta_m = 1 - 2\frac{Nu_{avg}}{Pe}\frac{z}{R},$$

(104-30)

is also shown.

It is occasionally stated that the Lévêque solution should be good for

$\zeta < 0.01$. At this value of $\zeta$, where 16 percent of the possible mass transfer has already occurred, the Lévêque solution predicts an average rate of mass transfer that is too high by 15.4 percent, while the three-term Lévêque series is accurate to 0.1 percent.

For a more detailed discussion of the Graetz problem, see reference 22.

## 105. The annulus

Axial flow in the annular space between two concentric cylinders provides a convenient situation for experimental studies of mass transfer. In the work of Lin et al.,[23] the electrode of interest formed part of the inner cylinder, and the outer cylinder formed the counter electrode. However, their experimental results and theoretical treatment have been severely criticized by Friend and Metzner.[24] Ross and Wragg[25] reviewed the problem and performed additional experiments with a similar arrangement. A circular tube with no inner cylinder is a limiting case of the annular geometry and has been studied by Van Shaw et al.[26] The theoretical treatment of this geometry in laminar flow constitutes the Graetz problem (see section 104). Another limiting case investigated by Tobias and Hickman[27] is the flow between two plane electrodes.

Let the radius of the outer cylinder be $R$, and the radius of the inner cylinder be $\kappa R$. The electrode of interest is of length $L$ and is located far enough downstream in the annulus that the velocity distribution is fully developed before this electrode is reached. A limiting current is reached at this electrode when the concentration of the reactant drops to zero over the entire surface.

For laminar flow in the annulus, the local, limiting current density should

[22]John Newman, "The Fundamental Principles of Current Distribution and Mass Transport in Electrochemical Cells," Allen J. Bard, ed., *Electroanalytical Chemistry* (New York: Marcel Dekker, Inc., to be published).

[23]C. S. Lin, E. B. Denton, H. S. Gaskill, and G. L. Putnam, "Diffusion-Controlled Electrode Reactions," *Industrial and Engineering Chemistry, 43* (1951), 2136–2143.

[24]W. L. Friend and A. B. Metzner, "Turbulent Heat Transfer Inside Tubes and the Analogy Among Heat, Mass, and Momentum Transfer," *A.I.Ch.E. Journal, 4* (1958), 393–402.

[25]T. K. Ross and A. A. Wragg, "Electrochemical Mass Transfer Studies in Annuli," *Electrochimica Acta, 10* (1965), 1093–1106.

[26]P. Van Shaw, L. Philip Reiss, and Thomas J. Hanratty, "Rates of Turbulent Transfer to a Pipe Wall in the Mass Transfer Entry Region," *A.I.Ch.E. Journal, 9* (1963), 362–364.

[27]Ch. W. Tobias and R. G. Hickman, "Ionic Mass Transport by Combined Free and Forced Convection," *Zeitschrift für physikalische Chemie* (Leipzig), *229* (1965), 145–166.

follow the theoretical expression

$$i_n = 0.8546 \frac{nFD_i c_\infty}{s_i} \left( \frac{\langle v \rangle \phi}{(1 - \kappa)RD_i x} \right)^{1/3}, \tag{105-1}$$

where $c_\infty$ is the bulk concentration, $\langle v \rangle$ is the average velocity in the annulus, $x$ is the distance from the upstream edge of the electrode, and $\phi^{1/3}$ is a function of the geometric parameter $\kappa$ and is shown in figure 1 for both the inner and the outer electrode.[24,25]

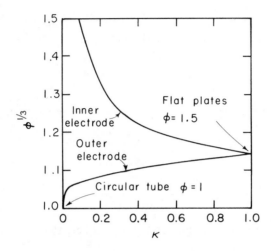

**Figure 105-1.** Coefficient for mass transfer in annuli.

Mass transfer in laminar flow in annuli is very similar to the classical Graetz problem, discussed in section 104. Equation 1 is analogous to the Lévêque solution, being useful for electrode lengths such that $L/2R \ll Sc \cdot Re$. Frequently, this covers the entire range of interest, particularly for electrolytic solutions where the Schmidt number is large. It is straightforward to apply the method of Lévêque to mass transfer in annular spaces by using the velocity derivative at the walls of the annulus instead of that for the tube. Thus,

$$\phi_0 = \frac{1 - \kappa}{2} \frac{1 - \kappa^2 - 2 \ln \dfrac{1}{\kappa}}{1 - \kappa^2 - (1 + \kappa^2) \ln \dfrac{1}{\kappa}} \tag{105-2}$$

for the outer electrode, and

$$\phi_i = \frac{\kappa - 1}{2\kappa} \frac{1 - \kappa^2 - 2\kappa^2 \ln \dfrac{1}{\kappa}}{1 - \kappa^2 - (1 + \kappa^2) \ln \dfrac{1}{\kappa}} \tag{105-3}$$

for the inner electrode. (See also problem 1.)

It might be estimated[25] that equation 1 is valid for

$$x < 0.005 \, ReSc \, d_e, \qquad (105\text{-}4)$$

where $d_e = 2(1 - \kappa)R$ is the equivalent diameter of the annulus, $Re = d_e \langle v \rangle / v$ is the Reynolds number, and $Sc = v/D_i$ is the Schmidt number (see the remark below equation 104-30). For $Sc = 2000$ and $Re = 500$, this condition yields

$$x < 5000 d_e, \qquad (105\text{-}5)$$

and is usually satisfied in experiments.

In order to facilitate comparison of results for different systems and with the standard correlations of heat and mass transfer, equation 1 is frequently written in dimensionless form:

$$Nu(x) = 1.0767 \left( \frac{\phi ReScd_e}{x} \right)^{1/3}, \qquad (105\text{-}6)$$

where the Nusselt number is a dimensionless mass-transfer rate:

$$Nu(x) = - \frac{N_i \, d_e}{c_\infty D_i}. \qquad (105\text{-}7)$$

The average value of the Nusselt number, corresponding to the average mass-transfer rate over the length $L$, is

$$Nu_{avg} = 1.6151 \left( \frac{\phi ReScd_e}{L} \right)^{1/3}. \qquad (105\text{-}8)$$

As $\kappa \longrightarrow 1$, these results apply to the flow between two flat plates, parts of which form plane electrodes. Then $\phi = 1.5$, and equations 1, 6, and 8 become

$$i_n = 0.9783 \frac{nFD_i c_\infty}{s_i} \left( \frac{\langle v \rangle}{hD_i x} \right)^{1/3}, \qquad (105\text{-}9)$$

where $h = (1 - \kappa)R$ is the distance between the planes,

$$Nu(x) = 1.2325 \left( \frac{ReScd_e}{x} \right)^{1/3}, \qquad (105\text{-}10)$$

$$Nu_{avg} = 1.8488 \left( \frac{ReScd_e}{L} \right)^{1/3}. \qquad (105\text{-}11)$$

In figure 105-2 the curve denoted *limited by convection and diffusion* depicts the local current density as a function of position along the electrode. The geometric arrangement, the electrodes, and the diffusion layer near the cathode are shown in figure 105-3. The mass-transfer rate is infinite at the upstream edge of the electrode where fresh solution is brought in contact with the electrode. The current decreases with increasing $x$, since the solution in the diffusion layer has already been depleted by the electrode reaction farther upstream. Later, it will be instructive to compare this current distribution with that which would be obtained when the ohmic potential drop in the solution is controlling.

The results of Lin *et al.*[23] for laminar flow fall roughly 17 percent below

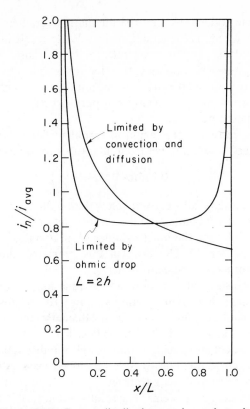

Figure 105-2. Current distribution on planar electrodes.

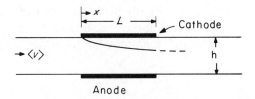

Figure 105-3. Plane electrodes in the walls of a flow channel.

the values predicted by equation 8. Part of this discrepancy can be attributed to the fact that some of the diffusion coefficients were determined by
fitting the experimental results to an erroneous equation. Ross and Wragg's[25]
laminar results are 9 to 13 percent below those predicted, while those of
Tobias and Hickman[27] scatter within 7 percent of the values predicted by
equation 11.

Turbulent flow is characterized by rapid and random fluctuations of the
velocity and pressure about their average values. The turbulence is greater

at a distance from solid walls, and the fluctuations gradually go to zero as the wall is approached. The fluctuations in velocity result in fluctuations in concentration and also in enhanced rates of mass transfer. Near the wall the fluctuations go to zero, and mass transfer at the wall is by diffusion. The details of the nature of the fluctuations are important in the region near the wall where diffusion and turbulent transport contribute roughly equally to the mass-transfer rate.

In the mass-transfer entry region in turbulent flow, Van Shaw et al.[26] expect the average Nusselt number in circular tubes to be given by

$$Nu_{avg} = 0.276 Re^{0.58} Sc^{1/3} \left(\frac{d_e}{L}\right)^{1/3}. \tag{105-12}$$

The experimental results fall 7 percent below these values but exhibit the same dependence upon the Reynolds number and the electrode length. The data of Ross and Wragg[25] for the inner cylinder of an annulus with $\kappa = 0.5$ are correlated by equation 12. However, in this geometry, those authors expect the coefficient to be 9 percent higher.

The mass-transfer entry region where equation 12 applies is much shorter in turbulent flow than in laminar flow. The results of Van Shaw et al.[26] indicate that this length ranges from 2 diameters to 0.5 diameter as the Reynolds number ranges from 5,000 to 75,000.

Beyond this short entry region, the Nusselt number rapidly approaches a constant value, corresponding to fully developed mass transfer. It is surprising that fully developed mass transfer has not been studied more extensively with electrochemical systems. The results of Lin et al.[23] agree well with the equation of Chilton and Colburn[28] for heat transfer:

$$Nu_{avg} = 0.023 Re^{0.8} Sc^{1/3}. \tag{105-13}$$

Friend and Metzner[24] discuss critically the applicability of such an equation for Schmidt numbers as large as those encountered in electrochemical systems. However, Hubbard and Lightfoot[29] also obtained agreement with this equation.

## 106. Two-dimensional diffusion layers in laminar forced convection

In 1942 Levich,[9] in treating electrolytic mass transfer to a rotating disk, remarked that in the case of diffusion, particularly the diffusion of ions, the

[28]T. H. Chilton and A. P. Colburn, "Mass Transfer (Absorption) Coefficients. Prediction from Data on Heat Transfer and Fluid Friction," *Industrial and Engineering Chemistry*, 26 (1934), 1183–1187.

[29]Davis W. Hubbard and E. N. Lightfoot, "Correlation of Heat and Mass Transfer Data for High Schmidt and Reynolds Numbers," *Industrial and Engineering Chemistry Fundamentals*, 5 (1966), 370–379.

Schmidt number reaches the value of several thousands. "Thus, in this case we deal with a peculiar limiting case of hydrodynamics, which may be called the hydrodynamics of Prandtl's [or Schmidt's] large numbers." Lighthill[30] developed a solution for the heat-transfer rate applicable when the region of temperature variation is thin compared to the region of velocity variation. Acrivos[31] realized that this method is applicable to a wide range of problems when the Schmidt number is large. Thus, for electrochemical systems where the Schmidt number is generally large, it is frequently possible to obtain the concentration distribution and the rate of mass transfer for steady problems when the velocity distribution near the electrode is known in advance. Many results for electrolytic mass transfer can be regarded as special cases of the application of this method.

The concentration distribution in a thin diffusion layer near an electrode is governed by the equation

$$v_x \frac{\partial c_i}{\partial x} + v_y \frac{\partial c_i}{\partial y} = D_i \frac{\partial^2 c_i}{\partial y^2}. \tag{106-1}$$

This equation applies to two-dimensional flow past an electrode, with $x$ measured along the electrode from its upstream end and $y$ measured perpendicularly from the surface into the solution.

Due to the thinness of the diffusion layer compared to the region of variation of the velocity, it is permissible to approximate the velocity components by their first terms in Taylor's expansions in the distance $y$ from the wall:

$$v_x = y\beta(x) \quad \text{and} \quad v_y = -\tfrac{1}{2}y^2\beta'(x), \tag{106-2}$$

where $\beta(x)$ is the velocity derivative $\partial v_x/\partial y$ evaluated at the wall ($y = 0$). These expressions for the velocity thus satisfy the applicable form of equation 100-6:

$$\frac{\partial v_x}{\partial x} + \frac{\partial v_y}{\partial y} = 0, \tag{106-3}$$

as well as the boundary conditions $v_x = v_y = 0$ at $y = 0$. With this approximation, equation 1 becomes

$$y\beta \frac{\partial c_i}{\partial x} - \frac{1}{2}y^2\beta' \frac{\partial c_i}{\partial y} = D_i \frac{\partial^2 c_i}{\partial y^2}. \tag{106-4}$$

If the concentration at the surface is a constant $c_0$, then the concentration profiles at different values of $x$ are similar and depend only on the combined variable

$$\xi = \frac{y\sqrt{\beta}}{\left(9D_i \int_0^x \sqrt{\beta}\, dx\right)^{1/3}}. \tag{106-5}$$

[30]M. J. Lighthill, "Contributions to the theory of heat transfer through a laminar boundary layer," *Proceedings of the Royal Society*, A202 (1950), 359–377.

[31]Andreas Acrivos, "Solution of the Laminar Boundary Layer Energy Equation at High Prandtl Numbers," *The Physics of Fluids*, 3 (1960), 657–658.

In terms of this similarity variable, the concentration profile is given by

$$\Theta = \frac{c_i - c_0}{c_\infty - c_0} = \frac{1}{\Gamma(4/3)} \int_0^\xi e^{-x^3}\, dx, \qquad (106\text{-}6)$$

where $\Gamma(4/3) = 0.89298$. This function is plotted in figure 102-1 and has been tabulated.[32]

The limiting current density (for $c_0 = 0$) is thus

$$i_n = \frac{\dfrac{nFD_i c_\infty \sqrt{\beta}}{s_i \Gamma(4/3)}}{\left( 9D_i \displaystyle\int_0^x \sqrt{\beta}\, dx \right)^{1/3}}. \qquad (106\text{-}7)$$

Equation 105-9 for flow between two plates is a special case of equation 7 for which $\beta$ is independent of $x$ and has the value $6\langle v \rangle / h$. Equation 7 gives the rate of mass transfer if $\beta$ is already known.

## 107. Axisymmetric diffusion layers in laminar forced convection

Equation 106-1 also applies to steady mass transfer in axisymmetric diffusion layers, that is, where the electrode forms part of a body of revolution. Examples would be the annulus and the disk electrode considered earlier and a sphere. The coordinates $x$ and $y$ have the same meaning; $x$ is measured along the electrode from its upstream end, and $y$ is measured perpendicularly from the surface into the solution. It is also necessary to specify the normal distance $\mathcal{R}(x)$ of the surface from the axis of symmetry. An axisymmetric body is sketched in figure 107-1.

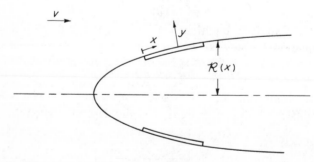

**Figure 107-1.** An electrode on an axisymmetric body with axisymmetric flow.

[32] Milton Abramowitz and Irene A. Stegun, eds., *Handbook of Mathematical Functions* (Washington: National Bureau of Standards, 1964), p. 320.

The applicable form of equation 100-6 now is[33]

$$\frac{\partial(\Re v_x)}{\partial x} + \Re \frac{\partial v_y}{\partial y} = 0. \tag{107-1}$$

Due to the thinness of the diffusion layer, it is still permissible to approximate the velocity components by their first terms in Taylor's expansions in $y$. However, in view of equation 1, these now take the form

$$v_x = y\beta(x) \quad \text{and} \quad v_y = -\frac{1}{2}y^2\frac{(\Re\beta)'}{\Re}, \tag{107-2}$$

and equation 106-1 becomes

$$y\beta\frac{\partial c_i}{\partial x} - \frac{1}{2}y^2\frac{(\Re\beta)'}{\Re}\frac{\partial c_i}{\partial y} = D_i\frac{\partial^2 c_i}{\partial y^2}. \tag{107-3}$$

The concentration profile is again given by equation 106-6, now in terms of the similarity variable

$$\xi = \frac{y\sqrt{\Re\beta}}{\left(9D_i\int_0^x \Re\sqrt{\Re\beta}\,dx\right)^{1/3}}, \tag{107-4}$$

and the limiting current density is

$$i_n = \frac{\dfrac{nFD_ic_\infty\sqrt{\Re\beta}}{s_i\Gamma(4/3)}}{\left(9D_i\int_0^x \Re\sqrt{\Re\beta}\,dx\right)^{1/3}}. \tag{107-5}$$

The Lévêque solution for a pipe in section 104 is an example of the application of this similarity transformation, and equation 105-1 for the annulus is a special case of equation 5 in which $\Re$ and $\beta$ are independent of $x$ and $\beta$ is equal to $4\langle v\rangle\phi/(1-\kappa)R$. For the rotating disk in section 103, $\Re = r = x$, and the value of the velocity derivative at the surface is (see equation 96-13 and the first of equations 96-10)

$$\beta = a\Omega x\sqrt{\frac{\Omega}{\nu}}. \tag{107-6}$$

Substitution into equation 5 yields

$$i_n = 0.62048\frac{nFc_\infty}{s_i}\sqrt{\Omega\nu}\left(\frac{D_i}{\nu}\right)^{2/3}, \tag{107-7}$$

which is seen to be the high-Schmidt-number limit obtained from figure 103-2 or equation 103-13.

## 108.  A flat plate in a free stream

The steady, laminar hydrodynamic flow parallel to a flat plate, beginning at $x = 0$ and extending along the positive $x$-axis, has been treated exten-

---

[33]Schlichting, *op. cit.*, p. 223.

sively. The value of the velocity derivative at the surface is[34]

$$\beta = 0.33206 v_\infty \sqrt{\frac{v_\infty}{vx}}, \qquad (108\text{-}1)$$

where $v_\infty$ is the value of $v_x$ far from the plate. Substitution into equation 106-7 yields

$$i_n = 0.3387 \frac{nFD_i c_\infty}{s_i} \left(\frac{v_\infty}{vx}\right)^{1/2} \left(\frac{v}{D_i}\right)^{1/3}. \qquad (108\text{-}2)$$

The average Nusselt number for an electrode of length $L$ is

$$Nu_{avg} = \frac{s_i L i_{avg}}{nFD_i c_\infty} = 0.6774 Re_L^{1/2} Sc^{1/3}, \qquad (108\text{-}3)$$

where $Re_L = L v_\infty / v$. These results apply for laminar flow. The flow becomes turbulent at a Reynolds number of about $10^5$.

Electrochemical systems for which these results are directly applicable are not frequently encountered. Unfortunately, the analysis for a flat plate in a free stream has been applied to annular geometries and the flow between two flat plates,[35,36,37] which should follow equations 105-1 and 105-9.

## 109. Rotating cylinders

Mass transfer between concentric cylinders, the inner of which is rotating with an angular speed $\Omega$, has been studied by Eisenberg *et al.*[38] and by Arvia and Carrozza.[39] If the flow between the electrodes is tangential and laminar, it does not contribute to the rate of mass transfer since the flow velocity is perpendicular to the mass flux. At higher rotation speeds, the flow is still laminar but no longer tangential, and so-called *Taylor vortices* are formed. Superimposed on the tangential motion is a radial and axial motion, outward at one point and inward at a different axial position (see figures 4-2 and 4-3). At still higher rotation speeds, the flow becomes turbulent. Mass transfer in

[34] *Ibid.*, p. 128.

[35] J. C. Bazán and A. J. Arvía, "Ionic Mass Transfer in the Electrolysis of Flowing Solutions. The Electrodeposition of Copper under Mass-Transfer Control on Tubular Electrodes," *Electrochimica Acta*, 9 (1964), 17–30.

[36] J. C. Bazán and A. J. Arvía, "Ionic Mass Transfer in Flowing Solutions. Electrochemical Reactions under Ionic Mass-Transfer Rate Control on Cylindrical Electrodes," *Electrochmica Acta*, 9 (1964), 667–684.

[37] G. Wranglén and O. Nilsson, "Mass Transfer under Forced Laminar and Turbulent Convection at Horizontal Plane Plate Electrodes," *Electrochimica Acta*, 7 (1962), 121–137.

[38] M. Eisenberg, C. W. Tobias, and C. R. Wilke, "Ionic Mass Transfer and Concentration Polarization at Rotating Electrodes," *Journal of the Electrochemical Society, 101* (1954), 306–319.

[39] A. J. Arvia and J. S. W. Carrozza. "Mass Transfer in the Electrolysis of CuSO₄– H₂SO₄ in Aqueous Solutions under Limiting Current and Forced Convection Employing a Cylindrical Cell with Rotating Electrodes," *Electrochimica Acta*, 7 (1962), 65–78.

this turbulent flow, which is achieved at lower rotation speeds if the inner cylinder rotates rather than the outer, has been studied and is reported in the above-mentioned works.

By the nature of the geometric arrangement, the current distribution is uniform. The results have been correlated by the equation

$$i_n = 0.0791 \frac{nFD_i c_\infty}{s_i d_R} \left(\frac{\Omega d_R^3}{2 v d_L}\right)^{0.70} \left(\frac{v}{D_i}\right)^{0.356}, \tag{109-1}$$

or, in dimensionless form,

$$Nu = 0.0791 \left(\frac{Re d_R}{d_L}\right)^{0.70} Sc^{0.356}, \tag{109-2}$$

where $d_R$ is the diameter of the inner, rotating cylinder, $d_L$ is the diameter of the cylinder with the limiting current, $Nu = i_n s_i d_R / nFD_i c_\infty$ is the Nusselt number, and $Re = \Omega d_R^2 / 2v$ is the Reynolds number.

In the work of Eisenberg et al., the limiting electrode was the inner, rotating electrode, and $d_R = d_L$. The results, for which the Reynolds number ranged from 112 to 162,000 and the Schmidt number from 2230 to 3650, agree with equations 1 and 2 within 8.3 percent. Arvia and Carrozza measured the limiting rates of mass transfer at the stationary, outer electrode.

Rotating cylinders where chosen to illustrate the behavior of electrochemical systems in chapter 1.

The behavior of the interface, particularly the electrode kinetics, is important in determining the behavior of an electrochemical system. In selecting a system for the study of electrode kinetics, care should be used to avoid complications not essential to the electrode kinetics.

The rotating disk electrode has been popular for the study of moderately fast electrode reactions because the hydrodynamic flow is well defined, concentration variations can be calculated, and the surface is uniformly accessible from the standpoint of diffusion and convection (see sections 96 and 103). However, it should be realized that the primary current distribution is not uniform; and this problem becomes more serious for faster reactions, larger current densities, and larger disks (see sections 116 and 117 and chapter 21).

Perhaps more attention should be devoted to the possibility of using rotating cylindrical electrodes. Here, both the primary and mass-transfer-limited current distributions are uniform on the electrodes, and both the ohmic potential drop and the concentration change at the electrodes can be accurately calculated even though the flow is turbulent. It might be more difficult to maintain cleanliness in such a system than with a rotating disk electrode.

Another way to avoid concentration variations in studies of the kinetics of moderately fast electrode reactions is to use a step change in current and follow the change in electrode potential in the time before the concentration can change significantly. For studies of the electrodeposition of copper by this method, Mattsson and Bockris[40] used small spherical electrodes, where the primary current distribution should be uniform and the ohmic potential drop can be calculated. Current-step methods should not be used if the primary current distribution is not uniform.

## 110. Growing mercury drops

The limiting diffusion current to a dropping mercury electrode finds important applications in the quantitative analysis of electrolytic solutions. Let the mercury flow at a constant rate from the capillary tube to the drop growing at the tip, so that the radius increases as

$$r_0 = \gamma t^{1/3}. \tag{110-1}$$

The diffusion layer on the drop has a thickness proportional to $\sqrt{t}$. Ilkovič[41,42] and also Mac Gillavry and Rideal[43] treated the problem with the assumption that the diffusion layer is thin compared to the radius of the drop.

For radial growth of the drop, without tangential surface motion, the limiting current density is

$$i_n = \frac{nFc_\infty}{s_i}\left(\frac{7D_i}{3\pi t}\right)^{1/2}\left(1 + 1.0302\frac{D_i^{1/2}t^{1/6}}{\gamma}\right). \tag{110-2}$$

This equation, without the correction term, was first derived by Ilkovič. The correction term, which accounts for the greater thickness of the diffusion layer and for which at least three different values of the coefficient can be found in the literature, was first derived correctly by Koutecký.[44] We have carried this slightly further[45] to express the coefficient in terms of gamma functions:

$$1.0302 = \frac{16}{11}\left(\frac{3}{7}\right)^{1/2}\frac{\Gamma(15/14)}{\Gamma(11/7)}. \tag{110-3}$$

[40]E. Mattsson and J. O'M Bockris, "Galvanostatic Studies of the Kinetics of Deposition and Dissolution in the Copper + Copper Sulphate System," *Transactions of the Faraday Society*, 55 (1959), 1586–1601.

[41]D. Ilkovič, "Polarographic Studies with the Dropping Mercury Kathode.—Part XLIV.—The Dependence of Limiting Currents on the Diffusion Constant, on the Rate of Dropping and on the Size of Drops," *Collection of Czechoslovak Chemical Communications*, 6 (1934), 498–513.

[42]D. Ilkovič, "Sur la valeur des courants de diffusion observés dans l'électrolyse à l'aide de l'électrode a gouttes des mercure. Étude polarographique," *Journal de Chimie Physique*, 35 (1938), 129–135.

[43]D. Mac Gillavry and E. K. Rideal, "On the Theory of Limiting Currents. I. Polarographic limiting currents," *Recueil des Travaux Chimiques des Pays-Bas*, 56 (1937), 1013–1021.

[44]Jaroslav Koutecký, "Correction for Spherical Diffusion to the Il'kovič Equation," *Czechoslovak Journal of Physics*, 2 (1953), 50–55.

[45]John Newman, "The Koutecký correction to the Ilkovič equation," *Journal of Electroanalytical Chemistry and Interfacial Electrochemistry*, 15 (1967), 309–312.

The total current to the drop, averaged over the life time $T$ of the drop, then takes the form

$$I_{avg} = 3.5723 \frac{nFc_\infty}{s_i} D_i^{1/2} m^{2/3} T^{1/6} \left[ 1 + 1.4530 \left( \frac{D_i^3 T}{m^2} \right)^{1/6} \right], \quad (110\text{-}4)$$

where $m$ is the volumetric flow rate of the mercury (cm$^3$/sec).

Since, in the absence of tangential surface motion, the convective flow is well defined, the dropping mercury electrode has frequently been used for the determination of diffusion coefficients.

## 111. Free convection

Free convection is a hydrodynamic flow which results from density variations in the solutions produced, in the cases of interest here, by concentration variations near the electrode. Free convection at a vertical plate electrode has been studied extensively. For deposition of a metal, the solution density is lower near the electrode than in the bulk, and an upward flow near the electrode occurs. This upward flow provides convective transport of the reactant to the electrode diffusion layer. Ibl[46] has reviewed the experimental work on this problem and reports the limiting current density to an electrode of length $L$.

$$i_{avg} = 0.66 \frac{nFD_ic_\infty}{s_i} \left[ \frac{g(\rho_\infty - \rho_0)}{\rho_\infty D_i \nu L} \right]^{1/4} \quad (111\text{-}1)$$

or

$$Nu_{avg} = \frac{s_i L i_{avg}}{nFD_ic_\infty} = 0.66(ScGr)^{1/4}, \quad (111\text{-}2)$$

where

$$Gr = \frac{g(\rho_\infty - \rho_0)L^3}{\rho_\infty \nu^2} \quad (111\text{-}3)$$

is the Grashof number. These results apply to values of $ScGr$ between $10^4$ and $10^{12}$.

The problem of free convection in a binary solution for a vertical plate has been treated theoretically. The coefficient in equation 2 is expressed as a function of the Schmidt number in table 111-1. Since the Schmidt number for electrolytic solutions is on the order of 1000, the agreement with equation 2 is good.

Free convection in solutions with an excess of supporting electrolyte is complicated by the fact that the concentration of the supporting electrolyte also varies in the diffusion layer and, therefore, contributes to the variation of the density. Approximate methods of estimating the interfacial density difference in the Grashof number consequently have been introduced, a popu-

[46]N. Ibl, "The Use of Dimensionless Groups in Electrochemistry," *Electrochimica Acta, 1* (1959), 117–129.

lar method being that of Wilke *et al.*[47] This subject is considered further in section 124.

TABLE 111-1. COEFFICIENT $C$ EXPRESSING THE RATE OF MASS TRANSFER FOR FREE CONVECTION AT A VERTICAL PLATE FROM A BINARY FLUID WITH A UNIFORM DENSITY DIFFERENCE BETWEEN THE VERTICAL SURFACE AND THE BULK SOLUTION (FROM REFERENCES 48, 49, and 50).

| $Sc$ | $C$ | $Sc$ | $C$ | $Sc$ | $C$ |
|------|------|------|------|------|------|
| 0.003 | 0.1816 | 0.72 | 0.5165 | 10 | 0.6200 |
| 0.01 | 0.2421 | 0.733 | 0.5176 | 100 | 0.6532 |
| 0.03 | 0.3049 | 1 | 0.5347 | 1000 | 0.6649 |
| | | | | $\infty$ | 0.670327 |

For turbulent natural convection at a vertical plate, Fouad and Ibl[51] obtained the relation

$$Nu_{avg} = 0.31(ScGr)^{0.28}, \tag{111-4}$$

applicable in the range $4 \times 10^{13} < ScGr < 10^{15}$.

Schütz[52] investigated experimentally free-convection mass transfer to spheres and horizontal cylinders and obtained for the average Nusselt number for spheres

$$Nu_{avg} = 2 + 0.59(ScGr)^{1/4} \tag{111-5}$$

in the range $2 \times 10^8 < ScGr < 2 \times 10^{10}$ and for cylinders

$$Nu_{avg} = 0.53(ScGr)^{1/4} \tag{111-6}$$

for $ScGr < 10^9$. In forming these dimensionless groups, $L = d$, the diameter of the sphere or cylinder. Schütz also measured local Nusselt numbers using a sectioned-electrode technique.

Acrivos[53] has obtained a solution of the laminar free-convection bound-

[47]C. R. Wilke, M. Eisenberg, and C. W. Tobias, "Correlation of Limiting Currents under Free Convection Conditions," *Journal of the Electrochemical Society, 100* (1953), 513–523.

[48]Simon Ostrach, "An Analysis of Laminar Free-Convection Flow and Heat Transfer about a Flat Plate parallel to the Direction of the Generating Body Force," Report 1111, *Thirty-Ninth Annual Report of the National Advisory Commitee for Aeronautics, 1953, Including Technical Reports Nos. 1111 to 1157* (Washington: U.S. Government Printing Office, 1955).

[49]E. M. Sparrow and J. L. Gregg, "Details of exact low Prandlt number boundary layer solutions for forced and for free convection," NASA Memo 2-27-59E (1959).

[50]E. J. LeFevre, "Laminar Free Convection from a Vertical Plane Surface," *Actes*, IX Congrès International de Mécanique Appliquée (Brussels), *4* (1957), 168–174.

[51]M. G. Fouad and N. Ibl, "Natural Convection Mass Transfer at Vertical Electrodes under Turbulent Flow Conditions," *Electrochimica Acta, 3* (1960), 233–243.

[52]G. Schütz, "Natural Convection Mass-transfer Measurements on Spheres and Horizontal Cylinders by an Electrochemical Method," *International Journal of Heat and Mass Transfer, 6* (1963), 873–879.

[53]Andreas Acrivos, "A Theoretical Analysis of Laminar Natural Convection Heat Transfer to Non-Newtonian Fluids," *A.I.Ch.E. Journal, 6* (1960), 584–590.

ary-layer equations for arbitrary two-dimensional and axisymmetric surfaces in the asymptotic limit $Sc \longrightarrow \infty$. These results should be of some interest here since the Schmidt number is large for electrolytic solutions. The local limiting current density for two-dimensional surfaces is predicted to be

$$i_n = 0.5029 \frac{nFD_ic_\infty}{s_i} \left[ \frac{g(\rho_\infty - \rho_0)}{\rho_\infty D_i v} \right]^{1/4} \frac{(\sin \epsilon)^{1/3}}{\left[ \int_0^x (\sin \epsilon)^{1/3} \, dx \right]^{1/4}}, \quad (111\text{-}7)$$

and the average limiting current density from $x = 0$ to $x = L$ is

$$i_{avg} = 0.6705 \frac{nFD_ic_\infty}{Ls_i} (ScGr)^{1/4} \left[ \frac{1}{L} \int_0^L (\sin \epsilon)^{1/3} \, dx \right]^{3/4}, \quad (111\text{-}8)$$

where $\epsilon(x)$ is the angle between the normal to the surface and the vertical. For a vertical electrode, $\sin \epsilon = 1$, and the coefficient 0.6705 of equation 8 can be compared directly with the experimental coefficient of equation 1 (or with the theoretical value 0.670327 in table 1).

For an axisymmetric surface, where $\Re(x)$ is again the distance of the surface from the axis of symmetry, the local limiting current density is

$$i_n = 0.5029 \frac{nFD_ic_\infty}{s_i} \left[ \frac{g(\rho_\infty - \rho_0)}{\rho_\infty D_i v} \right]^{1/4} \frac{(\Re \sin \epsilon)^{1/3}}{\left[ \int_0^x (\Re^4 \sin \epsilon)^{1/3} \, dx \right]^{1/4}}. \quad (111\text{-}9)$$

The axis of symmetry should coincide with the direction of the gravitational acceleration in order to assure an axisymmetric velocity distribution.

From the results of Acrivos, the predicted coefficients of $(ScGr)^{1/4}$, in the expressions for the average Nusselt number for the sphere and the horizontal cylinder, are 0.58 and 0.50, respectively, which can be compared with the experimental coefficients in equations 5 and 6.

Free convection at a horizontal plate is essentially different from that discussed above, since there is no chance for a laminar boundary layer to form and sweep fresh solution past the plate. At a horizontal electrode with a small density gradient, the solution at first remains stratified. With a higher density difference, a cellular flow pattern results; and for still higher density differences, the flow is turbulent. (This can be compared with the description of the flow behavior between rotating cylinders in section 4.) In the turbulent region, Fenech and Tobias[54] propose the relation

$$i_n = 0.19 \frac{nFD_ic_\infty}{s_i} \left[ \frac{g(\rho_\infty - \rho_0)}{\rho_\infty v D_i} \right]^{1/3}, \quad (111\text{-}10)$$

for electrodes with a minimum dimension greater than 2 cm.

[54] E. J. Fenech and C. W. Tobias, "Mass Transfer by Free Convection at Horizontal Electrodes," *Electrochimica Acta*, 2 (1960), 311–325.

## 112. Combined free and forced convection

When there is the possibility of effects of free convection superimposed on forced convection, the situation becomes essentially more complicated. Fortunately, it appears that one effect or the other predominates in the mass-transfer process, depending upon the values of the Reynolds and Grashof numbers. At horizontal electrodes, Tobias and Hickman[27] find that free convection predominates and the average rate of mass transfer is given by equation 111-10 if

$$\frac{L d_e g(\rho_\infty - \rho_0)}{\langle v \rangle \nu \rho_\infty} > 923, \tag{112-1}$$

where $L$ is the electrode length and $d_e$ is the equivalent diameter of the channel. Otherwise, forced convection predominates, and the average rate of mass transfer is given by equation 105-11. These results apply to laminar flow ($Re < 2100$). For turbulent flow, Tobias and Hickman find that forced convection predominates.

Acrivos[55] has analyzed the combined effect of free and forced convection for surfaces that are not horizontal and also finds that the transition region between predominance of free convection and predominance of forced convection is usually narrow.

The rule to follow is to calculate the mass-transfer rate separately for free convection and again for forced convection and to assume that the higher value applies.

## 113. Limitations of surface reactions

The work described above is restricted to processes at the limiting current where the concentration of the reactant at the surface has a constant value of zero. Most industrial processes are operated below the limiting current, and the kinetics of the surface reaction then influence the distribution of current. In this chapter on convective-transport problems, the ohmic potential drop is not considered. Thus, we must assume here that the ohmic potential drop is either negligible or constant for all parts of the electrode in question. The sum of the surface overpotential and the concentration overpotential is then constant, and the current distribution is determined by a balance of these overpotentials. The concentration and the current density at the surface vary with position on the electrode and must adjust themselves so that the total overpotential is constant. The more general problem involving the ohmic potential drop will be discussed in chapter 21.

[55]Andreas Acrivos, "On the combined effect of forced and free convection heat transfer in laminar boundary layer flows," *Chemical Engineering Science, 21* (1966), 343–352.

Under these conditions, the reaction rate at the electrode can be expressed in terms of the concentration at the surface, and the problem is similar to nonelectrolytic catalytic problems.[56,57,58,59] The convective-transport problem can then be reduced to an integral equation relating the reaction rate to an integral over the surface concentration at points upstream in the diffusion layer. Other, approximate methods have also been developed for calculating the surface concentration and reaction rate as a function of position on the electrode. These methods, including the integral-equation method, should also provide a useful starting point for attacking the more general problem involving the ohmic potential drop.

On the basis of the Lighthill transformation (see sections 106 and 107) one can express[16] the flux to the surface in terms of the (unknown) surface concentration

$$\frac{\partial c_i}{\partial y}\bigg|_{y=0} = -\frac{\sqrt{\Re\beta}}{\Gamma(4/3)} \int_0^x \frac{dc_0}{dx}\bigg|_{x=x_0} \frac{dx_0}{\left(9D_i \int_{x_0}^x \Re\sqrt{\Re\beta}\, dx\right)^{1/3}} \qquad (113\text{-}1)$$

or *vice versa*

$$c_0(x) - c_\infty = -\frac{\left(\dfrac{D_i}{3}\right)^{1/3}}{\Gamma(2/3)} \int_0^x \frac{\partial c_i}{\partial y}\bigg|_{\substack{y=0 \\ x=x_0}} \frac{\Re(x_0)\, dx_0}{\left(\int_{x_0}^x \Re\sqrt{\Re\beta}\, dx\right)^{2/3}}. \qquad (113\text{-}2)$$

In these equations, one can set $\Re = 1$ for a two-dimensional surface. These integrals contain the important part of the appropriate solutions of the diffusion-layer equations 106-4 and 107-3 since they relate the reaction rate and the surface concentration. The solution of the integral equation resulting from the introduction of the rate expression can then be carried out without further reference to the original partial differential equation.

## 114. Binary and concentrated solutions

It was shown in section 72 that the concentration of a binary electrolyte also obeys the equation of convective diffusion, even during the passage of current. The diffusion coefficient $D$ of the electrolyte then is to be used (see equations 72-5 and 72-6). This means that many of the results of this chapter

[56]D. A. Frank-Kamenetskiĭ, *Diffusion and Heat Exchange in Chemical Kinetics*, translated by N. Thon (Princeton, N.J.: Princeton University Press, 1955).

[57]Andreas Acrivos and Paul L. Chambré, "Laminar Boundary Layer Flows with Surface Reactions," *Industrial and Engineering Chemistry*, 49 (1957), 1025–1029.

[58]Daniel E. Rosner, "Reaction Rates on Partially Blocked Rotating Disks—Effect of Chemical Kinetic Limitations," *Journal of the Electrochemical Society*, 113 (1966), 624–625.

[59]Daniel E. Rosner, "Effects of convective diffusion on the apparent kinetics of zeroth order surface-catalysed chemical reactions," *Chemical Engineering Science*, 21 (1966), 223–239.

should be applicable to binary electrolytes. Two differences must be kept in mind. The first is that $D$ appears in the equation of convective diffusion, as noted above. The second is that migration makes a substantial contribution to the current density even at the limiting current. For deposition of the cation, this fact is reflected in the relationship 72-11 or 5-3 between the current density and the concentration derivative at the electrode (see also problems 2 and 3 of chapter 11).

Furthermore, the ohmic potential drop is much more important in a binary solution than in a solution with supporting electrolyte. This means that decomposition of the solvent may begin at one point on the electrode before a limiting current has been reached over all the remainder of the electrode. The limiting-current plateau on a current-potential plot is then difficult or impossible to discern.

Levich[9] originally treated the rotating disk for cation deposition from a binary electrolyte, his equation being

$$i_n = -0.62 \frac{z_+ \nu_+ F c_\infty}{1 - t_+} \frac{D^{2/3} \Omega^{1/2}}{\nu^{1/6}}. \tag{114-1}$$

The Ilkovič equation for a growing mercury drop has been extended to a binary electrolyte by Lingane and Kolthoff,[60] with the result

$$i_n = -\frac{z_+ \nu_+ F c_\infty}{1 - t_+} \left(\frac{7D}{3\pi t}\right)^{1/2}, \tag{114-2}$$

again for deposition of the cation.* For deposition of the cation from a binary electrolyte to a vertical electrode in free convection, the average limiting current density would be given by

$$i_{\mathrm{avg}} = -C \frac{z_+ \nu_+ F D c_\infty}{1 - t_+} \left[\frac{g(\rho_\infty - \rho_0)}{\rho_\infty D \nu L}\right]^{1/4}, \tag{114-3}$$

where $C$ is to be taken from table 111-1 with $Sc = \nu/D$. In all these equations, $c_\infty$ refers to the bulk concentration of the electrolyte, $\nu_+ c_\infty$ being the bulk concentration of the reacting cation.

Transport theory valid for dilute solutions has been applied fruitfully

---

*Note, however, that the formula given by Lingane and Kolthoff for anion reduction is not correct. For reduction of iodate ions from a solution of $KIO_3$ according to the reaction

$$IO_3^- + 3 H_2O + 6e^- \longrightarrow I^- + 6 OH^-,$$

we calculate (by the methods in chapter 19) the value $I_L/I_D = 0.6489$ for the ratio of the limiting current $I_L$ to the value $I_D$ prevailing in the absence of migration. This compares favorably with the experimental value of 0.65 reported by Lingane and Kolthoff. Neither their formula (yielding 0.84) nor one due to Heyrovský (yielding 0.74) works nearly as well.

For anion reduction, there is never a binary solution of $KIO_3$ near the electrode, since the product ions $I^-$ and $OH^-$ are always present. Consequently, the calculations are more complex and must be treated according to the development in chapter 19.

[60]James J. Lingane and I. M. Kolthoff, "Fundamental Studies with the Dropping Mercury Electrode. II. The Migration Current," *Journal of the American Chemical Society*, 61 (1939), 1045–1051.

to electrochemical systems. It should be pointed out, however, that equations valid for concentrated solutions and multicomponent transport are available and have been developed in chapter 12. Transport theory for solutions of a single salt is moderately simple and has been applied to electrodeposition on a rotating-disk electrode[61,62] and deposition from a stagnant solution.[63] Furthermore, transport properties for such solutions are frequently available in the literature (see chapter 14).

Multicomponent transport theory could be applied to certain simple geometries which would involve numerical solution of ordinary differential equations for the concentration profiles. However, in most cases, data for all the necessary transport properties are incomplete,[64] and a rigorous treatment is precluded.

The properties of the solution ($\rho$, $v$, $D_i$, etc.) have been treated largely as constant in this chapter. This is not completely valid since they depend on the composition. However, there is something to be said for such constant-property solutions. As soon as one accounts for variations of properties, one is faced with numerical solutions for each particular case; this means each particular concentration difference and temperature for each electrolytic system. One could produce an encyclopedia of results which would be of little general interest. The constant-property solutions, on the other hand, are much simpler, have approximately the correct behavior, and show more clearly the consequences of the physical phenomena. They illustrate the analogy between heat and mass transfer and allow the results to be used in both fields. On the whole, the constant-property solutions are superior from a pedagogic point of view. Empirical or theoretical corrections to constant-property solutions will have to be fairly simple in order to have any permanent value. (Of course, the determination and explanation of how the properties vary with composition are important matters of great interest.)

One then uses the constant-property solutions with the best average properties available. Fortunately, there is reason to believe that integral diffusion coefficients measured, say, with a rotating-disk electrode at the limiting current, would also be applicable to other geometries even though there is migration in the diffusion layer[65] and the transport properties vary with

[61]J. Newman and L. Hsueh, "The Effect of Variable Transport Properties on Mass Transfer to a Rotating Disk," *Electrochimica Acta*, *12* (1967), 417–427.

[62]L. Hsueh and J. Newman, "Mass Transfer and Polarization at a Rotating Disk Electrode," *Electrochimica Acta*, *12* (1967), 429–438.

[63]L. Hsueh and J. Newman, "Concentration Profile at the Limiting Current in a Stagnant Diffusion Cell," *Electrochimica Acta*, *16* (1971), 479–485.

[64]Donald G. Miller, "Application of Irreversible Thermodynamics to Electrolyte Solutions. II. Ionic Coefficients $\ell_{ij}$ for Isothermal Vector Transport Processes in Ternary Systems," *The Journal of Physical Chemistry*, *71* (1967), 616–632.

[65]John Newman, "The Effect of Migration in Laminar Diffusion Layers," *International Journal of Heat and Mass Transfer*, *10* (1967), 983–997.

composition in the diffusion layer.[31] Similarly, integral diffusion coefficients for polarography with growing mercury drops should be the same as those measured with an electrode at the limiting current at the end of a stagnant capillary. This possibility is discussed in section 92.

## PROBLEMS

1. The equation of convective diffusion for mass transfer in laminar flow in an annulus is given by equation 104-3. Neglect the axial diffusion term in the following analysis. The velocity profile is given by

$$v_z = Cr^2 + B \ln r + A.$$

The constants $A$ and $B$ can be selected so that the velocity vanishes at $r = \kappa R$ and $r = R$:

$$B = - CR^2 \frac{1 - \kappa^2}{\ln \frac{1}{\kappa}}, \qquad A = CR^2 \left( -1 + \frac{1 - \kappa^2}{\ln \frac{1}{\kappa}} \ln R \right).$$

The average velocity in the annulus is then given by

$$\langle v \rangle = \frac{CR^2}{2} \left( \frac{1 - \kappa^2}{\ln \frac{1}{\kappa}} - 1 - \kappa^2 \right).$$

(a) Show that the velocity profile near the outer wall $r = R$ can be expressed as

$$v_z = \beta_0 y \left[ 1 - \gamma_0 \frac{y}{R} + 0 \left( \frac{y^2}{R^2} \right) \right]$$

where
$$y = R - r$$
is the distance from the wall, and

$$\beta_0 = \frac{4 \langle v \rangle}{(1 - \kappa)R} \phi_0, \qquad \phi_0 = \frac{1 - \kappa}{2} \frac{1 - \kappa^2 - 2 \ln \frac{1}{\kappa}}{1 - \kappa^2 - (1 + \kappa^2) \ln \frac{1}{\kappa}},$$

$$2\gamma_0 = - \frac{1 - \kappa^2 + 2 \ln \frac{1}{\kappa}}{1 - \kappa^2 - 2 \ln \frac{1}{\kappa}}.$$

(b) Show that the velocity profile near the inner wall $r = \kappa R$ can be expressed as

$$v_z = \beta_i y \left[ 1 - \gamma_i \frac{y}{R} + 0 \left( \frac{y^2}{R^2} \right) \right]$$

where
$$y = r - \kappa R$$

is the distance from the wall, and

$$\beta_i = \frac{4\langle v\rangle}{(1-\kappa)R}\phi_i, \qquad \phi_i = \frac{\kappa-1}{2\kappa}\cdot\frac{1-\kappa^2-2\kappa^2\ln\dfrac{1}{\kappa}}{1-\kappa^2-(1+\kappa^2)\ln\dfrac{1}{\kappa}},$$

$$2\kappa\gamma_i = \frac{1-\kappa^2+2\kappa^2\ln\dfrac{1}{\kappa}}{1-\kappa^2-2\kappa^2\ln\dfrac{1}{\kappa}}.$$

(c)  By the substitution

$$\eta = y\left(\frac{\beta_i}{9D_iz}\right)^{1/3}, \quad Z = \left(\frac{9zD_i}{\beta_iR^3}\right)^{1/3}, \quad \Theta = \frac{c_i-c_0}{c_b-c_0},$$

transform the equation of convective diffusion for the diffusion layer near the inner electrode to

$$3\eta[1-\gamma_i\eta Z+0(\eta^2 Z^2)]\left(Z\frac{\partial\Theta}{\partial Z}-\eta\frac{\partial\Theta}{\partial\eta}\right)=\frac{\partial^2\Theta}{\partial\eta^2}+\frac{Z}{\kappa+\eta Z}\frac{\partial\Theta}{\partial\eta}.$$

The boundary conditions for this equation are

$$\Theta=0 \quad \text{at} \quad \eta=0, \qquad \Theta=1 \quad \text{at} \quad \eta=\infty.$$

(d)  Assume a solution of the form

$$\Theta=\Theta_0(\eta)+Z\Theta_1(\eta)+0(Z^2).$$

By substituting this form into the equation of part (c), expanding for small values of $Z$, and setting the coefficient of each power of $Z$ equal to zero, show that $\Theta_0$ and $\Theta_1$ satisfy the equations

$$\Theta_0''+3\eta^2\Theta_0'=0.$$

$$\Theta_1''+3\eta^2\Theta_1'-3\eta\Theta_1=\left(3\gamma_i\eta^3-\frac{1}{\kappa}\right)\Theta_0'.$$

(e)  Show that the appropriate solutions of the equations of part (d) are

$$\Theta_0=\frac{1}{\Gamma(4/3)}\int_0^\eta e^{-x^3}\,dx,$$

$$\Theta_1=-\frac{\dfrac{\gamma_i}{5}}{\Gamma(4/3)}\eta^2 e^{-\eta^3}-\frac{2\gamma_i-\dfrac{5}{\kappa}}{10\Gamma(4/3)}\eta\int_\eta^\infty e^{-x^3}\,dx,$$

so that

$$\Theta_0'(0)=\frac{1}{\Gamma(4/3)} \quad \text{and} \quad \Theta_1'(0)=-\frac{2\gamma_i-\dfrac{5}{\kappa}}{10}.$$

(f)  Show that the average Nusselt number for the inner electrode is

$$Nu_{\text{avg}}=1.6151\left(\frac{\phi_i ReScd_e}{L}\right)^{1/3}-(1-\kappa)\frac{2\gamma_i+\dfrac{5}{\kappa}}{5}-0\left(\frac{L}{d_e ReSc}\right)^{1/3}.$$

(g) In a similar manner, show that the average Nusselt number for the outer electrode is

$$Nu_{avg} = 1.6151 \left(\frac{\phi_0 ReScd_e}{L}\right)^{1/3} - (1 - \kappa)\frac{2\gamma_0 + 5}{5} + 0\left(\frac{L}{d_e ReSc}\right)^{1/3}.$$

(h) Show that as $\kappa$ approaches 1, both $\gamma_i$ and $\gamma_0$ approach $1/(1 - \kappa)$ and the average Nusselt number for flow between two plane electrodes can be expressed as

$$Nu_{avg} = 1.8488 \left(\frac{ReScd_e}{L}\right)^{1/3} - 0.4 + 0\left(\frac{L}{d_e ReSc}\right)^{1/3}.$$

Estimate the error in equation 105-11 when $L = 0.005 ReScd_e$.

(i) Show that $\gamma_0$ approaches $1/2$ as $\kappa$ approaches zero and that the average Nusselt number for a circular pipe can be expressed as

$$Nu_{avg} = 1.6151 \left(\frac{ReScd_e}{L}\right)^{1/3} - 1.2 + 0\left(\frac{L}{d_e ReSc}\right)^{1/3}.$$

(Compare equation 104-28.)

(j) Show that as $\kappa$ approaches zero, $\gamma_i$ approaches $1/2\kappa$ and $\phi_i$ approaches $1/2\kappa\ln(1/\kappa)$. Discuss how the error in equation 105-8 behaves for small values of $\kappa$.

2. Derive equation 103-13 by means of the velocity profile for distances close to the disk. Note that $\Gamma(4/3) = 0.89298$ and $\Gamma(5/3) = 0.90275$.

3. Show that the concentration profile in the diffusion layer near a rotating disk is given, at high Schmidt numbers, by figure 102-1 when the dimensionless variable $\xi$ is given by

$$\xi = z\left(\frac{av}{3D_i}\right)^{1/3}\sqrt{\frac{\Omega}{v}} = \zeta\left(\frac{aSc}{3}\right)^{1/3}.$$

Figure 102-1 is a plot of equation 106-6.

4. It was shown in section 72 that a binary electrolyte also obeys the equation of convective diffusion with the diffusion coefficient $D$ of the electrolyte. Show that figure 103-2 applies to metal ion deposition on a rotating disk from such a solution (see equation 5-3) if the ordinate is taken to represent

$$\frac{-i_n t_-}{z_+ \nu_+ F(c_\infty - c_0)\sqrt{v\Omega}}$$

and the abscissa is taken to represent $v/D$. The concentration $c$ of the electrolyte is taken to be $c_0$ at the electrode surface and $c_\infty$ in the bulk solution.

5. Use the development in section 107 to show that the limiting current density to a ring electrode, of inner radius $r_1$, embedded in a rotating, insulating disk is given by

$$i_n = 0.62048 \frac{nFc_\infty}{s_i}\sqrt{\Omega v}\left(\frac{D_i}{v}\right)^{2/3}\frac{r}{(r^3 - r_1^3)^{1/3}}.$$

**6.** Show that the average current density to a two-dimensional electrode of length $L$ and obeying equation 106-7 is

$$i_{avg} = \frac{nFc_\infty}{6Ls_i\Gamma(4/3)}\left[9D_i \int_0^L \sqrt{\beta}\, dx\right]^{2/3}.$$

Use this equation and $\beta = 6\langle v\rangle/h$ to obtain equation 105-11.

(b) Show that the total current to an axisymmetric electrode obeying equation 107-5 is

$$I = \frac{nFc_\infty}{s_i\Gamma(4/3)}\frac{\pi}{3}\left[9D_i \int_0^x \mathcal{R}\sqrt{\mathcal{R}\beta}\, dx\right]^{2/3}.$$

(c) Show that the total current to an axisymmetric electrode obeying equation 111-9 is

$$I = 1.341\frac{nFD_ic_\infty}{s_i}\left[\frac{g(\rho_\infty - \rho_0)}{\rho_\infty D_i v}\right]^{1/4}\pi\left[\int_0^x (\mathcal{R}^4 \sin \epsilon)^{1/3}\, dx\right]^{3/4}.$$

**7.** One can derive limiting-current expressions for the binary electrolyte by writing $(1/c_\infty)\, \partial c_i/\partial y$ at $y = 0$ from the results given for solutions with supporting electrolyte. With $D_i$ replaced by $D$, the same expression must apply to $(1/c_\infty)\, \partial c/\partial y$ at $y = 0$ for the binary electrolyte. Use this procedure to show that equation 114-3 is the correct expression for free convection from a binary electrolyte to a vertical electrode and to show that the limiting current density for cation deposition from a binary electrolyte in the rotating-cylinder system is

$$i_n = -\frac{z_+ v_+ FDc_\infty}{1 - t_+}\frac{0.0791}{d_R}\left(\frac{\Omega d_R^3}{2vd_L}\right)^{0.70}\left(\frac{v}{D}\right)^{0.356}.$$

**8.** Equation 113-1 involves a Stieltjes integral and should perhaps be written as

$$\left.\frac{\partial c_i}{\partial y}\right|_{y=0} = -\frac{\sqrt{\mathcal{R}\beta}}{\Gamma(4/3)}\int_{x_0=0}^{x_0=x}\frac{dc_0}{\left(9D_i \int_{x_0}^x \mathcal{R}\sqrt{\mathcal{R}\beta}\, dx\right)^{1/3}}.$$

Derive the Levich formula for the rotating disk by pretending that $c_0$ changes discontinuously from $c_\infty$ to 0 at $r = 0$. (For the rotating disk, $\mathcal{R} = r$ and $\beta = a\Omega r \sqrt{\Omega/v}$.) Derive the result of problem 5 by assuming that $c_0$ changes discontinuously from $c_\infty$ to 0 at $r = r_1$ and that $dc_0/dx = 0$ elsewhere.

## NOTATION

| | |
|---|---|
| $a$ | 0.51023 |
| $c_i$ | concentration of species $i$, mole/cm³ |
| $d_e$ | equivalent diameter of annulus, cm |
| $D$ | diffusion coefficient of electrolyte, cm²/sec |
| $D_i$ | diffusion coefficient of species $i$, cm²/sec |

| $F$ | Faraday's constant, 96,487 C/equiv |
|---|---|
| $g$ | acceleration of gravity, cm/sec$^2$ |
| $Gr$ | Grashof number |
| $h$ | distance between walls of a flow channel, cm |
| $H$ | dimensionless normal velocity for rotating disk |
| $i_n$ | normal component of current density at an electrode, A/cm$^2$ |
| $J$ | amount of material transferred to the wall, mole/sec |
| $L$ | length of electrode, cm |
| $m$ | volumetric flow rate of mercury, cm$^3$/sec |
| $M_k$ | coefficient in Graetz series |
| $n$ | number of electrons transferred in electrode reaction |
| $\mathbf{N}_i$ | flux of species $i$, mole/cm$^2$-sec |
| $Nu$ | Nusselt number |
| $Pe$ | Péclet number |
| $r$ | radial position coordinate, cm |
| $r_0$ | radius of growing mercury drop, cm |
| $R$ | radius of outer cylindrical electrode, cm |
| $R_k$ | Graetz function |
| $Re$ | Reynolds number |
| $\mathcal{R}$ | defines position of surface for an axisymmetric body, cm |
| $s_i$ | stoichiometric coefficient of species $i$ in electrode reaction |
| $Sc$ | Schmidt number |
| $t$ | time, sec |
| $t_+$ | cation transference number |
| $T$ | life time of drop, sec |
| $\mathbf{v}$ | fluid velocity, cm/sec |
| $\langle v \rangle$ | average velocity, cm/sec |
| $x$ | distance measured along an electrode surface, cm |
| $y$ | normal distance from surface, cm |
| $z$ | axial distance in cylindrical coordinates, cm |
| $z_i$ | charge number of species $i$ |
| $\beta$ | velocity derivative at the solid electrode, sec$^{-1}$ |
| $\gamma$ | constant in rate of growth of mercury drops, cm/sec$^{1/3}$ |
| $\Gamma(4/3)$ | 0.89298, the gamma function of 4/3 |
| $\epsilon$ | angle between the normal to a surface and vertical, radian |
| $\zeta$ | dimensionless axial distance for rotating disk or Graetz problem (see equation 96-9 or 104-6) |
| $\eta$ | similarity variable for Lévêque solution |
| $\Theta$ | dimensionless concentration |
| $\kappa$ | ratio of radii of inner to outer cylinder |
| $\lambda_k$ | eigenvalues for Graetz problem |
| $\nu$ | kinematic viscosity, cm$^2$/sec |

$v_+$          number of cations produced by dissociation of one molecule of electrolyte

$\xi$          dimensionless radial distance

$\zeta$          dimensionless similarity variable (see equations 106-5 and 107-4)

$\rho$          density, g/cm³

$\phi$          dimensionless velocity derivative at the surface

$\Omega$          rotation speed, radian/sec

subscripts

avg          average

$b$          inlet

$_0$          at the electrode surface

$\infty$          in the bulk solution

# Applications of Potential Theory

# 18

## 115. Simplifications for potential-theory problems

When concentration gradients in the solution can be ignored, substitution of equation 100-1 into equation 100-4 yields

$$\mathbf{i} = -\kappa \, \nabla\Phi, \tag{115-1}$$

where

$$\kappa = F^2 \sum_i z_i^2 u_i c_i \tag{115-2}$$

is the conductivity of the solution and where the convective transport terms sum to zero by the electroneutrality relation 100-3. Equation 100-2 when multiplied by $z_i$ and summed over $i$ yields

$$\nabla^2\Phi = 0, \tag{115-3}$$

that is, the potential satisfies Laplace's equation.

The boundary conditions are determined with equation 1. On insulators

$$\frac{\partial\Phi}{\partial y} = 0, \tag{115-4}$$

where $y$ is the normal distance from the surface. On electrodes, equation 1 relates this potential derivative to the surface overpotential through equation 101-3 or 101-4. If the potential $\Phi$ in the solution is measured with a reference electrode of the same kind as the working electrode,* then the surface overpotential can be eliminated with the relation

---

*Otherwise, an equilibrium potential difference must be included in equation 5. This difference is a constant here.

$$\eta_s = V - \Phi \quad \text{at} \quad y = 0, \tag{115-5}$$

where $V$ is the potential of the metal electrode. The resulting boundary condition is a nonlinear relationship between the potential and the potential derivative and is not commonly encountered in other applications of potential theory.

As formulated above, the potential-distribution problem is similar to the problem of the steady temperature distribution in solids, with the potential playing the rôle of the temperature, the current density that of the heat flux, and the electrical conductivity that of the thermal conductivity. Consequently, it is useful to be familiar with treatises on heat conduction, such as that of Carslaw and Jaeger.[1] A knowledge of electrostatics[2] and of the flow of inviscid fluids[3] is helpful since they are also involved with the solution of Laplace's equation.

Rousselot[4] presents an interesting discussion of potential-distribution problems. Kronsbein[5] has given an historical account of the literature, and Fleck[6] has reviewed the available analytic solutions of such problems.

## 116.  Primary current distribution

The so-called primary current and potential distributions apply when the surface overpotential can be neglected altogether, and the solution adjacent to an electrode is then taken to be an equipotential surface. Laplace's equation is not trivial to solve, even for relatively simple geometries.

Moulton[7] gave a classical solution for the primary current distribution for two electrodes placed arbitrarily on the boundary of a rectangle. This is an example of one way to solve Laplace's equation, that of conformal mapping,[8] using in this case the Schwarz-Christoffel transformation.

Consider a special case of this geometry, two plane electrodes placed opposite each other in the walls of a flow channel (see figure 105-3). The

[1] H. S. Carslaw and J. C. Jaeger, *Conduction of Heat in Solids* (Oxford: Clarendon Press, 1959).

[2] K. J. Binns and P. J. Lawrenson, *Analysis and Computation of Electric and Magnetic Field Problems* (New York: The Macmillan Company, 1963).

[3] L. M. Milne-Thomson, *Theoretical Hydrodynamics* (New York: The Macmillan Company, 1960).

[4] Robert H. Rousselot, *Répartition du potentiel et du courant dans les électrolytes* (Paris: Dunod, 1959).

[5] John Kronsbein, "Current and Metal Distribution in Electrodeposition. I. Critical Review of the Literature," *Plating*, 37 (1950), 851–854.

[6] R. N. Fleck, *Numerical Evaluation of Current Distribution in Electrochemical Systems*. M.S. Thesis, University of California, Berkeley, September, 1964 (UCRL-11612).

[7] H. Fletcher Moulton, "Current Flow in Rectangular Conductors," *Proceedings of the London Mathematical Society* (ser. 2), 3 (1905), 104–110.

[8] Ruel V. Churchill, *Complex Variables and Applications* (New York: McGraw-Hill Book Company, 1960).

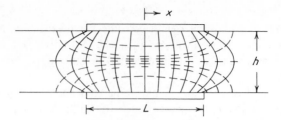

**Figure 116-1.** Two plane electrodes opposite each other in the walls of an insulating flow channel.

potential distribution is shown in figure 116-1 for $L = 2h$. Here, current lines are represented by solid curves and equipotential surfaces by dashed curves. These two sets of curves should be perpendicular to each other everywhere in the solution. The equipotential lines are close together near the edge of the electrode, and, at this point, the primary current density is infinite.

The primary current distribution on the electrode is shown in figure 116-2 for $L = 2h$ and is given by the equation

$$\frac{i_n}{i_{avg}} = \frac{\epsilon \cosh \epsilon / K(\tanh^2 \epsilon)}{\sqrt{\sinh^2 \epsilon - \sinh^2 (2x\epsilon/L)}}, \tag{116-1}$$

where $\epsilon = \pi L/2h$, $x$ is measured from the center of the electrode, and $K(m)$ is the complete elliptic integral of the first kind, tabulated in reference 9. From the complexity of this expression for the current density, one can perhaps appreciate the difficulty involved in obtaining the potential distribution. For contrast, the mass-transfer limiting current distribution for laminar flow is also shown on figure 2 (see section 105).

The primary current distribution shown in figure 2 is independent of the flow rate, since convection is great enough to eliminate concentration variations, and hence the distribution is symmetric. The current density is infinite at the ends of the electrodes since the current can flow through the solution beyond the ends of the electrodes (see figure 1). This is a general characteristic of primary current distributions. The current density where an electrode meets an insulator is either infinite or zero unless they form a right angle (see figure 116-3). When the electrode and the insulator lie in the same plane, the primary current density is inversely proportional to the square root of the distance from the edge for positions sufficiently close to the edge. This behavior is exhibited by equation 1 for the current density. Generally, the primary current distribution shows that the more inaccessible parts of an electrode receive a lower current density.

Primary current distribution is determined by geometric factors alone. Thus, only the geometric ratios of the cell enter into parameter $\epsilon$, but con-

[9] Milton Abramowitz and Irene A. Stegun, eds., *Handbook of Mathematical Functions* (Washington: National Bureau of Standards, 1964), p. 608.

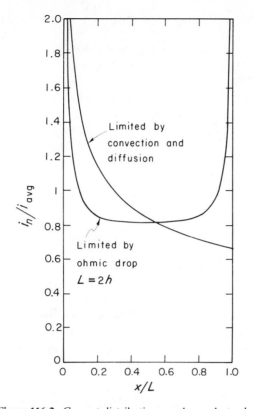

**Figure 116-2.** Current distribution on planar electrodes.

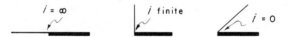

**Figure 116-3.** Behavior of the primary current distribution near the edge of an electrode.

ductivity of the solution does not enter. The resistance for this cell is

$$R = \frac{1}{\kappa W} \frac{K(1/\cosh^2 \epsilon)}{K(\tanh^2 \epsilon)},$$  (116-2)

where $W$ is width of electrodes perpendicular to the length of the channel.

Kasper[10] gives the primary current distribution for a point electrode and a plane electrode, for line electrodes parallel to plane electrodes and plane insulators, and for cylindrical electrodes in various configurations. These

[10]Charles Kasper, "The Theory of the Potential and the Technical Practice of Electrodeposition," *Transactions of the Electrochemical Society*, 77 (1940), 353–384; 78 (1940), 131–160; 82 (1942), 153–184.

systems illustrate the application of the method of images. Hine *et al.*[11] have described the primary current distribution for two plane electrodes of infinite length and finite width confined between two infinite, insulating planes perpendicular to, but not touching, the electrodes. Wagner[12] has given the primary current distribution for a two-dimensional slot in a plane electrode. These are further examples of the Schwarz-Christoffel transformation. Kojima[13] has collected various expressions for the resistance between two electrodes in various configurations. Analogous collections should be found for the resistance for heat conduction in solids[14] and for the capacitance of two electrodes.

For a disk electrode of radius $r_0$ embedded in an infinite insulating plane and with the counter electrode far away, the primary current distribution is given by[15]

$$\frac{i_n}{i_{\text{avg}}} = \frac{0.5}{\sqrt{1 - \dfrac{r^2}{r_0^2}}}, \tag{116-3}$$

and the equipotential and current lines in the solution are shown in figure 116-4. Again, the equipotential lines are close together near the edge of the electrode; and, at this point, the current density is infinite, being proportional to the reciprocal of the square root of the distance from the edge near this point. Only geometric factors enter into the current distribution, and the resistance to a hemispherical counter electrode at infinity is

$$R = \frac{1}{4\kappa r_0}. \tag{116-4}$$

This is a simple example of the application of the method of separation of the variables and the Fourier series and integrals.[16,17]

For two electrodes in the same plane with an insulator between, the inverse square root law applies near the edges of the electrodes. However,

[11] Fumio Hine, Shiro Yoshizawa, and Shinzo Okada, "Effect of Walls of Electrolytic Cells on Current Distribution," *Journal of the Electrochemical Society*, *103* (1956), 186–193.

[12] Carl Wagner, "Calculation of the Current Distributions at Electrodes Involving Slots," *Plating*, *48* (1961), 997–1002.

[13] Kaoru Kojima, "Engineering Analysis of Electrolytic Cells: Electric Resistance between Electrodes," *Research Reports of the Faculty of Engineering, Niigata University*, No. 13 (1964).

[14] Ulrich Grigull, *Die Grundgesetze der Wärmeübertragung* (Berlin: Springer-Verlag, 1955), p. 117.

[15] John Newman, "Resistance for Flow of Current to a Disk," *Journal of the Electrochemical Society*, *113* (1966), 501–502.

[16] Ruel V. Churchill, *Fourier Series and Boundary Value Problems* (New York: McGraw-Hill Book Company, 1963).

[17] Parry Moon and Domina Eberle Spencer, *Field Theory Handbook* (Berlin: Springer Verlag, 1961).

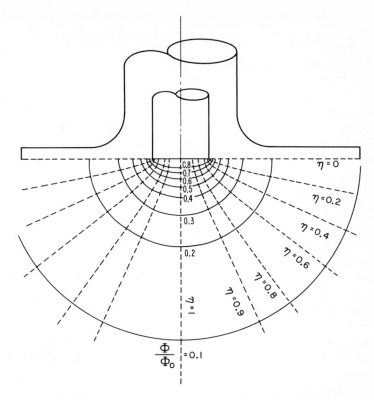

**Figure 116-4.** Current and potential lines for a disk electrode.

the coefficient becomes infinite, as the width of the separator approaches zero, in such a way that the total current flowing between the electrodes is infinite with no separation. Consequently, the solution for the primary distribution when two electrodes of different potentials meet has no physical meaning.

## 117. Secondary current distribution

When slow electrode kinetics are taken into account, the electrolytic solution near the electrode is no longer an equipotential surface, and the result of the calculations is the so-called *secondary current distribution*. The general effect of electrode polarization is to make the secondary current distribution more nearly uniform than the primary distribution, and an infinite current density at the edge of electrodes is eliminated. This can be regarded as the result of imposing an additional resistance at the electrode interface. The mathematical problem now involves the solution of Laplace's equation, subject to a more complicated, perhaps even nonlinear, boundary condition.

A variety of expressions for the electrode polarization has been used, which reflects the variety of electrode kinetics as well as a variety of approximations. In practice, the electrode kinetic equation is frequently replaced by a linear or a logarithmic (Tafel) relation between the surface overpotential and the potential derivative at the electrode. In any case, additional parameters besides geometric ratios are required to specify the current distribution. The advantage of the linear and logarithmic approximations is that they add only one new parameter. Thus, fairly realistic cases can be treated without excessive complication.

For sufficiently small surface overpotentials, equation 101-4 can be linearized to read

$$i_n = \eta_s \frac{di_n}{d\eta_s}\bigg|_{\eta_s=0} = (\alpha_a + \alpha_c)\frac{i_0 F}{RT}\eta_s = -\kappa \frac{\partial \Phi}{\partial y} \quad \text{at} \quad y = 0. \quad (117\text{-}1)$$

This provides a linear boundary condition for Laplace's equation and has been popular in the literature since there is some hope of solving the resulting linear problem. Furthermore, if the range of current densities at the electrode is sufficiently narrow, as one wants to achieve in electroplating, it is, of course, justified to linearize the polarization equation about some other, nonzero value of the surface overpotential. Finally, with linear polarization, one achieves an economy of parameters needed to determine the current distribution; and the calculation of a family of curves representing the current distribution for a particular geometry is justified.

The secondary current distribution $i_n/i_{avg}$ depends upon the same geometric ratios as the primary distribution and, in addition, for linear polarization, depends on the parameter $(L/\kappa)di_n/d\eta_s$, where $L$ is a length characteristic of the system. This parameter has been identified by Hoar and Agar[18] for the characterization of the influence of electrolytic resistance, polarization, and cell size on current distribution. When both electrodes are polarized, there are two such parameters involving the slope of the polarization curve on both the anode and the cathode.

For a disk electrode, the additional parameter for linear polarization is[19]

$$J = (\alpha_a + \alpha_c)\frac{i_0 F r_0}{RT\kappa}. \quad (117\text{-}2)$$

The secondary current distribution for linear polarization on a disk electrode is shown in figure 117-1, where $N = \infty$ means that the rotation speed is so high that concentration variations can be ignored.

[18]T. P. Hoar and J. N. Agar, "Factors in Throwing Power Illustrated by Potential-Current Diagrams," *Discussions of the Faraday Society*, No. 1 (1947), pp. 162–168.

[19]John Newman, "Current Distribution on a Rotating Disk below the Limiting Current," *Journal of the Electrochemical Society*, *113* (1966), 1235–1241.

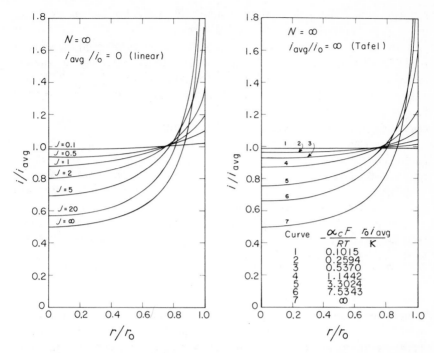

**Figure 117-1.** Secondary current distribution for linear polarization at a disk electrode.

**Figure 117-2.** Secondary current distribution for Tafel polarization at a disk cathode.

For $J \to \infty$, one obtains the primary current distribution. Then the ohmic resistance dominates over the kinetic resistance at the interface. For any finite value of $J$, the distribution is more nearly uniform and is finite at the edge of the disk. For $J \to 0$, the distribution is uniform, but the average current must then be small for the linear law still to apply. Except for this, current distribution $i_n/i_{avg}$ is independent of the magnitude of the current.

The Tafel polarization law, where one of the exponential terms in equation 101-4 is negligible, is also popular in the literature. For a cathodic reaction we have

$$\eta_s = -\frac{RT}{\alpha_c F}[\ln(-i_n) - \ln i_0]. \tag{117-3}$$

This is popular because, while being a fairly realistic polarization law, Tafel's equation introduces a minimum of additional parameters into the problem. In addition to depending on the same geometric ratios as the primary current distribution, the current distribution $i_n/i_{avg}$ now depends on the parameter $|i_{avg}|\alpha_c FL/RT\kappa$. The current distribution now depends on the magnitude of the current, but it is independent of the value of the exchange current density

$i_0$, insofar as Tafel polarization is applicable only for current densities appreciably above the exchange current density.

The secondary current distribution for Tafel polarization on a disk electrode[19] is shown in figure 117-2. This is similar to the secondary current distribution with linear polarization, but the parameter $|i_{avg}| \alpha_c Fr_0 / RT\kappa$ now plays the rôle of the parameter $J$ (the characteristic length $r_0$ being appropriate for the disk electrode). In particular, the primary current distribution is still approached as this parameter approaches infinity.

The parameter for polarization is always proportional to a characteristic length and inversely proportional to the conductivity $\kappa$ and involves the nature of the polarization. Therefore, we may state as a general rule that for large systems and small conductivities the primary current distribution will be approached, independent of the nature of the polarization law.

If we are unwilling to make either the linear or the Tafel approximation, then the secondary current distribution on a disk electrode depends on the parameters $\alpha_a / \alpha_c$, $J$ (defined by equation 2), and

$$\delta = \frac{(\alpha_a + \alpha_c) Fr_0}{RT\kappa} i_{avg}. \tag{117-4}$$

General plots of the current distribution now become unwieldy. Consequently, we select as a measure of the nonuniformity the ratio of the current density at the center of the disk to the average current density. This ratio has the value 1 for the uniform distribution and the value 0.5 for the extremely nonuniform primary distribution.

In order to illustrate how $J$ and $\delta$ jointly affect the nonuniformity, we plot[19] this ratio in figure 117-3 for $\alpha_a = \alpha_c$. For large currents ($|\delta| \gg J$),

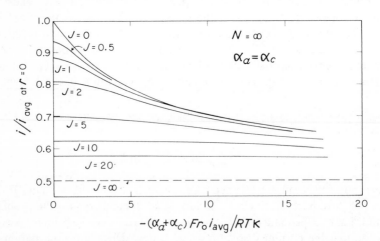

**Figure 117-3.** Current density at the center of the disk when concentration polarization is absent.

the Tafel results apply. As the current is decreased, the distribution becomes more nearly uniform but approaches at low currents the linear results for the given value of $J$ rather than a completely uniform current density. For larger values of $J$, the linear results apply to larger current densities.

Kasper[10] has treated the effect of linear polarization on some line-plane systems and for cylindrical electrodes. Wagner has treated the secondary current distribution for a plane electrode with a two-dimensional slot,[12] two cases of plane electrodes in the walls of an insulating channel, and a nonplanar electrode with a triangular profile.[20] One of the cases treated by Wagner for linear polarization, that of a plane electrode of finite width embedded in an insulating plane and with the counter electrode at infinity, has been treated by Gnusin et al.[21] for Tafel polarization. Parrish and Newman[22,23] discuss this case and the case of two plane electrodes opposite each other in the walls of a flow channel (see figure 105-3).

Some of these cases illustrate the use of current sources distributed along the electrode surface as a method of reducing the problem to an integral equation. This integral equation, which may be linear or nonlinear depending on the polarization law used, frequently requires a numerical solution.

The primary current distribution for a disk electrode, shown again in figure 117-4, can be contrasted with the uniform distribution found in section 103 when convection and diffusion are governing. The nonuniform ohmic potential drop to the disk spoils the uniform accessibility from the mass-transfer standpoint. The polarization of the electrode promotes a uniform distribution to a degree determined by the parameters $\delta$ and $J$. How these factors interact is treated in chapter 21. The potential distribution near the disk for a uniform current distribution[19] is also shown in figure 4 (see also Nanis and Kesselman[24]). This curve is normalized in such a way that it can be compared conveniently with the value $\Phi_0 4\kappa r_0 / I = 1$ for the primary current distribution. These extreme cases of uniform potential and uniform current density are clearly incompatible.

The uniform current density represents the extreme case of the variation of potential in the solution adjacent to the disk. The maximum potential

[20]Carl Wagner, "Theoretical Analysis of the Current Density Distribution in Electrolytic Cells," *Journal of the Electrochemical Society*, *98* (1951), 116–128.

[21]N. P. Gnusin, N. P. Poddubnyĭ, E. N. Rudenko, and A. G. Fomin, "Raspredelenie toka na katode v vide polosy v poluprostranstve elektrolita s polyarizatsionnoĭ krivoĭ vyrazhaemoĭ formuloĭ Tafelya," *Elektrokhimiya*, *1* (1965), 452–459.

[22]W. R. Parrish and John Newman, "Current Distribution on a Plane Electrode below the Limiting Current," *Journal of the Electrochemical Society*, *116* (1969), 169–172.

[23]W. R. Parrish and John Newman, "Current Distributions on Plane, Parallel Electrodes in Channel Flow," *Journal of the Electrochemical Society*, *117* (1970), 43–48.

[24]Leonard Nanis and Wallace Kesselman, "Engineering Applications of Current and Potential Distributions in Disk Electrode Systems," *Journal of the Electrochemical Society*, *118* (1971), 454–461.

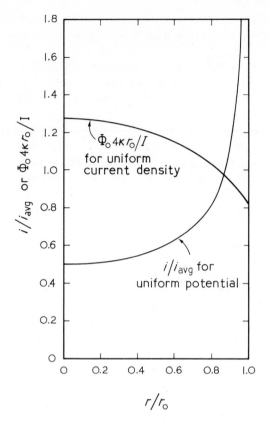

**Figure 117-4.** Primary current distribution and potential distribution for a uniform current density on a disk electrode.

difference between the center of the disk and the edge is

$$\Delta\Phi_0 = \frac{0.363 r_0 i_{avg}}{\kappa}. \tag{117-5}$$

This formula has important implications in regard to the shape of limiting current curves, the sizes of disks which can be protected from corrosion anodically or cathodically, the conditions for constant-potential electrolysis, and the determination of electrode kinetic parameters.[25] In each case, there is a maximum permissible variation of potential which governs the allowable values of $r_0$ and $i_{avg}$ for a solution of given conductivity.

[25]William H. Smyrl and John Newman. "Detection of Nonuniform Current Distribution on a Disk Electrode." *Journal of the Electrochemical Society, 119* (1972), 208–212.

## 118.  Numerical solution by finite differences

Analytic solutions of current-distribution problems are usually restricted to simple geometric arrangements and to no polarization or linear polarization. The use of some analytic solutions is facilitated by computer evaluation of certain integrals and infinite series. Some methods, such as Wagner's integral-equation method or solutions in infinite series with undetermined coefficients, require numerical evaluation of the current distribution on the electrodes or of the coefficients. When such methods can be used, the labor is less and the results more accurate than a numerical solution of Laplace's equation by finite-difference methods.

Finite-difference methods have been developed for heat conduction problems, for example, and extended to electrolytic cells. The applicability of this procedure is much less restricted than analytic or semi-analytic solutions.

Klingert et al.[26] have used a finite-difference method for the solution of Laplace's equation in an L-shaped region where the anode forms a right angle, one side of which is opposite an insulating surface. Fleck[6] has developed a general computer program for solving Laplace's equation by successive over-relaxation in arbitrarily shaped, two-dimensional regions, including those with curved boundaries, and for arbitrary polarization laws.


### NOTATION

$c_i$      concentration of species $i$, mole/cm$^3$
$F$      Faraday's constant, 96,487 C/equiv
$h$      distance between walls of flow channel, cm
$i_n$      normal component of current density at an electrode, A/cm$^2$
$i_0$      exchange current density, A/cm$^2$
$I$      total current, A
$J$      dimensionless exchange current density
$K$      complete elliptic integral of the first kind
$L$      length of electrodes, cm
$L$      characteristic length, cm
$r$      radial position coördinate, cm
$r_0$      radius of disk electrode, cm
$R$      universal gas constant, 8.3143 J/mole-deg
$R$      resistance, ohm


[26]J. A. Klingert, S. Lynn, and C. W. Tobias, "Evaluation of Current Distribution in Electrode Systems by High-speed Digital Computers," *Electrochmica Acta*, *9* (1964), 297–311.

| | |
|---|---|
| $T$ | absolute temperature, deg K |
| $u_i$ | mobility of species $i$, cm$^2$-mole/J-sec |
| $V$ | potential of an electrode, V |
| $W$ | width of electrodes, cm |
| $x$ | distance measured along an electrode surface, cm |
| $y$ | normal distance from the surface, cm |
| $z_i$ | charge number of species $i$ |
| $\alpha_a, \alpha_c$ | transfer coefficients |
| $\delta$ | dimensionless average current density |
| $\epsilon$ | $\pi L/2h$ |
| $\eta_s$ | surface overpotential, V |
| $\kappa$ | conductivity, mho/cm |
| $\Phi$ | electric potential, V |
| subscripts | |
| avg | average |
| 0 | at the electrode surface |

# Effect of Migration on Limiting Currents

**19**

Chapter 17 treated convective-transport problems, mostly at the limiting current and with an excess of supporting electrolyte. A relatively simple problem results if the current is maintained at a limiting value, but the concentration of supporting electrolyte is reduced relative to the concentration of the reacting ions. Since the current is at its limiting value, the ohmic potential drop in the bulk of the solution is still negligible,* and the current distribution is determined by mass transfer in the diffusion layer. However, the presence of an electric field in the diffusion layer can lead to an increase or a decrease in the limiting current due to migration of the reacting ions.

Let us regard figure 102-1 as the concentration profile of $CuSO_4$ for deposition at the limiting current. Within the diffusion layer, migration and diffusion contribute to mass transfer. The electric field is then very high at the electrode surface because the concentration is zero there. If we now add an inert electrolyte, such as $H_2SO_4$, the electric field will be greatly diminished, particularly at the electrode surface. The contribution of migration decreases, and the limiting current is reduced.

---

*However, a nonuniform ohmic potential drop can lead to a secondary reaction, such as decomposition of the solvent, on one part of the electrode before the limiting current can be attained on another part of the same electrode.

Because of the electroneutrality condition 100-3, solutions of only two ions also satisfy the equation of convective diffusion 102-2 but with $D_i$ replaced by the diffusion coefficient $D$ of the electrolyte (see section 72). Consequently, it is relatively simple to solve convective-transport problems at the limiting current for these solutions (see section 114). These results indicate an enhancement of the limiting current compared to the same discharging ions in a solution with excess inert electrolyte, and this can be attributed to the effect of migration in the diffusion layer.

There is some interest in calculating the limiting current for intermediate cases where there is some inert electrolyte but not a large excess. Eucken[1] gave the solution for three ion types in systems which could be represented by a stagnant Nernst diffusion layer (see also reference 2). Because experimental data[3] for the discharge of hydrogen ions on growing mercury drops did not agree with Eucken's formula, Heyrovský[4] rejected his method and introduced a correction factor involving the transference number of the discharging ion. This transference-number correction is not based on quantitative arguments, but it has become entrenched in the electrochemical literature.

Okada et al.[5] have considered the effect of ionic migration on limiting currents for a growing mercury drop, and Gordon et al.[6] for a rotating disk electrode. Newman[7] has treated the effect for four cases: the rotating disk, the growing mercury drop, penetration into a semi-infinite medium, and the stagnant Nernst diffusion layer. The ratio $I_L/I_D$ of the limiting current to the limiting diffusion current, calculated as in chapter 17 on convective-transport problems with excess supporting electrolyte, is a convenient measure of the effect of migration and depends on the ratios of concentrations in the bulk solution.

[1]Arnold Eucken, "Über den stationären Zustand zwischen polarisierten Wasserstoffelektroden," *Zeitschrift für physikalische Chemie*, 59 (1907), 72–117.

[2]Limin Hsueh and John Newman, "The Role of Bisulfate Ions in Ionic Migration Effects," *Industrial and Engineering Chemistry Fundamentals*, 10 (1971), 615–620.

[3]I. Šlendyk, "Polarographic Studies with the Dropping Mercury Kathode.—Part XXI.—Limiting Currents of Electrodeposition of Metals and of Hydrogen," *Collection of Czechoslovak Chemical Communications*, 3 (1931), 385–395.

[4]D. Ilkovič, "Polarographic Studies with the Dropping Mercury Kathode.—Part XLIV.—The Dependence of Limiting Currents on the Diffusion Constant, on the Rate of Dropping and on the Size of Drops," *Collection of Czechoslovak Chemical Communications*, 6 (1934), 498–513.

[5]Shinzo Okada, Shiro Yoshizawa, Fumio Hine, and Kameo Asada, "Effect of Migration on Polarographic Limiting Current," *Journal of the Electrochemical Society of Japan* (Overseas Edition), 27 (1959), E51–E52.

[6]Stanley L. Gordon, John S. Newman, and Charles W. Tobias, "The Role of Ionic Migration in Electrolytic Mass Transport; Diffusivities of $[Fe(CN)_6]^{3-}$ and $[Fe(CN)_6]^{4-}$ in KOH and NaOH Solutions," *Berichte der Bunsengesellschaft für physikalische Chemie*, 70 (1966), 414–420.

[7]John Newman, "Effect of Ionic Migration on Limiting Currents," *Industrial and Engineering Chemistry Fundamentals*, 5 (1966), 525–529.

The effect of migration on limiting currents is a simple example of a phenomenon which does not occur in nonelectrolytic systems, in contrast to the convective transport problems which have direct analogues in heat transfer and nonelectrolytic mass transfer.

## 119. Analysis

For the rotating disk, the normal component of the velocity depends only on $y$, the distance from the disk (see section 96). Consequently, $c_i$ and $\Phi$ also depend only on $y$ in the diffusion layer, and the limiting current density is uniform over the surface of the disk. Equations 100-1 and 100-2 can be combined to yield

$$D_i \frac{d^2 c_i}{dy^2} - v_y \frac{dc_i}{dy} + z_i u_i F\left(c_i \frac{d^2\Phi}{dy^2} + \frac{dc_i}{dy}\frac{d\Phi}{dy}\right) = 0. \tag{119-1}$$

We further approximate the velocity by the first term of its power-series expansion in $y$:

$$v_y = -a\Omega\sqrt{\frac{\Omega}{\nu}}\,y^2, \tag{119-2}$$

where $a = 0.51023$. This approximation should be valid within the diffusion layer at high Schmidt numbers. With the new variable

$$\xi = y\left(\frac{a\nu}{3D_R}\right)^{1/3}\sqrt{\frac{\Omega}{\nu}}, \tag{119-3}$$

where $D_R$ is the diffusion coefficient of the limiting reactant, equation 1 becomes

$$\frac{D_i}{D_R}c_i'' + 3\xi^2 c_i' + \frac{z_i u_i F}{D_R}(c_i\Phi'' + c_i'\Phi') = 0, \tag{119-4}$$

where primes denote differentiation with respect to $\xi$.

There is one equation of the form of equation 4 for each solute species. These equations are to be solved in conjunction with the electroneutrality equation 100-3 for the solute concentrations $c_i$ and the potential $\Phi$.

For boundary conditions we can state

$$c_i = c_{i\infty} \quad \text{at} \quad \xi = \infty, \qquad \Phi = 0 \quad \text{at} \quad \xi = \xi_{max}, \tag{119-5}$$

where $\xi_{max}$, the zero of potential, can be chosen arbitrarily. Let the electrode reaction be represented by equation 101-1. Then, the normal component of the flux of a species at the electrode is related to the normal component of the current density by equation 101-2. Since we do not know the current density in advance, we instead relate the flux of a species to the flux of the limiting reactant:

$$z_i u_i F c_i \frac{\partial\Phi}{\partial y} + D_i \frac{\partial c_i}{\partial y} = \frac{s_i}{s_R}\left(z_R u_R F c_R \frac{\partial\Phi}{\partial y} + D_R \frac{\partial c_R}{\partial y}\right) \tag{119-6}$$

at $y = 0$. In terms of the variable $\xi$, this becomes

$$z_i u_i F c_i \Phi' + D_i c_i' = \frac{s_i}{s_R}(z_R u_R F c_R \Phi' + D_R c_R') \quad \text{at} \quad \xi = 0. \quad (119\text{-}7)$$

The boundary condition for the limiting reactant at the limiting current is

$$c_R = 0 \quad \text{at} \quad \xi = 0. \quad (119\text{-}8)$$

These boundary conditions should be sufficient for the problem at hand. Equations 4 and 100-3 constitute a set of coupled, nonlinear, ordinary differential equations with boundary conditions at zero and infinity. These can be solved readily by the numerical method described in appendix C. In fact, the computer program for this particular problem is reproduced there. The results are discussed in the next few sections. After the concentration and potential profiles are calculated, the limiting current density can be obtained from the flux of the limiting reactant. The ratio $I_L/I_D$ of the limiting current to the limiting diffusion current is a convenient measure of the effect of migration.

The problem is similar for the effect of migration in other hydrodynamic situations. For a mercury drop growing in a solution which initially had a uniform composition, the transient transport equations can be reduced to

$$\frac{D_i}{D_R} c_i'' + 2\xi c_i' + \frac{z_i u_i F}{D_R}(c_i \Phi'' + c_i' \Phi') = 0 \quad (119\text{-}9)$$

if we make the same approximations as are used in the derivation of the Ilkovič equation 110-2 (without the correction term). Equation 9 replaces equation 4 for the disk. For a mercury drop growing at a constant volumetric rate, $\xi$ has the meaning

$$\xi = y\left(\frac{7}{12 D_R t}\right)^{1/2}, \quad (119\text{-}10)$$

although the analysis is not restricted to this case.

For mass transfer to a plane electrode from an infinite, stagnant medium, the transient transport equations also reduce to equation 9, where $\xi$ now has the meaning

$$\xi = \frac{y}{\sqrt{4 D_R t}}. \quad (119\text{-}11)$$

For steady mass transfer in a fictitious, stagnant, Nernst diffusion layer of thickness $\delta$, the transport equations are

$$\frac{D_i}{D_R} c_i'' + \frac{z_i u_i F}{D_R}(c_i \Phi'' + c_i' \Phi') = 0 \quad (119\text{-}12)$$

where $$\xi = \frac{y}{\delta}. \quad (119\text{-}13)$$

These several cases are very similar and can all be handled by the same computer program, particularly since the boundary conditions 5, 7, and 8 apply to all cases at the limiting current (except that condition 5 is applied at $\xi = 1$ for the Nernst diffusion layer).

Since the mathematical problems are identical for the growing mercury drops and penetration into an infinite stagnant medium, the correction factor $I_L/I_D$ is exactly the same for these two transient processes. One can also use the Lighthill transformation (see sections 106 and 107) to show[8] that the correction factor for steady transfer in arbitrary two-dimensional and axisymmetric diffusion layers is exactly the same as that calculated for the rotating disk. This means that the current density is distributed along the electrode in the same manner as when migration is neglected (see chapter 17), but the magnitude of the current density at all points is increased or diminished by a constant factor, $I_L/I_D$, which depends upon the bulk composition of the solution.

## 120. Correction factor for limiting currents

Figure 120-1 shows[7] the ratio $I_L/I_D$ for metal deposition on a rotating disk electrode. Consider the solid curve for the $CuSO_4$–$H_2SO_4$ system. The

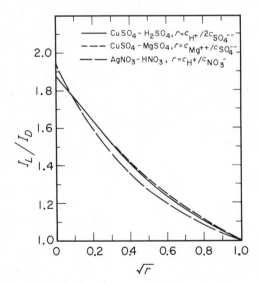

**Figure 120-1.** Effect of migration on limiting currents for metal deposition on a disk electrode.

so-called *diffusion limiting current* is that which exists when there is a small amount of $CuSO_4$ in a great excess of $H_2SO_4$, since migration should then make no contribution. Hence, the abscissa is the square root of the ratio of the normality of the added ion to that of the counter ion, and the ordinate

[8] John Newman, "The Effect of Migration in Laminar Diffusion Layers," *International Journal of Heat and Mass Transfer*, **10** (1967), 983–997.

shows how the limiting current is enhanced by the effect of migration. For $r = 0$, we have a solution of the single salt $CuSO_4$. We see that it makes little difference whether we add $MgSO_4$ or $H_2SO_4$. A curve for the deposition of silver from solutions of $AgNO_3$ and $HNO_3$ is also shown.

The square-root scale is used on the abscissa since the addition of only a small amount of supporting electrolyte to the solution of a single salt causes a considerable reduction of the limiting current, because it strongly affects the electric field at the electrode surface where the reactant concentration goes to zero.

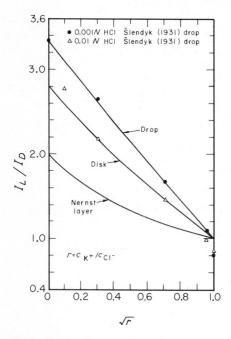

**Figure 120-2.** Effect of migration on limiting currents in discharge of hydrogen ions from KCl solutions. Lines represent values calculated with the present theory.

We have also been able to calculate this effect of migration for several other hydrodynamic situations; a growing mercury drop as encountered in polarography; a fictitious, stagnant, Nernst diffusion layer; and penetration into a semi-infinite stagnant medium. The differences in the effect of migration, among these situations, are pronounced only when the reactant ion has a diffusion coefficient considerably different from the other ions present, as shown in figure 120-2 for hydrogen ion discharge from HCl–KCl solutions. Here the polarographic data of Šlendyk[3] are shown for comparison. Theory and experiment agree well for the 0.001 $N$ HCl solutions. The discrepancy for the 0.01 $N$ HCl solutions can be attributed[9] to the fact that the solubility limit for the hydrogen produced in the electrode reaction is exceeded and the gas bubbles stir the solution in a way not accounted for in the theory.

The effect of migration does not always enhance the limiting current. For cathodic reduction of an anion, such as ferricyanide in KOH solutions, migration reduces the limiting current because the direction of the electric field is then such as to tend to drive

---

[9]John Newman and Limin Hsueh, "Currents Limited by Gas Solubility," *Industrial and Engineering Chemistry Fundamentals*, 9 (1970), 677–679.

the anions away from the electrode. This is shown in figure 120-3 for equimolar bulk concentrations of ferricyanide and ferrocyanide. The effect of ionic migration is always relatively small in redox systems because the product ion is always present at the electrode surface. Thus, in the absence of both the supporting · electrolyte and the product ion in the bulk solution, $I_L/I_D = 0.866$ for the cathodic process* and $I_L/I_D = 1.169$ for the anodic process on a rotating disk electrode.

We find that the correction factor for the effect of migration on limiting currents for unsteady transfer from a stagnant, semi-infinite fluid to a plane electrode follows exactly the curve for unsteady transfer to a growing mercury drop. Also, the effect is exactly the same[8] for steady transfer in arbitrary two-dimensional and axisymmetric diffusion layers as that shown for the rotating disk.

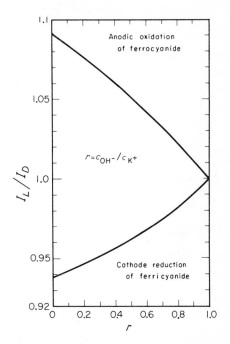

**Figure 120-3.** Effect of migration on limiting currents for a redox reaction. Equimolar potassium ferrocyanide and ferricyanide in KOH, for a disk electrode.

## 121.  Concentration variation of supporting electrolyte

For many of the discharge reactions, the concentration of supporting electrolyte is higher at the electrode surface than in the bulk solution. This difference is calculated as a by-product in the calculations of the effect of migration on limiting currents. The value, however, is of considerable interest in free-convection problems since the convective velocity is due to the density differences in the solution produced by the electrode reaction, and these density differences are affected by the concentration of the supporting electrolyte to roughly the same extent as by the concentration of the reactant (see section 111).

Figure 121-1 shows some of these concentrations at the electrode sur-

*See also footnote on page 334.

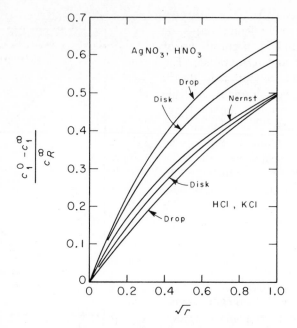

**Figure 121-1.** Concentration difference of the added ion divided by that of
the reactant. The abscissa scale is defined on figures 120-1
and 120-2.

face for two systems (discharge of $Ag^+$ from $AgNO_3$–$HNO_3$ solutions and
$H^+$ discharge from HCl–KCl solutions) and for several hydrodynamic situations. For these ions of valence 1, the concentration difference for the added
ion is roughly half that of the reactant ion when an excess of supporting
electrolyte is present. The difference, of course, goes to zero when there is
no supporting electrolyte, but it rises rapidly for even small amounts of impurities (note the square root scale on the abscissa).

For the redox systems, one must be concerned with both the added ion
and the product ion. Figure 121-2 shows these concentration differences for
the anodic oxidation of ferrocyanide ions, and figure 121-3 for the cathodic
reduction of ferricyanide. In the latter case, the added hydroxide ion is
depleted near the electrode. For these graphs, the ferrocyanide and ferricyanide ions have equal concentrations in the bulk solution, and the counter
ion is $K^+$. The abscissa scale is defined on figure 120-3.

Results for the copper sulfate, sulfuric acid system are reserved for
the next section.

In section 73 we treated systems with supporting electrolyte, showing
how one can calculate the concentration profile of the supporting electrolyte
as well as those of the minor species. Let us now use this method to calculate
the surface concentration of the added ion and a product ion in the limit

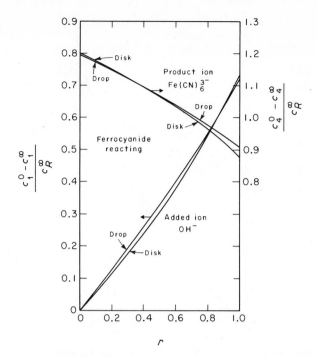

**Figure 121-2.** Surface concentrations for the anodic reaction in the $K_3Fe(CN)_6$–$K_4Fe(CN)_6$–$KOH$ system.

$r \to 1$, that is, with a large excess of supporting electrolyte. This procedure should yield the ordinate values on figures 1, 2, and 3 for $r = 1$.

With the use of the Nernst-Einstein relation 75-1, equation 119-4 for the rotating disk becomes

$$3\xi^2 \frac{dc_i}{d\xi} + \frac{D_i}{D_R}\left[\frac{d^2c_i}{d\xi^2} + z_i \frac{d}{d\xi}\left(c_i \frac{d\phi}{d\xi}\right)\right] = 0, \qquad (121\text{-}1)$$

where $\phi = F\Phi/RT$. Let the added ions and counter ions be species 1 and 2, the reactant be $R = 3$, and the product be species 4, also present in small amount.

For the reactant, we can neglect ionic migration in the limit $r \to 1$, and the concentration profile is given by

$$c_R = \frac{c_R^\infty}{\Gamma(4/3)} \int_0^\xi e^{-x^3}\, dx. \qquad (121\text{-}2)$$

For the product, migration can also be neglected, and equation 1 becomes

$$3\xi^2 \frac{dc_4}{d\xi} + \frac{D_4}{D_R}\frac{d^2c_4}{d\xi^2} = 0, \qquad (121\text{-}3)$$

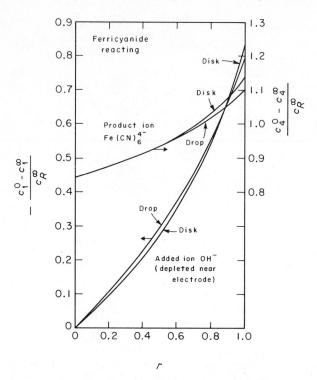

**Figure 121-3.** Surface concentrations for the cathodic reaction in the $K_3Fe(CN)_6$–$K_4Fe(CN)_6$–KOH system.

and the boundary condition 119-7 at the electrode reduces to

$$D_4 \frac{dc_4}{d\xi} = \frac{s_4}{s_R} D_R \frac{dc_R}{d\xi} \quad \text{at} \quad \xi = 0. \tag{121-4}$$

The solution therefore is

$$c_4 = c_4^\infty + \frac{s_4}{s_R} \frac{D_R}{D_4} \frac{c_R^\infty}{\Gamma(4/3)} \int_\infty^\xi e^{-x^3 D_R/D_4} \, dx, \tag{121-5}$$

from which we obtain

$$\frac{c_4^0 - c_4^\infty}{c_R^\infty} = -\frac{s_4}{s_R} \left( \frac{D_R}{D_4} \right)^{2/3}. \tag{121-6}$$

After linearization, the equations for the counter ion and the added ion are

$$3\xi^2 \frac{dc_1}{d\xi} + \frac{D_1}{D_R} \left[ \frac{d^2 c_1}{d\xi^2} + z_1 \frac{d}{d\xi} \left( c_1^0 \frac{d\phi}{d\xi} \right) \right] = 0, \tag{121-7}$$

$$3\xi^2 \frac{dc_2}{d\xi} + \frac{D_2}{D_R} \left[ \frac{d^2 c_2}{d\xi^2} + z_2 \frac{d}{d\xi} \left( c_2^0 \frac{d\phi}{d\xi} \right) \right] = 0, \tag{121-8}$$

where $c_1^0$ and $c_2^0$ are the uniform concentrations of these ions supposed to prevail in the absence of the reactant and product species (see section 73). The concentrations satisfy the electroneutrality relation in the forms

$$z_1 c_1 + z_2 c_2 + z_3 c_3 + z_4 c_4 = 0, \tag{121-9}$$

$$z_1 c_1^0 + z_2 c_2^0 = 0. \tag{121-10}$$

Elimination of $\phi$ and $c_2$ from equations 7 and 8 therefore yields

$$\left(1 - \frac{z_1}{z_2}\right)\left(\frac{d^2 c_1}{d\xi^2} + 3\xi^2 \frac{D_R}{D_e} \frac{dc_1}{d\xi}\right) = \frac{z_3}{z_2}\left(1 - \frac{D_R}{D_2}\right)\frac{d^2 c_3}{d\xi^2} + \frac{z_4}{z_2}\left(1 - \frac{D_4}{D_2}\right)\frac{d^2 c_4}{d\xi^2}, \tag{121-11}$$

where

$$D_e = \frac{D_1 D_2 (z_1 - z_2)}{z_1 D_1 - z_2 D_2}. \tag{121-12}$$

This is a generalization of equation 73-5 for the case where there are two minor species. Substitution of equations 2 and 5 gives

$$\left(1 - \frac{z_1}{z_2}\right)\left(\frac{d^2 c_1}{d\xi^2} + 3\xi^2 \frac{D_R}{D_e} \frac{dc_1}{d\xi}\right) = 3\xi^2 \frac{z_R}{z_2}\left(\frac{D_R}{D_2} - 1\right)\frac{c_R^\infty}{\Gamma(4/3)} e^{-\xi^3}$$
$$- 3\xi^2 \frac{z_4}{z_2} \frac{s_4}{s_R} \frac{D_R}{D_4}\left(\frac{D_R}{D_4} - \frac{D_R}{D_2}\right)\frac{c_R^\infty}{\Gamma(4/3)} e^{-\xi^3 D_R/D_4}. \tag{121-13}$$

The solution of this equation satisfying the boundary condition at infinity is

$$c_1 = c_1^\infty + \frac{z_R}{z_2 - z_1} \frac{\dfrac{D_R}{D_2} - 1}{\dfrac{D_R}{D_e} - 1} \frac{c_R^\infty}{\Gamma(4/3)} \int_\infty^\xi e^{-x^3}\, dx + B \int_\infty^\xi e^{-x^3 D_R/D_e}\, dx$$
$$+ \frac{z_4}{z_2 - z_1} \frac{s_4}{s_R} \frac{D_R}{D_4} \frac{\dfrac{D_4}{D_2} - 1}{\dfrac{D_4}{D_e} - 1} \frac{c_R^\infty}{\Gamma(4/3)} \int_\infty^\xi e^{-x^3 D_R/D_4}\, dx. \tag{121-14}$$

The boundary condition at the electrode is that the flux of ions 1 and 2 is zero and takes the form

$$\frac{dc_1}{d\xi} + z_1 c_1^0 \frac{d\phi}{d\xi} = 0 \quad \text{and} \quad \frac{dc_2}{d\xi} + z_2 c_2^0 \frac{d\phi}{d\xi} = 0. \tag{121-15}$$

With equation 10 this becomes

$$\frac{dc_1}{d\xi} + \frac{dc_2}{d\xi} = 0, \tag{121-16}$$

and with equation 9 we have

$$\frac{dc_1}{d\xi}\left(1 - \frac{z_1}{z_2}\right) = \frac{z_3}{z_2} \frac{dc_3}{d\xi} + \frac{z_4}{z_2} \frac{dc_4}{d\xi} = \frac{z_R}{z_2} \frac{c_R^\infty}{\Gamma(4/3)} + \frac{z_4}{z_2} \frac{s_4}{s_R} \frac{D_R}{D_4} \frac{c_R^\infty}{\Gamma(4/3)} \tag{121-17}$$

at $\xi = 0$. This allows the determination of the constant $B$ in equation 14:

$$B = \frac{c_R^\infty}{\Gamma(4/3)} \frac{\dfrac{D_R}{D_e} - \dfrac{D_R}{D_2}}{z_2 - z_1}\left[\frac{z_R}{\dfrac{D_R}{D_e} - 1} + \frac{s_4}{s_R} \frac{z_4}{\dfrac{D_4}{D_e} - 1}\right]. \tag{121-18}$$

Finally, we can calculate the concentration change of species 1 between the bulk and the electrode surface:

$$\frac{c_1^0 - c_1^\infty}{c_R^\infty} = \frac{-z_R}{z_2 - z_1}\left\{\frac{\frac{D_R}{D_2} - 1}{\frac{D_R}{D_e} - 1}\left[1 - \left(\frac{D_e}{D_R}\right)^{1/3}\right] + \left(\frac{D_e}{D_R}\right)^{1/3}\right\}$$

$$- \frac{z_4}{z_2 - z_1}\frac{s_4}{s_R}\left\{\frac{\frac{D_4}{D_2} - 1}{\frac{D_4}{D_e} - 1}\left[1 - \left(\frac{D_e}{D_4}\right)^{1/3}\right] + \left(\frac{D_e}{D_4}\right)^{1/3}\right\}\left(\frac{D_R}{D_4}\right)^{2/3}. \quad (121\text{-}19)$$

The case where there is no product ion can be treated by setting $s_4 = 0$ or $z_4 = 0$. The corresponding equation for $c_2$ can be obtained from equation 19 by reversing the subscripts 1 and 2.

For a growing mercury drop, the expressions for the concentration differences should become

$$\frac{c_1^0 - c_1^\infty}{c_R^\infty} = \frac{-z_R}{z_2 - z_1}\left\{\frac{\frac{D_R}{D_2} - 1}{\frac{D_R}{D_e} - 1}\left[1 - \left(\frac{D_e}{D_R}\right)^{1/2}\right] + \left(\frac{D_e}{D_R}\right)^{1/2}\right\}$$

$$- \frac{z_4}{z_2 - z_1}\frac{s_4}{s_R}\left\{\frac{\frac{D_4}{D_2} - 1}{\frac{D_4}{D_e} - 1}\left[1 - \left(\frac{D_e}{D_4}\right)^{1/2}\right] + \left(\frac{D_e}{D_4}\right)^{1/2}\right\}\left(\frac{D_R}{D_4}\right)^{1/2} \quad (121\text{-}20)$$

and

$$\frac{c_4^0 - c_4^\infty}{c_R^\infty} = -\frac{s_4}{s_R}\left(\frac{D_R}{D_4}\right)^{1/2}. \quad (121\text{-}21)$$

## 122. The rôle of bisulfate ions

Bisulfate ions do not completely dissociate in sulfuric acid solutions (see section 33). A simple, dramatic example is found in the conductivity of solutions of copper sulfate and sulfuric acid, as shown in figure 122-1. When copper sulfate is added to a solution of sulfuric acid, the conductivity is found to decrease. If the conductivity is predicted from limiting ionic mobilities (see table 75-1), the result is in accord with this observation if bisulfate ions are assumed to be undissociated. Predictions based on sulfate and hydrogen ions are, on the other hand, in qualitative and quantitative discord with the experimental values.

The incomplete dissociation of bisulfate ions should also have dramatic consequences for the effect of ionic migration on limiting currents. When copper sulfate is added to sulfuric acid solutions, the electric field increases not only because the current increases but also because hydrogen ions combine to form bisulfate ions and the conductivity decreases. This is shown

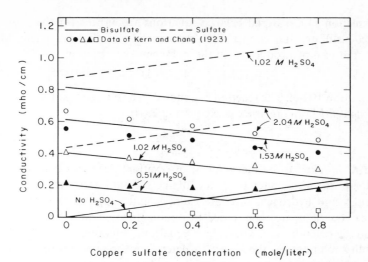

**Figure 122-1.** Conductivity of aqueous solutions of copper sulfate and sulfuric acid at 25°C (data from Kern and Chang, *Trans. Am. Electrochem. Soc.*, *41*, 181-196 (1923)).

on figure 122-2. The parameter $r$ in the abscissa is still based on the ratio of the stoichiometric concentrations of sulfuric acid and copper sulfate.

The partial dissociation of bisulfate ions can also be taken into account.[2]

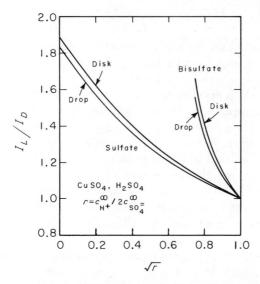

**Figure 122-2.** Effect of migration in the $CuSO_4$–$H_2SO_4$ system with no dissociation and with complete dissociation of bisulfate ions.

As in section 119, material balance equations are written for the hydrogen, sulfate, bisulfate, and copper ions, the equations including, where appropriate, the rate of production in the homogeneous reaction

$$HSO_4^- \rightleftharpoons H^+ + SO_4^= \tag{122-1}$$

(see equation 100-2). The reaction is assumed to be fast, so that the concentrations also satisfy the relation (see equation 33-4)

$$K' = \frac{c_{H^+}^* c_{SO_4^=}^*}{c_{HSO_4^-}^*}, \tag{122-2}$$

where $K'$ is taken to be independent of position. Asterisks denote the fact that these quantities refer to a view of the solution as composed of water molecules and hydrogen, bisulfate, sulfate, and copper ions. The material balance equations are then added to obtain three equations which do not include the reaction rate. These three equations, equation 2, and the electroneutrality relation are then used to determine the concentration distributions of the four ions and the potential distribution by the numerical method described in appendix C.

Let $c_A^\infty$ and $c_B^\infty$ be the bulk stoichiometric concentrations of copper sulfate and sulfuric acid, and let $I$ be the bulk ionic strength based on a convention of complete dissociation:

$$I = 4c_A^\infty + 3c_B^\infty. \tag{122-3}$$

The two important parameters will then be the relative amounts of reactant and supporting electrolyte, expressed as

$$r = \frac{c_B^\infty}{c_A^\infty + c_B^\infty}, \tag{122-4}$$

and the ratio $I/K'$ of the ionic strength to the dissociation constant.

The effect of migration on limiting current is shown in figures 122-3 and 122-4 for the rotating-disk electrode and the growing mercury drop. The ordinate, $I_L/I_D$, is the ratio of the limiting current to the limiting diffusion current of a well-supported solution when the effect of viscosity variations is excluded. The abscissa is the ratio $r$ of equation 4, and values of $I/K'$ are given as a parameter. The two solid lines indicate the two extreme cases of complete ($I/K' = 0$) and no dissociation ($I/K' = \infty$) of bisulfate ions.

The concentration difference of sulfuric acid between the electrode surface and the bulk solution is shown in figures 122-5 and 122-6 for the two electrochemical systems. One may notice that the concentration of sulfuric acid would even decrease near the electrode surface for some values of $r$ when bisulfate ions are not completely dissociated. Qualitatively speaking, the bisulfate ions, containing hydrogen, are driven away from the electrode because of their negative charge. For no dissociation, one should consider that $r = 0.5$ corresponds to a binary solution of copper bisulfate.

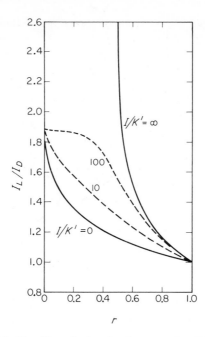

**Figure 122-3.** The effect of migration for a rotating disk electrode.

An analytic solution can be obtained for a stagnant Nernst diffusion layer. Although this system is not of great physical interest, the results shown in figures 122-7 and 122-8 do serve to complete the picture that is presented in figures 122-3 to 122-6.

In practical applications, the dissociation constant $K'$ can be related to the true ionic strength $I_r$ of the bulk solution

$$I_r = \tfrac{1}{2} \sum_i z_i^2 c_i^{*\infty}, \qquad (122\text{-}5)$$

the correlation having been given in figure 33-1 and equation 33-6.

Figure 122-9 shows the wide range of concentration differences which are conceivable in the copper sulfate, sulfuric acid system. Results for free convection, from section 124, are shown in addition to the rotating disk, the growing mercury drop, and the stagnant Nernst diffusion layer. For comparison, the values calculated by Wilke et al.[10] and by Fenech and Tobias[11] are also shown. For the quantity plotted on figure 9, the value 0.71 can be

[10]C. R. Wilke, M. Eisenberg, and C. W. Tobias, "Correlation of Limiting Currents under Free Convection Conditions," *Journal of the Electrochemical Society*, *100* (1953), 513–523.

[11]E. J. Fenech and C. W. Tobias, "Mass Transfer by Free Convection at Horizontal Electrodes," *Electrochimica Acta*, *2* (1960), 311–325.

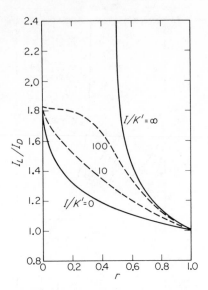

**Figure 122-4.** The effect of migration for a growing mercury drop or in a stagnant diffusion cell.

deduced from the results of one of Brenner's experiments.[12] Hsueh and Newman[2] obtained four values in the range from 0.50 to 0.57 and one value of 0.75.

## 123. Paradoxes with supporting electrolyte

The use of a supporting electrolyte raises a number of questions, such as:

1. Which species is carrying the current?

2. If the reactant is carrying the current in the diffusion layer, how does the supporting electrolyte have an effect?

3. If the supporting electrolyte is motionless in the diffusion layer, is it also motionless in the bulk of the solution? Again, what species is carrying the current?

We shall endeavor to answer such questions in this section.

[12]A. Brenner, "A Method for Studying Cathode Films by Freezing," *Proc. Amer. Electroplaters' Soc.* (1940), pp. 95–98.

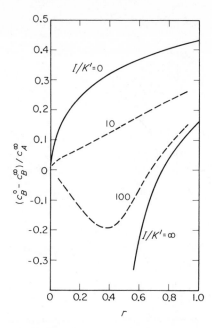

**Figure 122-5.** The surface concentration change for a rotating disk
electrode.

The nature of the problem can perhaps be seen more clearly from the
fundamental transport relations. If there are no concentration gradients,
then equation 115-1 applies

$$\mathbf{i} = -\kappa\, \nabla\Phi, \tag{123-1}$$

where

$$\kappa = F^2 \sum_i z_i^2 u_i c_i. \tag{123-2}$$

This is the conductivity we measure with a conductivity cell, using alternating current, and is the usual basis for defining transference numbers (equation 70-5) and deciding how the various species contribute to the current.
These concepts sometimes clash with what we find on more detailed analysis
of the effect of supporting electrolyte.

Consider an electrolytic cell with two electrodes and a solution between.
Near each electrode there is a stagnant diffusion layer in which mass-transfer
effects are important, and the bulk of the solution is well mixed.

At the electrode, the flux of all species is zero except for a reactant. In
a stagnant diffusion layer, this implies that the added ions and the counter
ions are motionless, having adopted concentration distributions so that the
forces of migration and diffusion cancel. How then can the supporting elec-

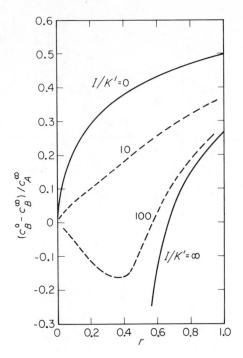

**Figure 122-6.** The surface concentration change for a growing mercury drop or in a stagnant diffusion cell.

trolyte act to reduce the electric field strength when it carries no current? In this regard, the effect of adding supporting electrolyte is essentially no different from adding more reacting electrolyte. This excess electrolyte is not moving either.

There are several reasons why we might use a supporting electrolyte instead of adding more reacting electrolyte. Without the supporting electrolyte, the conductivity of the solution may be limited by the solubility of the reacting electrolyte. The current distribution is likely to be more uniform with a higher conductivity (see section 117), and the ohmic potential drop will be smaller. We may want to keep the inventory of working material down and use a cheaper material which we can discard when it gets dirty. We may want to adjust the pH. (See also section 10.)

Now suppose that the region outside the stagnant diffusion layer is *well stirred* so that there are no concentration gradients. (It may be useful to consider the rotating cylinder system treated in chapter 1. A concentration profile was shown in figures 5-1 and 7-1.) In this region, equation 1 should apply, so that the supporting electrolyte is apparently moving and carrying

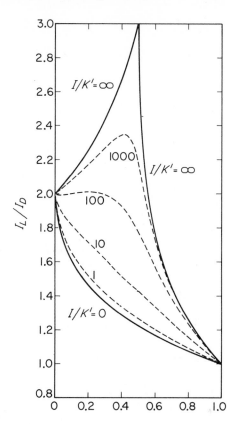

**Figure 122-7.** The effect of migration in a Nernst diffusion layer.

a current. How can we have a flux which suddenly becomes zero at the edge of the diffusion layer? Actually, the net flux of supporting electrolyte must be zero throughout the solution. The velocity of stirring cannot be one dimensional. A sufficient concentration gradient exists across the bulk solution so that convection cancels the migration flux of the supporting electrolyte and augments that of the reactant; this is the purpose of stirring in the first place. The faster the stirring, the lower the gradients; but the fluid motion carries no current since the solution is electrically neutral. However, the potential drop in the bulk is still given by equation 1 and can be considerably reduced by the addition of supporting electrolyte.

A statement of how much current each ionic species is carrying thus depends upon one's reference frame. Some people base their answer on

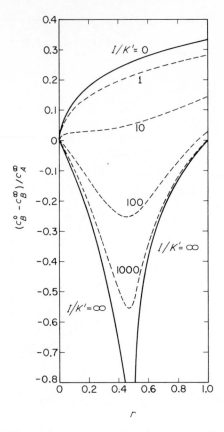

**Figure 122-8.** The surface concentration change in a Nernst diffusion layer.

the net flux. Others prefer to ignore the convective flux, saying that it does not contribute to the current. Still others give an answer based only on migration fluxes and transference numbers.

Finally, one may be puzzled by the fact that the addition of a supporting electrolyte reduces the ohmic potential drop and yet can lead to a reduction of the limiting current by a factor of two. The reason is that the potential drop in the bulk solution is not relevant at the limiting current. By lowering the electric field in the diffusion layer, so that migration makes no contribution to the flux of the reactant, the supporting electrolyte reduces the limiting current density. At this point one can contrast figure 10-2 with figure 9-1. To pass 0.5 amp with no supporting electrolyte requires about 0.6 volt, whereas with supporting electrolyte only about 0.26 volt is required.

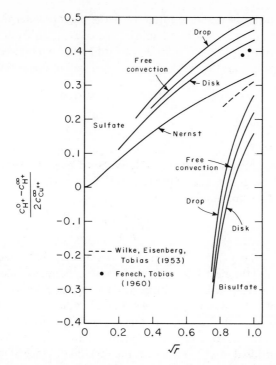

**Figure 122-9.** Concentration differences of sulfuric acid possible in the copper sulfate, sulfuric acid system with complete and with no dissociation of bisulfate ions and for several hydrodynamic situations.

At higher potentials, mass-transfer in the diffusion layer becomes a more severe restriction with the supporting electrolyte.

## 124.  Limiting currents for free convection

The addition of supporting electrolyte to a solution does not make the free-convection problem directly comparable to that of heat transfer and nonelectrolytic mass transfer in a binary fluid because, while it does reduce the effect of ionic migration, the concentration variation of the supporting electrolyte affects the density variation to roughly the same extent as the reactant and thus influences the velocity profile. Since the mass-transfer rate depends upon the velocity profile, the limiting current density is also affected.

The quantities of practical interest are the mass transfer to the wall and the shear stress at the wall. For laminar free convection from a solution to a vertical electrode with a constant density difference $\Delta\rho$ between the surface and the bulk solution, the results can be expressed in dimensionless form as

$$Nu_{avg} = \frac{s_R i_{avg} L}{nFD_R c_{R\infty}} = C(ScGr)^{1/4} \qquad (124\text{-}1)$$

and

$$\frac{\tau_0}{Lg\,\Delta\rho} = B(ScGr)^{-1/4}, \qquad (124\text{-}2)$$

where $\tau_0$ is the shear stress at the wall averaged over the length $L$, $g$ is the magnitude of the gravitational acceleration, $Sc = \nu/D_R$ is the Schmidt number, $Gr$ is the Grashof number

$$Gr = \frac{gL^3\,\Delta\rho}{\rho_\infty\nu^2}, \qquad (124\text{-}3)$$

and $\Delta\rho = |\rho_\infty - \rho_0|$ is the magnitude of the density difference between the bulk solution and the electrode surface. $C$ and $B$ are dimensionless coefficients which depend on the Schmidt number and the composition of the bulk solution. Values of $C$ for a binary fluid were given in table 111-1. For free convection to a vertical surface with a constant density difference $\Delta\rho$, the local rate of mass transfer is inversely proportional to $x^{1/4}$, and the local shear stress is proportional to $x^{1/4}$, where $x$ is the vertical distance along the surface measured from the beginning of the boundary layer.

We treat[13] this problem in the limit of infinite Schmidt number and express the results in the form of $C/C_b$ and $B/B_b$, where $C_b$ and $B_b$ are the values appropriate to a binary fluid and have the values $C_b = 0.670327$ and $B_b = 0.932835$.

The copper sulfate, sulfuric acid system was treated first. In view of the low value of the dissociation constant of bisulfate ions, the calculations were carried out for no dissociation of bisulfate ions as well as for complete dissociation to sulfate and hydrogen ions (see also section 122). Results are shown in figure 124-1 for $B/B_b$ for complete dissociation of bisulfate ions and the ratio $C/C_b$ for both no dissociation and complete dissociation. Dashed lines show the corresponding values of $I_L/I_D$ for a rotating disk.

For metal deposition from a binary electrolytic solution, one can show that

$$C = \frac{(D/D_R)^{3/4}}{1 - t_R} C_b(Sc_e) \qquad (124\text{-}4)$$

---

[13]Jan Robert Selman and John Newman, "Free-Convection Mass Transfer with a Supporting Electrolyte," *Journal of the Electrochemical Society, 118* (1971), 1070–1078.

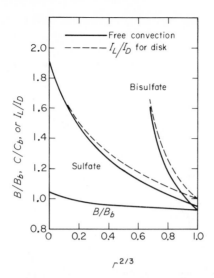

**Figure 124-1.** Coefficients for shear stress (complete dissociation only) and mass transfer in the $CuSO_4$–$H_2SO_4$ system. Dashed curves show for comparison the values of $I_L/I_D$ for the rotating disk.

and
$$B = \left(\frac{D}{D_R}\right)^{1/4} B_b(Sc_e) \tag{124-5}$$

(see equation 114-3), where $Sc_e = \nu/D$ is based on the diffusion coefficient $D$ of the salt and $t_R$ is the transference number of the reacting cation (see equations 72-6 and 72-12).

As $r$ approaches unity, one would expect $C/C_b$ and $B/B_b$ to approach unity if the appearance of $\Delta\rho$ in the Grashof number were sufficient to correlate the effect of the supporting electrolyte. The contrary behavior emphasizes the fact that these ratios express not only the effect of ionic migration but also the effect of the density profile not being similar to that for a binary fluid.

To be specific, the diffusion layer thickness is greater for $H_2SO_4$ than for $CuSO_4$ because of the larger value of the diffusion coefficient of hydrogen ions. Thus, the density difference in the outer part of the diffusion layer is positive while it is negative near the electrode. Consequently, the value of $\Delta\rho$ does not, by itself, give sufficient information about the density profile. In fact, with added $H_2SO_4$, the velocity profile shows a maximum within the diffusion layer. This is shown for excess sulfuric acid in figure 124-2. Since these phenomena occur in a more drastic fashion in some redox systems,

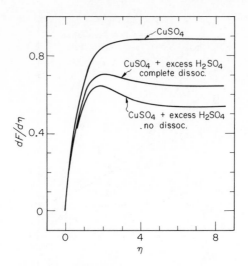

**Figure 124-2.** Velocity profiles for binary salt solution (CuSO₄) and for
CuSO₄ with excess H₂SO₄ ($r = 0.99998$) completely
dissociated and undissociated.

we shall postpone their further discussion. In figure 2, the abscissa is the similarity variable

$$\eta = y\left(\frac{3g\,\Delta\rho}{4\nu D_R\rho_\infty x}\right)^{1/4},$$
(124-6)

and the velocity profile is related to $dF/d\eta$ by

$$\frac{dF}{d\eta} = \left(\frac{3\nu\rho_\infty}{4g\,\Delta\rho D_R x}\right)^{1/2} v_x.$$
(124-7)

The redox reaction

$$Fe(CN)_6^{3-} + e^- \rightleftharpoons Fe(CN)_6^{4-}$$
(124-8)

is popular in mass-transfer studies and has been used in free convection, although it is not common. The densification in this system is much weaker than in copper sulfate solutions since the excess of product ion largely compensates for the deficit of the reactant.

The ratio $C/C_b$ is shown in figure 124-3 as a function of

$$r = \frac{c_{OH^-}^\infty}{c_{K^+}^\infty + c_{Na^+}^\infty}.$$
(124-9)

The solutions have equal bulk concentrations of potassium ferrocyanide and potassium ferricyanide, with either sodium hydroxide or potassium hydroxide added as a supporting electrolyte. On figure 3, values of $I_L/I_D$ for the rotating disk with KOH supporting electrolyte are plotted for comparison.

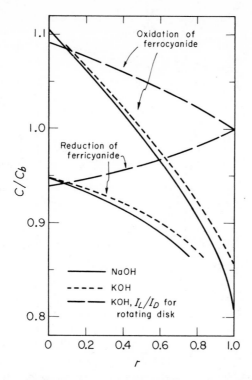

**Figure 124-3.** Coefficient for mass-transfer rate in the supported ferricyanide-ferrocyanide systems, for equal bulk concentrations
of $K_3Fe(CN)_6$ and $K_4Fe(CN)_6$.

Figure 3 shows a conspicuous deviation of the values of $C/C_b$ from the
values of $I_L/I_D$ for the rotating disk. In contrast, the concentration ratios
shown in figure 124-4 are essentially independent of the hydrodynamic situation, almost coinciding with the results for the rotating disk (which are not
shown). Figure 3 reflects the strong dissimilarity of the density profile in the
supported solutions compared to that in a binary solution. A dramatic consequence of this is shown in the velocity profiles in figure 124-5. There is a
velocity maximum which becomes more pronounced as KOH is added,
and the magnitude of the velocities becomes smaller. The profile for $r = 0.95$
yields a converged but physically unreasonable solution, since the velocity
far from the electrode has reversed sign. Reasonable solutions were not
obtained in the cathodic case for $r$ greater than 0.85 for KOH, and 0.75 for
NaOH, supporting electrolyte.

The situation is different only in degree from the one encountered in the
case of supported $CuSO_4$. Normalized density profiles for the two cases are

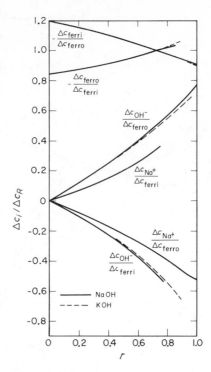

**Figure 124-4.** Surface concentrations in the supported ferricyanide-ferrocyanide systems, for equal bulk concentrations of $K_3Fe(CN)_6$ and $K_4Fe(CN)_6$.

compared in figure 124-6. The ferricyanide-ferrocyanide system has a weaker densification than $CuSO_4$, and, consequently, the addition of supporting electrolyte can have relatively a much greater effect on the density profile, as we see in figure 6.

Many of the phenomena reported here can be attributed to the large diffusion coefficient of the supporting electrolyte used. For a system where the diffusion coefficients of the solutes are roughly the same, one could estimate the value of $I_L/I_D$ from calculations for other hydrodynamic situations and then assume that this is equal to the value of $C/C_b$ for free convection with little error.

The analysis applies to large Schmidt numbers. In this limit, the present results can be applied to other geometries by using the transformation of Acrivos (see section 111). This means that the coefficient 0.6705 in equation 111-8 is replaced by $C$ or that the coefficient 0.5029 in equations 111-7 and 111-9 is replaced by $3C/4$.

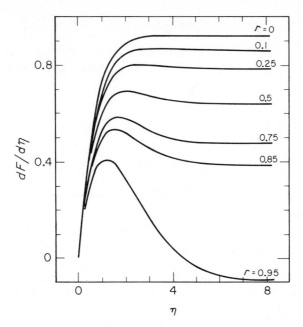

**Figure 124-5.** Velocity profiles for various values of $r$ for cathodic reduction of ferricyanide ions with KOH supporting electrolyte.

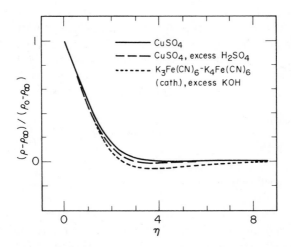

**Figure 124-6.** Normalized density profiles for binary salt solution (CuSO$_4$), for CuSO$_4$ with excess H$_2$SO$_4$ ($r = 0.99998$), and for equimolar ferricyanide-ferrocyanide with excess KOH (cathodic reaction, $c_{OH^-}/c_{K^+} = 0.95$).

## PROBLEMS

**1.** Use the results of section 114 to show that the correction factor for migration for the discharge of cations from a binary salt solution is

$$\frac{I_L}{I_D} = \frac{\left(\dfrac{D}{D_R}\right)^{2/3}}{1 - t_R}$$

for a disk electrode and

$$\frac{I_L}{I_D} = \frac{\left(\dfrac{D}{D_R}\right)^{1/2}}{1 - t_R}$$

for a growing mercury drop. Show that the corresponding expression for a stagnant Nernst diffusion layer is

$$\frac{I_L}{I_D} = \frac{\dfrac{D}{D_R}}{1 - t_R}.$$

**2.** Verify equations 121-20 and 121-21.

**3.** Show that, for a stagnant Nernst diffusion layer, equations 121-20 and 121-21 should be replaced by

$$\frac{c_4^0 - c_4^\infty}{c_R^\infty} = -\frac{s_4}{s_R}\frac{D_R}{D_4}$$

and
$$\frac{c_1^0 - c_1^\infty}{c_R^\infty} = \frac{-z_R}{z_2 - z_1} - \frac{z_4}{z_2 - z_1}\frac{s_4}{s_R}\frac{D_R}{D_4}.$$

**4.** The latter part of section 121 deals with concentration profiles at the limiting current with a large excess of supporting electrolyte. Use these results to reexamine the questions raised in problem 4 of chapter 11.

## NOTATION

| | |
|---|---|
| $a$ | 0.51023 |
| $B$ | coefficient for shear stress in free convection |
| $B_b$ | 0.932835 |
| $c_i$ | concentration of species $i$, mole/cm$^3$ |
| $C$ | coefficient for mass transfer in free convection |
| $C_b$ | 0.670327 |
| $D$ | diffusion coefficient for a binary electrolyte, cm$^2$/sec |
| $D_e$ | diffusion coefficient of supporting electrolyte, cm$^2$/sec |
| $D_i$ | diffusion coefficient of species $i$, cm$^2$/sec |
| $F$ | Faraday's constant, 96,487 C/equiv |
| $F$ | dimensionless stream function for free convection |
| $g$ | magnitude of the gravitational acceleration, cm/sec$^2$ |
| $Gr$ | Grashof number |

| $i$ | current density, $A/cm^2$ |
|---|---|
| $I$ | ionic strength, mole/liter |
| $I_D$ | limiting diffusion current |
| $I_L$ | limiting current |
| $I_r$ | *true* ionic strength, mole/liter |
| $K'$ | dissociation constant, mole/liter |
| $L$ | height of vertical electrode, cm |
| $n$ | number of electrons transferred in electrode reaction |
| $Nu$ | Nusselt number |
| $r$ | ratio of supporting electrolyte to total electrolyte |
| $R$ | universal gas constant, 8.3143 J/mole-deg |
| $s_i$ | stoichiometric coefficient of species $i$ in electrode reaction |
| $Sc$ | Schmidt number |
| $t$ | time, sec |
| $t_i$ | transference number of species $i$ |
| $T$ | absolute temperature, deg K |
| $u_i$ | mobility of species $i$, $cm^2$-mole/J-sec |
| $v_x$ | velocity component parallel to the electrode, cm/sec |
| $v_y$ | velocity component perpendicular to the electrode, cm/sec |
| $x$ | distance along the electrode measured from the beginning of the boundary layer, cm |
| $y$ | normal distance from the electrode, cm |
| $z_i$ | charge number of species $i$ |
| $\Gamma(4/3)$ | 0.89298, the gamma function of 4/3 |
| $\delta$ | thickness of Nernst stagnant diffusion layer, cm |
| $\eta$ | similarity variable for free convection |
| $\kappa$ | conductivity, mho/cm |
| $\nu$ | kinematic viscosity, $cm^2$/sec |
| $\xi$ | dimensionless distance from the electrode |
| $\rho$ | density, $g/cm^3$ |
| $\Delta\rho$ | $\lvert \rho_\infty - \rho_0 \rvert$ |
| $\tau_0$ | shear stress averaged over the electrode, dyne/$cm^2$ |
| $\phi$ | $F\Phi/RT$ |
| $\Phi$ | electric potential, V |
| $\Omega$ | rotation speed, radian/sec |

subscripts and superscripts

| avg | average |
|---|---|
| $R$ | limiting reactant |
| 0 | at the electrode surface |
| 0 | zero-order concentration of supporting ions |
| $\infty$ | in the bulk solution |
| * | see equation 122-2 |

# Concentration
# Overpotential

# 20

## 125. Definition

The concentration overpotential was defined in section 7 as the potential difference between a reference electrode adjacent to the surface, just outside the diffuse double layer, and another reference electrode in the bulk of the solution, minus the potential difference which would exist between these reference electrodes if the current distribution were unchanged but there were no concentration variations between the electrode surface and the bulk solution. These reference electrodes were to involve the same electrode reaction as that under consideration at the working electrode.

This idealized definition is based on the concept of a diffusion layer near the electrode, where the concentration variations occur, and a bulk solution, where the composition is uniform. The ohmic potential is subtracted from the measurement, so that the resulting concentration overpotential is independent of the precise placement of the reference electrode in the bulk solution. The ohmic potential which is subtracted is not the actual ohmic potential drop but is that which would prevail in a solution of uniform composition (with the same current distribution). This has the advantage that this ohmic potential drop can be calculated by solving Laplace's equation, as in chapter 18, without explicit consideration of the concentration variations near the electrode, which would destroy the validity of Laplace's equation in this region. We shall see how this works in the next chapter, where we want to treat the current distribution below, but at an appreciable fraction of, the limiting current.

In section 7 we also considered another possible decomposition of the potential variation in the solution, in which the ohmic portion is that which would disappear immediately if the current density were to become zero everywhere. This decomposition has the practical disadvantage that in most geometries interruption of the external current to an electrode does not automatically ensure that the local current density is everywhere equal to zero,[1] even in the absence of concentration variations near electrodes. It has the theoretical disadvantage that the calculation of the ohmic potential drop would then include directly some of the effect of the variation of composition.

According to the development in chapter 2, the potential $V_r$ of a movable reference electrode (relative to a fixed reference electrode) varies with position as

$$\nabla V_r = -\sum_i \frac{s_i}{nF} \nabla \mu_i, \tag{125-1}$$

where the electrode reaction for the reference electrode is given by equation 12-6. By selecting an ionic species $n$, we can write this as

$$\nabla V_r = \frac{1}{z_n F} \nabla \mu_n - \sum_i \frac{s_i}{nF} \left( \nabla \mu_i - \frac{z_i}{z_n} \nabla \mu_n \right), \tag{125-2}$$

since    $$\sum_i z_i s_i = -n. \tag{125-3}$$

Substitution of equation 16-3 yields

$$\nabla V_r = -\frac{i}{\kappa} - \sum_i \frac{s_i}{nF} \left( \nabla \mu_i - \frac{z_i}{z_n} \nabla \mu_n \right)$$
$$- \sum_j \frac{t_j^0}{z_j F} \left( \nabla \mu_j - \frac{z_j}{z_n} \nabla \mu_n \right). \tag{125-4}$$

The first two terms on the right are the ohmic potential drop and the terms relating to the specific electrode reaction. The last term represents the diffusion potential (see equation 70-7). The last two terms are expressed in terms of the gradients of the electrochemical potentials of neutral combinations of ions and are zero in the absence of concentration variations, in which case $\kappa$ is a constant.

Let us introduce the concentrations into equation 4 by means of equation 77-7. Then we have

$$\nabla V_r = -\frac{i}{\kappa} - \frac{RT}{nF} \sum_i s_i \nabla \ln c_i f_{i,n} - \frac{RT}{F} \sum_j \frac{t_j^0}{z_j} \nabla \ln c_j f_{j,n}, \tag{125-5}$$

where $f_{i,n}$ is the molar activity coefficient of species $i$ referred to the ionic species $n$ (see equation 77-9).

[1] John Newman, "Ohmic Potential Measured by Interrupter Techniques," *Journal of the Electrochemical Society, 117* (1970), 507–508.

If we subtract the ohmic drop which would exist in the absence of concentration variations and integrate across the diffusion layer, we obtain the concentration overpotential as defined above

$$\eta_c = \int_0^\infty i_y \left( \frac{1}{\kappa} - \frac{1}{\kappa_\infty} \right) dy + \frac{RT}{nF} \sum_i s_i \ln \frac{(c_i f_{i,n})_\infty}{(c_i f_{i,n})_0}$$
$$+ \frac{RT}{F} \int_0^\infty \sum_j \frac{t_j^0}{z_j} \frac{\partial \ln c_j f_{j,n}}{\partial y} dy, \qquad (125\text{-}6)$$

where $\infty$ denotes the bulk solution and $0$ denotes the electrode surface (outside the diffuse double layer). The current density $i_y$ is approximately constant in the diffusion layer and can be taken to be equal to $i_n$, the value at the electrode. For dilute solutions, we can neglect the activity coefficients and let the transference numbers be given by equation 70-5, with the result

$$\eta_c = i_n \int_0^\infty \left( \frac{1}{\kappa} - \frac{1}{\kappa_\infty} \right) dy + \frac{RT}{nF} \sum_i s_i \ln \frac{c_{i\infty}}{c_{i0}}$$
$$+ F \int_0^\infty \sum_j \frac{z_j D_j}{\kappa} \frac{\partial c_j}{\partial y} dy, \qquad (125\text{-}7)$$

where the Nernst-Einstein relation 75-1 has been used. Equation 7 can be compared with equation 30 of reference 2.

Subtraction of $i_y/\kappa_\infty$ in the integrals in equations 6 and 7 corresponds to subtracting the ohmic contribution which would exist in the absence of concentration variations. The concentration overpotential is thus the potential difference of a concentration cell plus an ohmic contribution due to the variation of conductivity within the diffusion layer, which can logically be associated with concentration variations near electrodes.

## 126. Binary electrolyte

The potentials of concentration cells involving solutions of a single electrolyte were treated in section 17. On the basis of equation 17-17, the concentration overpotential in this case can be expressed as

$$\eta_c = i_n \int_0^\infty \left( \frac{1}{\kappa} - \frac{1}{\kappa_\infty} \right) dy$$
$$- \frac{\nu RT}{F} \int_0^\infty \left( \frac{t_-^0}{z_+ \nu_+} - \frac{s_-}{n\nu_-} + \frac{s_0 c}{n c_0} \right) \frac{\partial \ln c f_{+-}}{\partial y} dy. \qquad (126\text{-}1)$$

For dilute solutions, this reduces to

$$\eta_c = \frac{i_n}{z_+ \nu_+ F^2(z_+ u_+ - z_- u_-)} \int_0^\infty \left( \frac{1}{c} - \frac{1}{c_\infty} \right) dy$$
$$+ \frac{z_+ - z_-}{z_+} \frac{RT}{F} \left( \frac{t_-^0}{z_-} + \frac{s_-}{n} \right) \ln \frac{c_\infty}{c_0}. \qquad (126\text{-}2)$$

[2]John Newman, "The Effect of Migration in Laminar Diffusion Layers," *International Journal of Heat and Mass Transfer, 10* (1967), 983–997.

(In equation 1, $c_0$ is the solvent concentration; in equation 2 it is the value of $c$ at the electrode.)

It is tempting to try to simplify this expression for the concentration overpotential even further. For the purpose of evaluating the integral in equation 2, the concentration profile could be approximated as (compare figure 102-1)

$$c = c_0 + (c_\infty - c_0)\frac{y}{\delta} \quad \text{for} \quad y < \delta \left. \right\}$$
$$c = c_\infty \qquad\qquad \text{for} \quad y > \delta, \left. \right\} \tag{126-3}$$

where $\delta$ is given by

$$\frac{\partial c}{\partial y} = \frac{c_\infty - c_0}{\delta} \quad \text{at} \quad y = 0. \tag{126-4}$$

Then

$$\int_0^\infty \left(\frac{1}{c} - \frac{1}{c_\infty}\right) dy = \frac{\delta}{c_\infty - c_0} \ln \frac{c_\infty}{c_0} - \frac{\delta}{c_\infty}$$
$$= \frac{\ln \dfrac{c_\infty}{c_0} - \dfrac{c_\infty - c_0}{c_\infty}}{\dfrac{\partial c}{\partial y}\Big|_{y=0}}. \tag{126-5}$$

Now

$$\frac{i_n}{F}\left(\frac{s_-}{n} + \frac{t_-^0}{z_-}\right) = v_- D \frac{\partial c}{\partial y}\Big|_{y=0}. \tag{126-6}$$

Consequently, with the Nernst-Einstein relation 75-1, the concentration overpotential becomes

$$\eta_c = \frac{RT}{z_+ z_- F} \frac{z_+ - z_-}{1 + \dfrac{z_- s_-}{n t_-^0}}$$
$$\times \left\{\left[1 + \frac{z_- s_-}{n}\left(2 + \frac{z_- s_-}{n t_-^0}\right)\right] \ln \frac{c_\infty}{c_0} - t_+^0\left(1 - \frac{c_0}{c_\infty}\right)\right\}, \tag{126-7}$$

and, for a metal deposition reaction ($s_- = 0$), this reduces to

$$\eta_c = \frac{(z_+ - z_-)RT}{z_+ z_- F}\left[\ln \frac{c_\infty}{c_0} - t_+^0\left(1 - \frac{c_0}{c_\infty}\right)\right]. \tag{126-8}$$

This is the basis for equation 7-6.

## 127. Supporting electrolyte

For solutions with an excess of supporting electrolyte, it should be possible to neglect conductivity variations in the diffusion layer. Then equation 125-7 for the concentration overpotential becomes

$$\eta_c = \frac{RT}{nF}\sum_i s_i \ln \frac{c_{i\infty}}{c_{i0}} + \frac{F}{\kappa_\infty}\sum_j z_j D_j(c_{j\infty} - c_{j0}). \tag{127-1}$$

Now, the last term is also on the order of the reactant concentration divided by the supporting electrolyte concentration and can be neglected, with the result

$$\eta_c = \frac{RT}{nF} \sum_i s_i \ln \frac{c_{i\infty}}{c_{i0}}. \qquad (127\text{-}2)$$

This is the basis for equation 10-1.

## 128.  Calculated values

The most salient feature of the concentration overpotential is that it becomes infinite when the concentration of one of the reactants becomes zero at the electrode, corresponding to the limiting current. The concentration overpotential also allows us to calculate the current-potential relationship for a complete cell, as presented in figures 9-1 and 10-2. (There we see that the surface overpotential also becomes infinite at the limiting current, because the exchange current density goes to zero.) For many cell geometries, the calculations are more difficult. This will be treated in the next chapter.

The nature of concentration overpotentials can be revealed by some examples. Figure 7-2 was calculated for a rotating disk electrode[3] by using equation 125-6 or 126-1, that is, without approximations for dilute solutions. The values of $i$ and $\eta_c$ are those for the center of the disk, in case the current distribution is nonuniform.

Tables 1 and 2 give values of $\eta_c$ calculated according to equation 125-7

TABLE 128-1. VALUES OF CONCENTRATION OVERPOTENTIAL $\eta_c$ (in mV) FOR
COPPER DEPOSITION ON A ROTATING DISK FROM SOLUTIONS OF COPPER
SULFATE AND SULFURIC ACID, WITH COMPLETE DISSOCIATION OF BISULFATE
IONS.

| $c_{R0}/c_{R\infty} \longrightarrow$ | 0.1 | 0.25 | 0.5 | 0.7 | 0.9 | 0.95 |
|---|---|---|---|---|---|---|
| $r$ | | | | | | |
| $0^a$ | −49.84 | −27.85 | −12.63 | −6.06 | −1.67 | −0.800 |
| 0 | −50.52 | −28.27 | −12.79 | −6.10 | −1.67 | −0.798 |
| 0.25 | −37.35 | −23.94 | −12.65 | −6.69 | −2.01 | −0.984 |
| 0.5 | −34.23 | −21.62 | −11.38 | −6.04 | −1.83 | −0.897 |
| 0.7 | −32.18 | −19.96 | −10.33 | −5.43 | −1.63 | −0.799 |
| 0.9 | −30.39 | −18.48 | − 9.35 | −4.85 | −1.44 | −0.704 |
| 0.99 | −29.66 | −17.87 | − 8.95 | −4.61 | −1.36 | −0.663 |
| $1^b$ | −29.58 | −17.81 | − 8.90 | −4.58 | −1.35 | −0.659 |

$^a$Equation 126-8
$^b$Equation 127-2

$r = c_{H^+}^\infty / 2c_{SO_4^=}^\infty$

[3]J. Newman and L. Hsueh, "The Effect of Variable Transport Properties on Mass Transfer to a Rotating Disk," *Electrochimica Acta*, 12 (1967), 417–427.

TABLE 128–2. VALUES OF CONCENTRATION OVERPOTENTIAL $\eta_c$ (in mV) FOR
REDUCTION OF FERRICYANIDE IONS ON A ROTATING DISK FROM SOLUTIONS
EQUIMOLAR IN POTASSIUM FERRICYANIDE AND POTASSIUM FERROCYANIDE
AND WITH VARIOUS AMOUNTS OF ADDED POTASSIUM HYDROXIDE.

| $c_{R0}/c_{R\infty} \longrightarrow$ | 0.1 | 0.25 | 0.5 | 0.7 | 0.9 | 0.95 |
|---|---|---|---|---|---|---|
| $r$ | | | | | | |
| 0 | −74.22 | −48.67 | −27.15 | −15.14 | −4.84 | −2.41 |
| 0.25 | −74.85 | −49.22 | −27.57 | −15.42 | −4.95 | −2.46 |
| 0.5 | −75.47 | −49.79 | −28.02 | −15.73 | −5.08 | −2.53 |
| 0.7 | −76.04 | −50.32 | −28.44 | −16.03 | −5.20 | −2.59 |
| 0.9 | −76.78 | −51.01 | −29.00 | −16.44 | −5.36 | −2.68 |
| 0.99 | −77.21 | −51.41 | −29.34 | −16.68 | −5.46 | −2.73 |
| 1 | −77.27 | −51.47 | −29.38 | −16.71 | −5.47 | −2.74 |

$r = c_{OH^-}^\infty / c_{K^+}^\infty$.

for a rotating disk electrode. The concentration profiles were calculated with the computer program in appendix C for the effect of migration on limiting currents, but with a nonzero value for the reactant concentration at the electrode. Thus, approximations for dilute solutions are already introduced: variations of activity coefficients are neglected, equations 70-3 and 70-5 are used for the conductivity and transference numbers, and the Nernst-Einstein relation 75-1 is used.

In table 1 for copper deposition, comparison with equations 126-8 and 127-2 is made, thus providing a check on the error involved in the additional approximations used to derive these equations. The presence of the diffusion potential shows up in this table. For example, the value $-6.69$ mV for $c_{R0}/c_{R\infty} = 0.7$ and $r = 0.25$ is greater in magnitude than the corresponding values for $r = 0$ and $r = 1$. In table 2 for the reduction of ferricyanide ions, equation 127-2 works well throughout the range of the table, being exact for $r = 1$.

## PROBLEMS

**1.** For the electrode reaction
$$Pb(s) + SO_4^= \rightleftharpoons PbSO_4(s) + 2e^-$$
in a sulfuric acid solution supposed to be dissociated into hydrogen and sulfate ions, show that the concentration overpotential can be approximated as
$$\eta_c = -\frac{(z_+ - z_-)RT}{z_+ z_- F}\left[\ln\frac{c_\infty}{c_0} - t_-^0\left(1 - \frac{c_0}{c_\infty}\right)\right].$$

**2.** For the electrode reaction
$$PbO_2(s) + SO_4^= + 4\,H^+ + 2e^- \rightleftharpoons PbSO_4(s) + 2\,H_2O$$
in a sulfuric acid solution supposed to be dissociated into hydrogen and sulfate ions, show that the concentration overpotential can be approximated

as

$$\eta_c = \frac{(z_+ - z_-)RT}{z_+ z_- F}\left[\frac{1 + 3t^0_-}{1 + t^0_-}\ln\frac{c_\infty}{c_0} - \frac{t^0_+ t^0_-}{1 + t^0_-}\left(1 - \frac{c_0}{c_\infty}\right)\right].$$

**3.** For an anode, equation 126-8 can give negative values for $\eta_c$. Explain how this situation arises and discuss whether any basic physical laws are violated by having a negative overpotential at an anode.

**4.** Using the ferrous-ferric redox reaction

$$Fe^{++} \rightleftharpoons Fe^{+++} + e^-$$

in an excess of supporting electrolyte as an example, show that equations 57-15 and 57-16 can be rearranged to yield the rate equation

$$i = i^\infty_0\left\{\frac{c_{10}}{c_{1\infty}}\exp\left[\frac{(1 - \beta)nF}{RT}\eta\right] - \frac{c_{20}}{c_{2\infty}}\exp\left[-\frac{\beta nF}{RT}\eta\right]\right\},$$

where species 1 is the ferrous ion, species 2 is the ferric ion, $\eta$ is the total overpotential $\eta_s + \eta_c$, and $i^\infty_0$ is the exchange current density at the bulk composition, having the composition dependence

$$i^\infty_0 = nFk^{1-\beta}_c k^\beta_a c^\beta_{1\infty} c^{1-\beta}_{2\infty}$$

if the rules for reaction orders after equation 57-12 are followed.

## NOTATION

| | |
|---|---|
| $c$ | concentration of binary electrolyte, mole/cm³ |
| $c_i$ | concentration of species $i$, mole/cm³ |
| $D$ | diffusion coefficient of binary electrolyte, cm²/sec |
| $D_i$ | diffusion coefficient of species $i$, cm²/sec |
| $f_{i,n}$ | molar activity coefficient of species $i$ referred to species $n$ |
| $f_{+-}$ | mean molar activity coefficient of binary electrolyte |
| $F$ | Faraday's constant, 96,487 C/equiv |
| $i$ | current density, A/cm² |
| $n$ | number of electrons involved in the electrode reaction |
| $R$ | universal gas constant, 8.3143 J/mole-deg |
| $s_i$ | stoichiometric coefficient of species $i$ in electrode reaction |
| $t^0_i$ | transference number of species $i$ with respect to the velocity of species 0 |
| $T$ | absolute temperature, deg K |
| $u_i$ | mobility of species $i$, cm²-mole/J-sec |
| $V_r$ | potential of a reference electrode, V |
| $y$ | normal distance from an electrode, cm |
| $z_i$ | charge number of species $i$ |
| $\delta$ | equivalent diffusion-layer thickness, cm |
| $\eta_c$ | concentration overpotential, V |
| $\kappa$ | conductivity, mho/cm |

$\mu_i$        electrochemical potential of species $i$, J/mole

$\nu$         $\nu_+ + \nu_-$

$\nu_+, \nu_-$    numbers of cations and anions produced by dissociation of one molecule of electrolyte

subscripts

0         at the electrode surface

$\infty$       in the bulk solution

# Currents below the
# Limiting Current

**21**

At currents below, but at an appreciable fraction of, the limiting current, it is necessary to consider concentration variations near electrodes, the surface overpotential associated with the electrode reaction, and the ohmic potential drop in the bulk of the solution. These problems are inherently of greater complexity than either the convective-transport problems or the potential-theory problems, treated in chapters 17 and 18, in which one or more of these factors could be ignored.

In many electrolytic cells, the concentration variations are still restricted to thin diffusion layers near the electrodes, and Laplace's equation still applies to the bulk of the solution outside these diffusion layers. This means that one can devote separate attention to these different regions. Since the diffusion layers are thin, the bulk region essentially fills the region of the electrolytic solution bounded by the walls of the cell and the electrodes. In this region, the potential is determined so as to satisfy Laplace's equation and agree with the current density distribution on the boundaries of the region. In the diffusion layers, the concentrations are determined so as to satisfy the appropriate form of the transport equations, with a mass flux at the wall appropriate to the current density distribution on the electrodes and with the concentration approaching the bulk concentrations far from the electrode. The current distribution and concentrations at the electrode surface must adjust themselves so as to agree with the overpotential variation determined from the calculation of the potential in the bulk region. We are thus faced with a singular-perturbation problem, and the treatment of the two regions is coupled through the boundary conditions.

The thinness of the diffusion layers also allows one to separate the irreversible part of the cell potential into the sum of the surface overpotentials, the concentration overpotentials, and the ohmic potential drop in the solution (see sections 9 and 10). The surface overpotential has been defined and discussed in sections 8 and 101 and chapter 8. It is related to the concentrations and current density at the electrode surface by the polarization equation 101-3. The surface overpotential varies with position on the electrode unless the concentrations and current density are uniform on the electrode. The concentration overpotential was discussed in section 7 and chapter 20. In general, the concentration overpotential also depends upon position along the electrode surface.

Asada et al.[1] have used a separate treatment of the diffusion layers and the bulk solution to treat free convection in a rectangular cell with a vertical electrode at each end for currents below the limiting current. Newman has given a detailed justification for such a procedure for systems with laminar, forced convection[2] and has applied the method to the rotating disk electrode.[3,4,5] References 6 to 11 reflect on the experimental verification of the results. The problem for two electrodes of length L placed opposite each other

[1] Kameo Asada, Fumio Hine, Shiro Yoshizawa, and Shinzo Okada, "Mass Transfer and Current Distribution under Free Convection Conditions," *Journal of the Electrochemical Society*, *107* (1960), 242–246.

[2] John Newman, "The Effect of Migration in Laminar Diffusion Layers," *International Journal of Heat and Mass Transfer*, *10* (1967), 983–997.

[3] John Newman, "Current Distribution on a Rotating Disk below the Limiting Current," *Journal of the Electrochemical Society*, *113* (1966), 1235–1241.

[4] John Newman, "The Diffusion Layer on a Rotating Disk Electrode," *Journal of the Electrochemical Society*, *114* (1967), 239.

[5] W. R. Parrish and John Newman, "Current Distribution on a Plane Electrode below the Limiting Current," *Journal of the Electrochemical Society*, *116* (1969), 169–172.

[6] D. H. Angell, T. Dickinson, and R. Greef, "The Potential Distribution near a Rotating-Disk Electrode," *Electrochimica Acta*, *13* (1968), 120–123.

[7] W. J. Albery and J. Ulstrup, "The Current Distribution on a Rotating Disk Electrode," *Electrochimica Acta*, *13* (1968), 281–284.

[8] Vinay Marathe and John Newman, "Current Distribution on a Rotating Disk Electrode," *Journal of the Electrochemical Society*, *116* (1969), 1704–1707.

[9] Stanley Bruckenstein and Barry Miller, "An Experimental Study of Nonuniform Current Distribution at Rotating Disk Electrodes," *Journal of the Electrochemical Society*, *117* (1970), 1044–1048.

[10] William H. Smyrl and John Newman, "Ring-Disk and Sectioned Disk Electrodes," *Journal of the Electrochemical Society*, *119* (1972), 212–219.

[11] William H. Smyrl and John Newman, "Detection of Nonuniform Current Distribution on a Disk Electrode," *Journal of the Electrochemical Society*, *119* (1972), 208–212.

at a distance $h$, embedded in the walls of a flow channel with steady, laminar flow, is formulated in reference 12 and has been worked out by Parrish and Newman.[5,13]

## 129. The bulk medium

We deal here with forced-convection systems where the hydrodynamic velocity distribution can be assumed to be known. When the Péclet number $Pe = UL/D_R$ (where $U$ is a characteristic velocity and $L$ is a characteristic length) is large, mass transfer by convection predominates over diffusion except in a thin diffusion layer near an electrode surface. Outside these diffusion layers, in the bulk solution, the concentrations are uniform; and the potential satisfies Laplace's equation (see section 115). We use a tilde to denote the potential and current distributions in this region. Thus, we have

$$\nabla^2 \tilde{\Phi} = 0. \qquad (129\text{-}1)$$

In the bulk medium, the current density is related to the potential gradient by Ohms' law

$$\tilde{\mathbf{i}} = -\kappa_\infty \nabla \tilde{\Phi}. \qquad (129\text{-}2)$$

In this singular-perturbation problem, the diffusion layers approach zero thickness as the Péclet number approaches infinity. Consequently, we solve Laplace's equation in the region confined by the electrodes and the insulating walls of the cell, as though the diffusion layers were not present. At the walls, the current density in the bulk region must match with the current density in the outer limit of the diffusion layer. In section 131, we argue that the normal component of the current density changes very little in the thin diffusion layer and is essentially equal to the value at the wall. Therefore, the boundary condition for Laplace's equation is

$$\frac{\partial \tilde{\Phi}}{\partial y} = -\frac{i_n}{\kappa_\infty} \quad \text{at} \quad y = 0, \qquad (129\text{-}3)$$

where $y$ is the distance from the wall and $i_n$ is the $y$-component of $\mathbf{i}$ at the wall. Thus, $i_n$ represents the contribution to the external current flowing to an electrode and is zero on insulating surfaces. On electrodes, $i_n$ is not known until we have solved simultaneously for the bulk medium and the diffusion layers.

Laplace's equation is to be solved for the bulk medium in much the same way as in chapter 18, where there were no concentration variations; the same methods can be used, and geometric arrangements which proved

[12]John Newman, "Engineering Design of Electrochemical Systems," *Industrial and Engineering Chemistry, 60* (no. 4) (April, 1968), 12–27.

[13]W. R. Parrish and John Newman, "Current Distributions on Plane, Parallel Electrodes in Channel Flow," *Journal of the Electrochemical Society, 117* (1970), 43–48.

intractable there would be equally difficult to treat here. For plane electrodes in the walls of a flow channel,[5,12,13] an integral equation can be used to relate the potential and the normal component of the potential gradient at the wall, for the solution of Laplace's equation. For the disk electrode, the same thing can be accomplished by using the coefficients of an infinite series[3] obtained by the method of separation of variables, although an integral equation could also be used.[14] Where these methods can be used, they are superior, in terms of computational effort and accuracy, to a numerical solution of Laplace's equation in the bulk medium by finite differences.

If $V_{met}$ is the potential of the electrode metal and $\tilde{\Phi}$ in the bulk solution is that measured by a reference electrode of the same kind as the working electrode, then the total overpotential at the electrode is

$$\eta = V_{met}(x) - \tilde{\Phi}_0(x), \qquad (129\text{-}4)$$

where $x$ is distance measured along the electrode and $\tilde{\Phi}_0$ is the value of $\tilde{\Phi}$ evaluated at $y = 0$. This is the sum of the concentration overpotential $\eta_c$, associated with concentration changes in the diffusion layer, and the surface overpotential $\eta_s$, associated with the heterogeneous electrode reaction,

$$\eta = \eta_c + \eta_s. \qquad (129\text{-}5)$$

We can see that this conforms to our previous definitions of $\eta_c$ and $\eta_s$ in terms of reference electrodes located outside the diffuse double layer and in the bulk solution. Since $\tilde{\Phi}_0$ is the value of $\tilde{\Phi}$ at $y = 0$, subtracting it from $V_{met}$ in equation 4 corresponds to subtracting the ohmic potential drop in the bulk solution, calculated with the actual current distribution but extrapolated to the electrode surface with a constant conductivity $\kappa_\infty$, as though there were no concentration variations in the diffusion layer. Hence $\eta$, and $\eta_c$ in particular, includes only the ohmic potential drop associated with concentration variations in the diffusion layer.

## 130.  The diffusion layers

Because of the thinness of the diffusion layer, effects of curvature can be neglected, and we adopt the usual boundary layer coordinates: $x$ measured along the electrode from its upstream end and $y$, the normal distance from the surface. In the diffusion layer, the transport equation simplifies to

$$\frac{\partial c_i}{\partial t} + v_x \frac{\partial c_i}{\partial x} + v_y \frac{\partial c_i}{\partial y} = D_i \frac{\partial^2 c_i}{\partial y^2} + z_i u_i F \left( c_i \frac{\partial^2 \Phi}{\partial y^2} + \frac{\partial c_i}{\partial y} \frac{\partial \Phi}{\partial y} \right). \qquad (130\text{-}1)$$

On the right side, derivatives with respect to $x$ have been ignored compared to the derivatives with respect to $y$.

---

[14]John Newman, "The Fundamental Principles of Current Distribution and Mass Transport in Electrochemical Cells," Allen J. Bard, ed., *Electroanalytical Chemistry* (New York: Marcel Dekker, Inc.), to be published.

We shall also assume that the Schmidt number $Sc = \nu/D_R$ is large. This means that the diffusion layer is thin even when compared with any hydrodynamic boundary layer that may be present, and, within the diffusion layer, the velocity components can be represented as (see equations 106-2 and 107-2)

$$v_x = y\beta(x) \quad \text{and} \quad v_y = -\frac{1}{2}y^2 \frac{(\Re\beta)'}{\Re}, \tag{130-2}$$

where $\beta(x)$ is the velocity derivative at the solid wall, $\beta = \partial v_x/\partial y$ at $y = 0$, and the prime denotes the derivative with respect to $x$. These equations apply to two-dimensional and axisymmetric diffusion layers; for a two-dimensional diffusion layer, $\Re(x)$ is to be set equal to 1.

In the equations 1 for mass transfer in the diffusion layer, only derivatives of potential with respect to $y$ appear and not $\Phi$ itself or the $x$ derivative of $\Phi$. Consequently, we can introduce a new potential $\phi$ in the diffusion layer, defined as

$$\phi = \Phi(x, y) - \tilde{\Phi}_0(x), \tag{130-3}$$

or we can even assign the zero of $\phi$ arbitrarily at each value of $x$. Then $\tilde{\Phi}_0(x)$ is important only in the determination of the total overpotential $\eta$.

Matters are simplified considerably if we are further willing to neglect migration in the diffusion layer, so that equation 1 becomes

$$\frac{\partial c_i}{\partial t} + y\beta \frac{\partial c_i}{\partial x} - \frac{y^2(\Re\beta)'}{2\Re} \frac{\partial c_i}{\partial y} = D_i \frac{\partial^2 c_i}{\partial y^2}. \tag{130-4}$$

Even though the ohmic potential drop in the bulk solution has an important effect on the variation of the total overpotential along the electrode surface, migration within the diffusion layer does not have a crucial effect on the current distribution. We have seen that, at the limiting current, the effect of migration is to change the magnitude but not the distribution of the current.

Equation 4 applies if there is an excess of supporting electrolyte (see section 73). However, the importance of the ohmic drop in the bulk solution depends on the ratio of a characteristic length to the conductivity $\kappa_\infty$, as brought out in chapter 18, and this ratio can be large even in the presence of supporting electrolyte. Equation 4 also applies to the other extreme case, solutions of a binary electrolyte (see section 72), where $D_i$ is to be replaced by $D$, the diffusion coefficient of the electrolyte.

Equations 1, one for each species, are to be solved along with the electroneutrality equation 100-3 for the concentrations and the potential. For the simplified case, equation 4 need be solved only for those species that participate in the electrode reaction.

The concentrations approach their bulk values as $y$ approaches infinity. At the electrode surface, the fluxes are related to the current density by equation 101-2:

$$N_{in} = -\frac{s_i}{nF}i_n \quad \text{at} \quad y = 0. \tag{130-5}$$

When migration can be ignored, this becomes

$$\frac{\partial c_i}{\partial y} = \frac{s_i i_n}{nFD_i} \quad \text{at} \quad y = 0. \tag{130-6}$$

The equations of section 113 then allow us to solve the diffusion-layer equations 4 (in the steady state) and to relate the surface concentration to the concentration derivative at the surface. Further reference to the diffusion-layer equations is then unnecessary. Substitution of equation 6 into equation 113-1 gives

$$\frac{s_i i_n(x)}{nFD_i} = -\frac{\sqrt{\Re\beta}}{\Gamma(4/3)} \int_0^x \frac{dc_{i0}}{dx}\bigg|_{x=x_0} \frac{dx_0}{\left(9D_i \int_{x_0}^x \Re\sqrt{\Re\beta}\,dx\right)^{1/3}}, \tag{130-7}$$

where $c_{i0}(x)$ is the surface concentration of species $i$. In applications, the fact that this is a Stieltjes integral should be borne in mind (see problem 8 of chapter 17).

## 131.  Boundary conditions and method of solution

Certain boundary conditions have already been discussed in connection with the diffusion layers and the bulk medium. The solution for the potential $\Phi$ in the bulk solution must satisfy the condition 129-3, relating the potential derivative to the external current density. It also provides the total overpotential through equation 129-4. The solution in the diffusion layer requires matching the fluxes and the current density through equation 130-5. This is already incorporated into equation 130-7.

The current densities for the diffusion layer and the bulk medium must match. The current density satisfies equation 71-2:

$$\nabla \cdot \mathbf{i} = 0. \tag{131-1}$$

In a two-dimensional diffusion layer, this can be written as

$$\frac{\partial i_x}{\partial x} + \frac{\partial i_y}{\partial y} = 0 \tag{131-2}$$

or

$$i_y = i_n - \int_0^y \frac{\partial i_x}{\partial x}\,dy. \tag{131-3}$$

Since the diffusion layer is thin, $i_y$ is approximately constant throughout the thickness of the diffusion layer, and, therefore, the value at the wall is appropriate to use in the boundary condition 129-3 for the solution of Laplace's equation in the bulk medium.

It remains to adopt expressions for the concentration overpotential $\eta_c$ and the surface overpotential $\eta_s$. The former can be calculated from equation 125-7. However, equations 126-8 and 127-2 have the advantage of involving only the concentrations at the surface and not the concentration profiles in the diffusion layer. They are therefore appropriate to use if equation 130-7 has been adopted. For many electrodes, the surface overpotential can be related to the current density through equation 101-4:

$$i_n = i_0 \left[ \exp\left(\frac{\alpha_a F}{RT}\eta_s\right) - \exp\left(-\frac{\alpha_c F}{RT}\eta_s\right) \right], \qquad (131\text{-}4)$$

where the exchange current density $i_0$ depends on the composition of the solution adjacent to the electrode.

The situation at this point may be confusing, particularly with regard to the diffusion layer, because we have presented several alternative equations. Let us suppose that we have only one reactant species and that we have adopted equation 130-7 as the diffusion-layer equation and either equation 126-8 or 127-2 for the concentration overpotential. The principal unknowns then are the current density and the concentration at the electrode surface. These must adjust themselves so as to agree with the total overpotential $\eta$ available after subtracting the ohmic potential drop from the electrode potential.

Suppose we know the distribution of $\eta$ along the electrode. Then $\eta_s$ in equation 4 can be replaced by $\eta - \eta_c$, where $\eta_c$ is related to the surface concentration by equation 126-8 or 127-2. Substitution of equation 4 into equation 130-7 then gives an integral equation for the surface concentration. The numerical solution of this integral equation is actually fairly simple since there is no upstream propagation of effects in the diffusion layer, and a nonlinear equation for $c_{i0}$ need be solved only once at each value of $x$.

The following procedure might be suggested for solving the problem:

1. Assume a distribution of $i_n(x)$ along the surfaces of the electrodes.

2. Calculate the potential in the bulk medium from Laplace's equation and boundary condition 129-3. There is an arbitrary, additive constant in the solution.

3. Calculate the distribution of total overpotential $\eta$ along an electrode.

4. Solve the integral equation for the surface concentration along this electrode. This integral equation is formed as described above from equations 130-7, 131-4, 126-8 or 127-2, and 129-5. In addition to the surface concentration, this calculation also yields the current distribution $i_n$ and the split of the total overpotential into $\eta_c$ and $\eta_s$.

5. In calculating the total overpotential in step 3, there is an uncertainty in the additive constant which is removed by specifying the electrode potential and the additive constant in step 2. In the case of a single electrode whose behavior is not influenced by the placement of the counter electrode, a trial-and-error calculation can be avoided at this point by specifying the electrode potential relative to a suitably placed reference electrode, or the current density or total overpotential at the beginning of the diffusion layer. However, if the total current to the electrode is specified, it is now necessary to adjust the constant in step 3 until this current is reached. Thus, steps 3 and 4 must be repeated until this condition is satisfied. If there are two

electrodes which directly influence each other, it is necessary to specify the total current (so that it will be the same on both electrodes), and repetition of steps 3 and 4 is necessary.

6. If there are two electrodes, steps 3 to 5 must now be carried out for the second electrode.

7. Steps 3 to 6 yield a new current distribution $i_n$ on the electrodes, which may be different from that used in step 2. Steps 2 to 6 must be repeated with a new current density distribution, and this must be continued until the distribution obtained from steps 3 to 6 agrees sufficiently well with that used in step 2. Convergence of this procedure can usually be accomplished if the new current distribution chosen is some average of the previous distribution used in step 2 and that produced by steps 3 to 6.

## 132.  Results for the rotating disk

Let us look at the current distribution on a rotating disk electrode[3,5] (see figure 103-1) embedded in a larger insulating disk, both of which rotate about their axis in an electrolytic solution. The counter electrode is supposed to be far enough away that it does not affect the current distribution on the disk electrode. The limiting current was treated in sections 103, 114, and 120 and is distributed uniformly over the surface of the electrode. The primary and secondary current distributions were discussed in sections 116 and 117.

It is assumed that there is one reactant whose concentration is important, and this concentration in the diffusion layer obeys the equation

$$ay\Omega\sqrt{\frac{\Omega}{v}}\left(r\frac{\partial c_R}{\partial r} - y\frac{\partial c_R}{\partial y}\right) = D\frac{\partial^2 c_R}{\partial y^2}, \tag{132-1}$$

a form of equation 130-4. For a solution of a single salt, $D$ is the diffusion coefficient of the salt. For reaction of a minor component in a solution with excess supporting electrolyte, $D$ denotes the diffusion coefficient of the reactant. In both cases, the normal current density at the electrode surface is given by

$$\frac{s_R i_n}{nF} = \frac{D}{1-t}\frac{\partial c_R}{\partial y} \quad \text{at} \quad y = 0, \tag{132-2}$$

where $t$ is the transference number of the reactant.

For the concentration overpotential we used the expression

$$\eta_c = -\frac{RT}{ZF}\left[\ln\left(\frac{c_\infty}{c_0}\right) - t\left(1 - \frac{c_0}{c_\infty}\right)\right]. \tag{132-3}$$

This corresponds to metal deposition from a solution of a single salt (see equation 126-8) if we take $t$ to be the transference number of the reacting cation and set

$$Z = -\frac{z_+ z_-}{z_+ - z_-}. \tag{132-4}$$

Equation 1 applies approximately to the reaction of a minor component from a solution with excess indifferent electrolyte (see equation 127-2) if we set $t = 0$ and $Z = -n/s_R$.

The electrode kinetics are expressed by equation 101-4 where we have set $\alpha_a = \alpha Z$ and $\alpha_c = \beta Z$ and have given the concentration dependence of the exchange current density the form

$$i_0(c_0) = \left(\frac{c_0}{c_\infty}\right)^\gamma i_0(c_\infty). \tag{132-5}$$

The current distribution is then determined by seven dimensionless parameters. These are a dimensionless exchange current density $J$, a dimensionless average current density $\delta$, a dimensionless limiting current density $N$, the transference number $t$, the exponent $\gamma$ in the concentration dependence of the exchange current density, and $\alpha$ and $\beta$ in the kinetic equation. $J$, $\delta$, and $N$ are given by

$$J = \frac{ZFr_0}{RT\kappa_\infty} i_0(c_\infty), \qquad \delta = \frac{ZFr_0}{RT\kappa_\infty} i_{\text{avg}}, \tag{132-6}$$

and

$$N = -\left(\frac{r_0^2\Omega}{\nu}\right)^{1/2}\left(\frac{a\nu}{3D}\right)^{1/3}\frac{nZF^2Dc_\infty}{s_RRT(1-t)\kappa_\infty}, \tag{132-7}$$

where $r_0$ is the radius of the disk electrode.

Figure 132-1 shows the distribution of the reactant concentration at the electrode surface. Due to the ohmic potential drop, there tends to be a higher current density near the edge of the disk, and this produces a decrease in the concentration. The distribution is more nonuniform for higher rotation

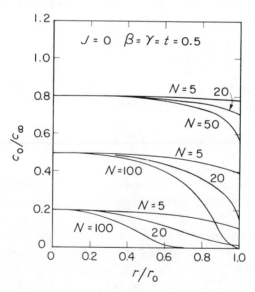

**Figure 132-1.** Surface concentration for Tafel kinetics.

speeds $N^2$, but the concentration cannot become negative. The disk becomes mass-transfer limited first near the edge. Figure 132-2 shows the corresponding current distribution, expressed as $i/i_{\text{lim}}$. The local current density can rise above the average limiting current density because of the nonuniform poten-

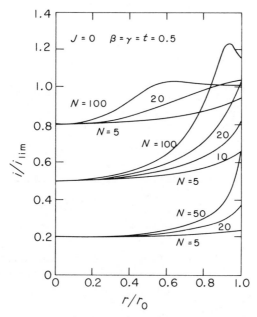

**Figure 132-2.** Current distribution for Tafel kinetics with an appreciable fraction of the limiting current.

tial drop, but it is likely to decrease again toward the edge due to the limited rate of mass transfer by convection and diffusion. Higher values of $N$ lead again to a more nonuniform distribution.

The corresponding curves on figures 1 and 2 can be identified by the fact that

$$\frac{i}{i_{\text{lim}}} = 1 - \frac{c_0}{c_\infty} \tag{132-8}$$

at the center of the disk.

Figure 132-3 shows how the parameters of the system affect the non-uniformity of the current distribution (compare figure 117-3), but this time the effect of concentration variations is included ($N$) along with the current level ($\delta$), while $J$ is restricted to the reversible and Tafel cases. Notice that the mass-transfer limitations do not ensure a uniform current density except very close to the limiting current.

Figure 132-4 shows the polarization curve for copper deposition from

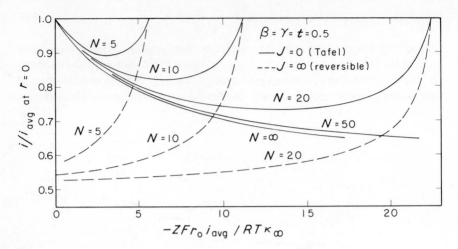

**Figure 132-3.** Current density at the center of the disk.

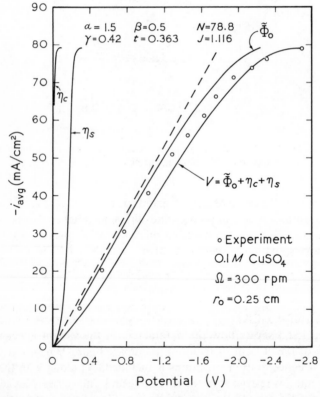

**Figure 132-4.** Overpotentials for copper deposition on a rotating disk. Dashed line is ohmic drop for the primary current distribution. $\tilde{\Phi}_0$, $\eta_c$, and $\eta_s$ are evaluated at the center of the disk.

a 0.1 *M* cupric sulfate solution.[15] The concentration and surface overpotentials at the center of the disk contribute relatively little compared to the ohmic potential drop in this solution of low conductivity. The ohmic potential is not linear since the current distribution changes near the limiting current. Additional parameters for this system are $i_0 = 1$ mA/cm² and $\kappa_\infty = 0.00872$ mho/cm for a 0.1 *M* CuSO₄ solution at 25°C.

Figure 132-5 shows the total overpotential $\eta$ at the center of the disk,

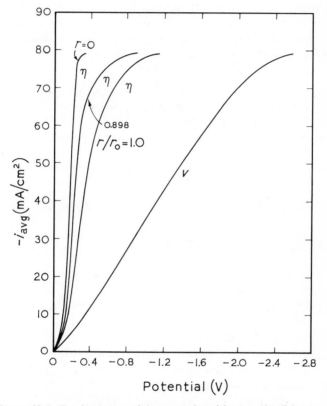

**Figure 132-5.** Total overpotential at several positions on the disk.

at the edge of the disk, and at $r = 0.898\ r_0$. The ohmic drop $\Phi_0$ also depends on radial position in such a way that the electrode potential $V = \Phi_0 + \eta$ is uniform. Figure 5 shows that the overpotential at the edge can be 0.8 V, while that at the center is only 0.25 V. Thus, hydrogen may begin to be evolved at the edge before the center has reached the limiting current.

[15]L. Hsueh and J. Newman, "Mass Transfer and Polarization at a Rotating Disk Electrode," *Electrochimica Acta*, *12* (1967), 429–438.

## PROBLEMS

**1.** Suppose that we wish to treat the current distribution in an electro-chemical system involving a redox reaction and an excess of supporting electrolyte. Use equation 113-2 to show that the product concentration at the surface can be related to that of the reactant according to

$$\frac{c_{40} - c_{4\infty}}{c_{R0} - c_{R\infty}} = \frac{s_4}{s_R}\left(\frac{D_R}{D_4}\right)^{2/3}$$

(compare equation 121-6), where the product of the reaction is labeled as species 4. Use this result to show that the concentration overpotential according to equation 127-2 can be expressed, for ferricyanide reduction from a solution equimolar in ferricyanide and ferrocyanide, as

$$\eta_c = -\frac{RT}{F}\ln\left[\frac{c_{R\infty}}{c_{R0}} + \left(\frac{D_R}{D_4}\right)^{2/3}\left(\frac{c_{R\infty}}{c_{R0}} - 1\right)\right].$$

These results mean that only one diffusion-layer equation of the form 130-7 need be solved for each diffusion layer.

**2.** For the rotating disk electrode, show that the parameter $N$ can be given as

$$N = -\Gamma(4/3)\frac{ZFr_0}{RT\kappa_\infty}i_{\lim},$$

thus substantiating that it is a dimensionless limiting current density.

## NOTATION

| | |
|---|---|
| $a$ | 0.51023 |
| $c_i$ | concentration of species $i$, mole/cm$^3$ |
| $D$ | diffusion coefficient of reactant or of binary electrolyte, cm$^2$/sec |
| $D_i$ | diffusion coefficient of species $i$, cm$^2$/sec |
| $F$ | Faraday's constant, 96,487 C/equiv |
| $h$ | distance between walls of a flow channel, cm |
| $i_n$ | normal current density at electrode surface, A/cm$^2$ |
| $i_0$ | exchange current density, A/cm$^2$ |
| $J$ | dimensionless exchange current density |
| $L$ | electrode length, cm |
| $L$ | characteristic length, cm |
| $n$ | number of electrons transferred in electrode reaction |
| $N$ | dimensionless limiting current density |
| $N_{in}$ | normal component of the flux of species $i$, mole/cm$^2$-sec |
| $Pe$ | Péclet number |
| $r$ | radial distance, cm |
| $r_0$ | radius of disk electrode, cm |
| $R$ | universal gas constant, 8.3143 J/mole-deg |

| | |
|---|---|
| $\mathcal{R}$ | distance of axisymmetric surface from axis of symmetry, cm |
| $s_i$ | stoichiometric coefficient of species $i$ in electrode reaction |
| $Sc$ | Schmidt number |
| $t$ | time, sec |
| $t$ | transference number of reactant |
| $T$ | absolute temperature, deg K |
| $u_i$ | mobility of species $i$, cm²-mole/J-sec |
| $U$ | characteristic velocity, cm/sec |
| $\mathbf{v}$ | fluid velocity, cm/sec |
| $V$, $V_{met}$ | electrode potential, V |
| $x$ | distance along electrode from its upstream edge, cm |
| $y$ | distance from electrode surface, cm |
| $z_i$ | charge number of species $\mathbf{i}$ |
| $Z$ | $-z_+ z_-/(z_+ - z_-)$ for binary electrolyte |
| | $-n/s_R$ for excess supporting electrolyte |
| $\alpha_a$, $\alpha_c$ | transfer coefficients |
| $\alpha$, $\beta$ | transfer coefficients |
| $\beta(x)$ | velocity derivative $\partial v_x/\partial y$ at the wall, sec⁻¹ |
| $\gamma$ | exponent in concentration dependence of $i_0$ |
| $\Gamma(4/3)$ | 0.89298, the gamma function of 4/3 |
| $\delta$ | dimensionless average current density |
| $\eta$ | total overpotential, V |
| $\eta_c$ | concentration overpotential, V |
| $\eta_s$ | surface overpotential, V |
| $\kappa$ | conductivity, mho/cm |
| $\nu$ | kinematic viscosity, cm²/sec |
| $\phi$ | potential (see equation 130-3), V |
| $\Phi$ | electric potential, V |
| $\Omega$ | rotation speed, radian/sec |

subscripts and superscripts

| | |
|---|---|
| $\sim$ | in the bulk medium |
| 0 | at the electrode surface |
| $\infty$ | in the bulk solution |
| $R$ | reactant |

*Appendices*

# Partial Molar Volumes

Partial molar volumes are occasionally used as properties of solutions. However, the density of the solution is the quantity which is measured experimentally. For $n$ components, the density is a function of $n - 1$ independent concentrations $c_i$ at a given temperature and pressure. Let us suppose that the density $\rho$ of the solution is given by the experimental correlation in terms of these $n - 1$ concentrations:

$$\rho = \rho(c_1, c_2, \ldots, c_{n-1}). \tag{A-1}$$

We want to show here how to calculate the partial molar volumes of the components from the experimental density data. However, we are not concerned here with the partial molar volumes of individual ions, only with those of the neutral components of the solution. The method is thus equally applicable to nonelectrolytic solutions.

The partial molar volume of component $j$ is given by

$$\bar{V}_j = \left(\frac{\partial V}{\partial n_j}\right)_{\substack{T, p, n_i \\ i \neq j}}, \tag{A-2}$$

where $V$ is the volume of the solution and $n_i$ is the number of moles of component $i$ in the solution. The density of the solution is also given by

$$\rho = \sum_{i=1}^{n} c_i M_i = \frac{1}{V} \sum_{i=1}^{n} n_i M_i, \tag{A-3}$$

where $M_i$ is the molecular weight of component $i$ and $n_i = c_i V$.

From equations 2 and 3, the partial molar volume of component $j$ is

$$\bar{V}_j = \frac{1}{\rho} \frac{\partial}{\partial n_j} \sum_{i=1}^{n} n_i M_i + \left( \sum_{i=1}^{n} n_i M_i \right) \frac{\partial \frac{1}{\rho}}{\partial n_j}$$

$$= \frac{M_j}{\rho} - \frac{V}{\rho} \frac{\partial \rho}{\partial n_j}. \tag{A-4}$$

From the density correlation 1 we obtain

$$\frac{\partial \rho}{\partial n_j} = \sum_{k=1}^{n-1} \frac{\partial \rho}{\partial c_k} \frac{\partial c_k}{\partial n_j} = \sum_{k=1}^{n-1} \frac{\partial \rho}{\partial c_k} \frac{\partial \frac{n_k}{V}}{\partial n_j}$$

$$= \sum_{k=1}^{n-1} \frac{\partial \rho}{\partial c_k} \left( \frac{1}{V} \frac{\partial n_k}{\partial n_j} - \frac{n_k}{V^2} \frac{\partial V}{\partial n_j} \right) = \sum_{k=1}^{n-1} \frac{\partial \rho}{\partial c_k} \left( \frac{\delta_{jk}}{V} - \frac{c_k}{V} \bar{V}_j \right), \tag{A-5}$$

where $\delta_{jk}$ is the Kronecker delta, $\delta_{jk} = 1$ for $j = k$ and $\delta_{jk} = 0$ for $j \neq k$.

Substitution of equation 5 into equation 4 yields

$$\bar{V}_j = \frac{M_j}{\rho} - \frac{1}{\rho} \sum_{k=1}^{n-1} \frac{\partial \rho}{\partial c_k} \delta_{jk} + \frac{\bar{V}_j}{\rho} \sum_{k=1}^{n-1} c_k \frac{\partial \rho}{\partial c_k}. \tag{A-6}$$

Solving for $\bar{V}_j$, we obtain

$$\bar{V}_j = \frac{M_j - \sum_{k=1}^{n-1} \frac{\partial \rho}{\partial c_k} \delta_{jk}}{\rho - \sum_{k=1}^{n-1} c_k \frac{\partial \rho}{\partial c_k}} = \frac{M_j - \frac{\partial \rho}{\partial c_j} (1 - \delta_{jn})}{\rho - \sum_{k=1}^{n-1} c_k \frac{\partial \rho}{\partial c_k}}. \tag{A-7}$$

In particular, for a single salt solution where the density is given by $\rho = \rho(c)$, $c$ being the concentration of the salt, the partial molar volume of the electrolyte is

$$\bar{V}_e = \frac{M_e - \dfrac{d\rho}{dc}}{\rho - c \dfrac{d\rho}{dc}}, \tag{A-8}$$

and the partial molar volume of the solvent is

$$\bar{V}_0 = \frac{M_0}{\rho - c \dfrac{d\rho}{dc}}. \tag{A-9}$$

It the density is a linear function of the concentration, the partial molar volumes of the electrolyte and the solvent are constant. Conversely, if the partial molar volume of one component is constant, the partial molar volume of the other component is also constant, and the density is a linear function of concentration.

# Vectors and Tensors     **B**

Transport equations can become quite lengthy, and this frequently leads one to introduce vector notation, which has several advantages:

1. The equations become considerably more compact when written in vector notation.

2. The equations have significance independent of any particular coordinate system.

3. It is easier to grasp the meaning of an equation (after the vector notation becomes familiar).

You may regard vector notation as a form of shorthand writing, but it would be a good idea for you to develop an intuitive feel for the significance of some of the more common vector operations.

A vector has both magnitude and direction and can be decomposed into components in three rectangular directions:

$$\mathbf{v} = \mathbf{e}_x v_x + \mathbf{e}_y v_y + \mathbf{e}_z v_z. \tag{B-1}$$

Here, $\mathbf{e}_x$, $\mathbf{e}_y$, and $\mathbf{e}_z$ denote unit vectors in the $x$, $y$, and $z$ directions, respectively.

The *divergence* of a vector field is

$$\nabla \cdot \mathbf{v} = \frac{\partial v_x}{\partial x} + \frac{\partial v_y}{\partial y} + \frac{\partial v_z}{\partial z}. \tag{B-2}$$

(These operations have different forms in other coordinate systems; see reference 1, pp. 738-741). This quantity is a scalar whose physical significance can be seen most easily from the continuity equation 93-2

$$\frac{\partial \rho}{\partial t} = -\nabla \cdot (\rho \mathbf{v}). \tag{B-3}$$

The mass flux is $\rho \mathbf{v}$, showing the direction and magnitude of mass transfer per unit area, and $\nabla \cdot (\rho \mathbf{v})$ represents the *rate of mass flowing away from a point*. Hence the name *divergence*. We might call $-\nabla \cdot (\rho \mathbf{v})$ the *convergence* of the mass flux $\rho \mathbf{v}$. Then, the equation of continuity says

$$\frac{\partial \rho}{\partial t} = -\nabla \cdot (\rho \mathbf{v})$$

rate of accumulation = rate of convergence or net input.

Similar conservation or continuity equations have appeared in other places, for example, in equations 69-3 and 71-1.

The *curl* of a vector field yields another vector defined as

$$\Omega = \nabla \times \mathbf{v} = \begin{vmatrix} \mathbf{e}_x & \mathbf{e}_y & \mathbf{e}_z \\ \partial/\partial x & \partial/\partial y & \partial/\partial z \\ v_x & v_y & v_z \end{vmatrix}$$

$$= \mathbf{e}_x \left( \frac{\partial v_z}{\partial y} - \frac{\partial v_y}{\partial z} \right) + \mathbf{e}_y \left( \frac{\partial v_x}{\partial z} - \frac{\partial v_z}{\partial x} \right) + \mathbf{e}_z \left( \frac{\partial v_y}{\partial x} - \frac{\partial v_x}{\partial y} \right). \tag{B-4}$$

When $\mathbf{v}$ is the fluid velocity, $\Omega$ is known as the *vorticity*, which can be regarded as proportional to the angular velocity (radian/sec) of a fluid element. This vector operation is encountered in fluid mechanics and in electromagnetic theory (see equation 22-3), but electrochemists may find little use for it.

The *gradient* of a scalar field is a vector:

$$\nabla \Phi = \mathbf{e}_x \frac{\partial \Phi}{\partial x} + \mathbf{e}_y \frac{\partial \Phi}{\partial y} + \mathbf{e}_z \frac{\partial \Phi}{\partial z}. \tag{B-5}$$

The gradient of $\Phi$ shows the change of electric potential with position and is the negative of the electric field. The direction of the gradient shows the direction of the greatest change, and the magnitude is the rate of change in this direction. The gradient of a vector field, on the other hand, is a *tensor*. It has nine components because it is necessary to describe the rate of change of each component of the vector in each of three directions.

$$\nabla \mathbf{v} = \begin{pmatrix} \dfrac{\partial v_x}{\partial x} & \dfrac{\partial v_y}{\partial x} & \dfrac{\partial v_z}{\partial x} \\[2mm] \dfrac{\partial v_x}{\partial y} & \dfrac{\partial v_y}{\partial y} & \dfrac{\partial v_z}{\partial y} \\[2mm] \dfrac{\partial v_x}{\partial z} & \dfrac{\partial v_y}{\partial z} & \dfrac{\partial v_z}{\partial z} \end{pmatrix}. \tag{B-6}$$

[1]R. Byron Bird, Warren E. Stewart, and Edwin N. Lightfoot, *Transport Phenomena* (New York: John Wiley & Sons, Inc., 1960).

A tensor is an operator for vectors. The result of a tensor operating on a vector is another vector:

$$\mathbf{\tau \cdot a} = \mathbf{e}_x(\tau_{xx}a_x + \tau_{xy}a_y + \tau_{xz}a_z)$$
$$+ \mathbf{e}_y(\tau_{yx}a_x + \tau_{yy}a_y + \tau_{yz}a_z)$$
$$+ \mathbf{e}_z(\tau_{zx}a_x + \tau_{zy}a_y + \tau_{zz}a_z). \tag{B-7}$$

(The result of a tensor operating on a vector can also be written $\mathbf{a \cdot \tau}$, but this is not the same as $\mathbf{\tau \cdot a}$; see entry 1(d) of table B-1.)

The stress $\mathbf{\tau}$ due to viscous forces is a tensor (see section 94). Its nine components tell us the force acting on surfaces with various orientations.

$$\mathbf{f = n \cdot \tau}, \tag{B-8}$$

where $\mathbf{n}$ is a unit vector normal to a surface and $\mathbf{f}$ is the stress on the surface. The equation of motion for a Newtonian fluid of constant density and viscosity (equation 94-4) is a vector equation involving the tensor $\nabla\mathbf{v}$. The components of this equation in rectangular coordinates are

$$\frac{\partial v_x}{\partial t} + v_x\frac{\partial v_x}{\partial x} + v_y\frac{\partial v_x}{\partial y} + v_z\frac{\partial v_x}{\partial z} = -\frac{1}{\rho}\frac{\partial p}{\partial x}$$
$$+ \nu\left(\frac{\partial^2 v_x}{\partial x^2} + \frac{\partial^2 v_x}{\partial y^2} + \frac{\partial^2 v_x}{\partial z^2}\right) + g_x.$$

$$\frac{\partial v_y}{\partial t} + v_x\frac{\partial v_y}{\partial x} + v_y\frac{\partial v_y}{\partial y} + v_z\frac{\partial v_y}{\partial z} = -\frac{1}{\rho}\frac{\partial p}{\partial y}$$
$$+ \nu\left(\frac{\partial^2 v_y}{\partial x^2} + \frac{\partial^2 v_y}{\partial y^2} + \frac{\partial^2 v_y}{\partial z^2}\right) + g_y.$$

$$\frac{\partial v_z}{\partial t} + v_x\frac{\partial v_z}{\partial x} + v_y\frac{\partial v_z}{\partial y} + v_z\frac{\partial v_z}{\partial z} = -\frac{1}{\rho}\frac{\partial p}{\partial z}$$
$$+ \nu\left(\frac{\partial^2 v_z}{\partial x^2} + \frac{\partial^2 v_z}{\partial y^2} + \frac{\partial^2 v_z}{\partial z^2}\right) + g_z.$$

$$\tag{B-9}$$

This equation and others of frequent use to us can be found written out in several coordinate systems in reference 1.

A few definitions and identities are given in table B-1. Vectors are denoted by boldface Latin characters, and tensors by boldface Greek characters. The directions $x$, $y$, and $z$ in rectangular coordinates are denoted 1, 2, and 3, so that $x_2 = y$, $\mathbf{e}_2 = \mathbf{e}_y$, etc., and sums extend over the indices 1, 2, and 3.

TABLE B-1. VECTOR AND TENSOR ALGEBRA AND CALCULUS

---

1. Definitions.
   (a) Dyadic product.     $(\mathbf{ac})_{ij} = a_ic_j$.     ($\mathbf{ac}$ is a tensor.)
   (b) Double dot product.

$$\mathbf{\sigma : \tau} = \sum_i \sum_j \sigma_{ij}\tau_{ji}.$$

   (c) A tensor operating on a vector from the right yields a vector.

$$\mathbf{a \cdot \tau} = \sum_i \sum_j \mathbf{e}_i a_j \tau_{ji}.$$

(d) Transpose of a tensor.     $(\tau^*)_{ij} = \tau_{ji}$   or   $\mathbf{\tau} \cdot \mathbf{a} = \mathbf{a} \cdot \mathbf{\tau}^*$.

(e) Product of two tensors.     $(\mathbf{\tau} \cdot \mathbf{\sigma}) \cdot \mathbf{v} = \mathbf{\tau} \cdot (\mathbf{\sigma} \cdot \mathbf{v})$ or

$$(\mathbf{\tau} \cdot \mathbf{\sigma})_{ij} = \sum_k \tau_{ik}\sigma_{kj}.$$

(f) The divergence of a tensor is a vector.

$$\nabla \cdot \mathbf{\tau} = \sum_i \sum_j \mathbf{e}_i \left( \frac{\partial \tau_{ji}}{\partial x_j} \right).$$

(g) Laplacian of a scalar.

$$\nabla^2 \Phi = \nabla \cdot \nabla \Phi = \sum_i \left( \frac{\partial^2 \Phi}{\partial x_i^2} \right).$$

(h) Gradient of a vector.     $(\nabla \mathbf{v})_{ij} = \partial v_j / \partial x_i$.

(i) Laplacian of a vector.     $\nabla^2 \mathbf{v} = \nabla \cdot \nabla \mathbf{v} = \nabla(\nabla \cdot \mathbf{v}) - \nabla \times \nabla \times \mathbf{v}$.

2. Algebra.

(a) $\mathbf{\tau} : (\mathbf{ab}) = \mathbf{b} \cdot (\mathbf{\tau} \cdot \mathbf{a})$.

(b) $(\mathbf{uv}):(\mathbf{wz}) = (\mathbf{uw}):(\mathbf{vz}) = (\mathbf{u} \cdot \mathbf{z})(\mathbf{v} \cdot \mathbf{w})$.

(c) $\mathbf{a} \cdot (\mathbf{bc}) = (\mathbf{a} \cdot \mathbf{b})\mathbf{c}$.

(d) $(\mathbf{ab}) \cdot \mathbf{c} = \mathbf{a}(\mathbf{b} \cdot \mathbf{c})$.

(e) $\mathbf{a} \times (\mathbf{b} \times \mathbf{c}) = \mathbf{b}(\mathbf{a} \cdot \mathbf{c}) - \mathbf{c}(\mathbf{a} \cdot \mathbf{b})$.

(f) $\mathbf{u} \cdot (\mathbf{v} \times \mathbf{w}) = \mathbf{v} \cdot (\mathbf{w} \times \mathbf{u})$.

(g) $(\mathbf{u} \times \mathbf{v}) \cdot (\mathbf{w} \times \mathbf{z}) = (\mathbf{u} \cdot \mathbf{w})(\mathbf{v} \cdot \mathbf{z}) - (\mathbf{u} \cdot \mathbf{z})(\mathbf{v} \cdot \mathbf{w})$.

(h) $\mathbf{v} \cdot (\mathbf{\tau}^* \cdot \mathbf{w}) = \mathbf{w} \cdot (\mathbf{\tau} \cdot \mathbf{v})$.

3. Differentiation of products.

(a) $\nabla \phi \psi = \phi \nabla \psi + \psi \nabla \phi$ (a vector).

(b) $\nabla \phi \mathbf{v} = \phi \nabla \mathbf{v} + (\nabla \phi)\mathbf{v}$ (a tensor).

(c) $\nabla(\mathbf{a} \cdot \mathbf{c}) = \mathbf{a} \cdot \nabla \mathbf{c} + \mathbf{c} \cdot \nabla \mathbf{a} + \mathbf{a} \times \nabla \times \mathbf{c} + \mathbf{c} \times \nabla \times \mathbf{a}$

$\qquad\qquad = (\nabla \mathbf{c}) \cdot \mathbf{a} + (\nabla \mathbf{a}) \cdot \mathbf{c}$ (a vector).

(d) $\nabla \cdot (\phi \mathbf{v}) = \phi \nabla \cdot \mathbf{v} + \mathbf{v} \cdot \nabla \phi$ (a scalar).

(e) $\nabla \cdot (\mathbf{v} \times \mathbf{w}) = \mathbf{w} \cdot (\nabla \times \mathbf{v}) - \mathbf{v} \cdot (\nabla \times \mathbf{w})$ (a scalar).

(f) $\nabla \times (\phi \mathbf{v}) = \phi \nabla \times \mathbf{v} + (\nabla \phi) \times \mathbf{v}$ (a vector).

(g) $\nabla \times (\mathbf{b} \times \mathbf{c}) = \mathbf{b}(\nabla \cdot \mathbf{c}) - \mathbf{c}(\nabla \cdot \mathbf{b}) + \mathbf{c} \cdot \nabla \mathbf{b} - \mathbf{b} \cdot \nabla \mathbf{c}$ (a vector).

(h) $\nabla \cdot (\mathbf{ab}) = (\nabla \cdot \mathbf{a})\mathbf{b} + \mathbf{a} \cdot \nabla \mathbf{b}$ (a vector).

(i) $\nabla \cdot (\phi \mathbf{\tau}) = \phi \nabla \cdot \mathbf{\tau} + (\nabla \phi) \cdot \mathbf{\tau}$ (a vector).

(j) $\nabla \cdot (\mathbf{u} \cdot \mathbf{\tau}) = \mathbf{\tau} : \nabla \mathbf{u} + \mathbf{u} \cdot \nabla \cdot \mathbf{\tau}^*$ (a scalar).

4. Various forms of Gauss's law (divergence theorem) and Stokes's law. ($dS$ = area element, $d\ell$ = line element, $dv$ = volume element. Integration over a closed surface or a closed curve is denoted by a circle through the integral sign. In the first case, $\mathbf{dS}$ is normally outward from the surface; in the second case, $\mathbf{d\ell}$ and $\mathbf{dS}$ are related by a right-hand screw rule, that is, a right-hand screw turned in the direction of $\mathbf{d\ell}$ advances in the direction of $\mathbf{dS}$.)

(a) $\oint \mathbf{dS} \cdot \mathbf{F} = \int dv \, \nabla \cdot \mathbf{F}$.

(b) $\oint \mathbf{dS} \, \phi = \int dv \, \nabla \phi$.

(c) $\oint (\mathbf{dS} \cdot \mathbf{G})\mathbf{F} = \int dv \, \mathbf{F} \nabla \cdot \mathbf{G} + \int dv \, \mathbf{G} \cdot \nabla \mathbf{F}$.

(d) $\oint \mathbf{dS} \times \mathbf{F} = \int dv \, \nabla \times \mathbf{F}$.

(e) $\oint \mathbf{dS} \cdot \mathbf{\tau} = \oint dv \, \nabla \cdot \mathbf{\tau}$.

(f) $\oint \mathbf{dS} \cdot (\Psi \nabla \phi - \phi \nabla \Psi) = \int dv \, (\Psi \nabla^2 \phi - \phi \nabla^2 \Psi)$.

(g) $\oint d\boldsymbol{\ell} \cdot \mathbf{F} = \int d\mathbf{S} \cdot \nabla \times \mathbf{F}.$

(h) $\oint d\boldsymbol{\ell} \, \phi = \int d\mathbf{S} \times \nabla \phi.$

5. Miscellaneous.

(a) $\nabla \cdot \nabla \times \mathbf{E} = 0.$

(b) $\nabla \times \nabla \phi = 0.$

(c) $\mathbf{w} \cdot \nabla \mathbf{v} = \sum_i \sum_j \mathbf{e}_i w_j \, \partial v_i / \partial x_j.$

(d) $D/Dt = \partial/\partial t + \mathbf{v} \cdot \nabla.$

(e) $D\mathbf{v}/Dt = \partial \mathbf{v}/\partial t + \frac{1}{2}\nabla v^2 - \mathbf{v} \times \nabla \times \mathbf{v}.$  where $\mathbf{v}$ is the mass-average velocity.

# Numerical Solution of Coupled, Ordinary Differential Equations

**C**

The mathematical modeling of physical phenomena usually finds expression in partial differential equations. Often these reduce to ordinary differential equations, either because only one independent variable is pertinent or because of the applicability of a special technique such as a similarity transformation or the method of separation of variables. The availability of high-speed digital computers and a generalized method of solution allows such problems to be treated without the drastic approximations frequently needed to obtain analytic solutions. The original problems are often nonlinear and involve several dependent variables, but by a proper linearization of such problems a convergent iteration scheme usually results, although convergence cannot generally be assured. Consequently, it is appropriate to discuss first the solution of coupled, *linear* differential equations, followed by illustrations of the linearization method.

Since other, very different techniques (such as the Runge-Kutta method) work well with initial-value problems, attention is restricted here to boundary-value problems—that is, with boundary conditions at $x = 0$ and $x = L$ or $x = \infty$. The procedure used here has been found to be quite useful in a variety of problems, and it seems appropriate to report it[1,2] so that other workers can implement it with ease.

Boundary-value problems of interest in the present context arise, for example, in the following situations:

1. Mass transfer into a semi-infinite, stagnant medium (*penetration model*).

2. Mass transfer in a stagnant film or a porous solid, as encountered with heterogeneous catalysis or porous electrodes.

3. Mass transfer in boundary layers possessing profiles similar in the distance along a surface. This can include both free and forced convection, and for large Schmidt numbers the similarity of the hydrodynamics ceases to be essential.

4. Velocity distributions in similar boundary layers (see, for example, the hydrodynamics of the rotating disk in section 96).

5. Distribution of charge and mass in diffuse, electric double layers.

The concentration distributions of several species may be coupled among themselves or with the velocity and temperature fields for a number of reasons.

1. The diffusion coefficients, viscosity, and other physical properties depend upon the composition, temperature, and pressure.

2. The interfacial velocity at the surface is related to the rate of mass transfer.

3. The species may be charged and interact with each other through the electric potential.

4. The components may be involved in heterogeneous or homogeneous reactions described by equilibrium or rate expressions.

5. For free convection, the fluid motion results from density differences created by nonuniform composition or temperature.

The calculation procedure was first generalized to an arbitrary number of coupled equations for treating the effect of ionic migration on limiting currents (see chapter 19), where an arbitrary number of species may be involved.

### Coupled, linear, difference equations

A set of $n$ coupled, linear, second-order differential equations is represented as

$$\sum_{k=1}^{n} a_{i,k}(x)\frac{d^2c_k}{dx^2} + b_{i,k}(x)\frac{dc_k}{dx} + d_{i,k}(x)c_k = g_i(x), \qquad \text{(C-1)}$$

[1] John Newman, "Numerical Solution of Coupled, Ordinary Differential Equations," *Industrial and Engineering Chemistry Fundamentals*, 7 (1968), 514–517.

[2] John Newman, *Numerical Solution of Coupled, Ordinary Differential Equations* (UCRL-17739), Lawrence Radiation Laboratory, University of California, Berkeley, August, 1967.

where $c_k(x)$ are the $n$ unknown functions. The subscript $i$ denotes the equation number, and each of the equations can involve all of the unknowns, $c_k$, through the sum. The equations are linear; that is, the coefficients $a_{i,k}$, $b_{i,k}$, and $d_{i,k}$ are independent of the unknowns.

For central difference approximations of the derivatives with a mesh distance $h$,

$$\frac{d^2 c_k}{dx^2} = \frac{c_k(x_j + h) + c_k(x_j - h) - 2c_k(x_j)}{h^2} + 0(h^2), \qquad \text{(C-2)}$$

$$\frac{dc_k}{dx} = \frac{c_k(x_j + h) - c_k(x_j - h)}{2h} + 0(h^2), \qquad \text{(C-3)}$$

one obtains the following difference equations

$$\sum_{k=1}^{n} A_{i,k}(j)C_k(j - 1) + B_{i,k}(j)C_k(j) + D_{i,k}(j)C_k(j + 1) = G_i(j), \qquad \text{(C-4)}$$

where
$$C_k(j) = c_k(x_j), \qquad \text{(C-5)}$$

$$A_{i,k}(j) = a_{i,k}(x_j) - \frac{h}{2}b_{i,k}(x_j), \qquad \text{(C-6)}$$

$$B_{i,k}(j) = -2a_{i,k}(x_j) + h^2 d_{i,k}(x_j), \qquad \text{(C-7)}$$

$$D_{i,k}(j) = a_{i,k}(x_j) + \frac{h}{2}b_{i,k}(x_j), \qquad \text{(C-8)}$$

$$G_i(j) = h^2 g_i(x_j). \qquad \text{(C-9)}$$

At $j = 1$, the equations are

$$\sum_{k=1}^{n} B_{i,k}(1)C_k(1) + D_{i,k}(1)C_k(2) + X_{i,k}C_k(3) = G_i(1). \qquad \text{(C-10)}$$

There is no point for $j = 0$, so $A_{i,k}$ does not appear. However, in order to allow the treatment of complex boundary conditions, the third term involving the unknowns at $j = 3$ has been added. For example, general boundary conditions for equation 1 at $x = 0$ would read

$$\sum_{k=1}^{n} p_{i,k}\frac{dc_k}{dx} + e_{i,k}c_k = f_i \quad \text{at} \quad x = 0. \qquad \text{(C-11)}$$

If we use the finite-difference form of equation 3, we can expedite matters by the introduction of an *image point* at $x = -h$, outside the domain of interest (see figure 1). Then the coefficients in equation 10 become

$$X_{i,k} = -B_{i,k}(1) = \frac{p_{i,k}}{2}, \qquad D_{i,k}(1) = he_{i,k}, \qquad G_i(1) = hf_i. \qquad \text{(C-12)}$$

In this case of boundary conditions involving the first derivative, it is desirable to use the difference approximation to the differential equation at the boundary point ($x = 0$ and $j = 2$), even though this involves an imaginary point outside the domain of interest. The finite-difference form of the boundary conditions is then written as an extra set of equations which should be sufficient to eliminate the values at the image point. There seems to be no

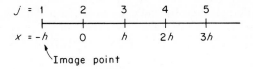

**Figure C-1.** Image point for treatment of boundary conditions with derivatives.

advantage in eliminating the values at the image point in an operation separate from the solution of the difference equations themselves.

If the boundary conditions do not involve derivatives ($p_{i,k} = 0$), then it is not necessary to use the image point, and $x = 0$ corresponds to $j = 1$.

For similar reasons, the difference equations at $j = j_{\max}$ are written

$$\sum_{k=1}^{n} Y_{i,k} C_k(j-2) + A_{i,k}(j)C_k(j-1) + B_{i,k}(j)C_k(j) = G_i(j), \quad \text{(C-13)}$$

where the coefficients $Y_{i,k}$ again allow the introduction of complex boundary conditions at the upper limit ($x = L$) of the domain of interest, in the same way that the coefficients $X_{i,k}$ do at $x = 0$.

### Solution of coupled, linear, difference equations

We turn now to the method of solution of the difference equations 4, 10, and 13. For $j = 1$, let $C_k(j)$ take the form

$$C_k(1) = \xi_k(1) + \sum_{\ell=1}^{n} E_{k,\ell}(1)C_\ell(2) + x_{k,\ell}C_\ell(3). \quad \text{(C-14)}$$

Substitution into equation 10 shows that $\xi_k$, $E_{k,\ell}$, and $x_{k,\ell}$ satisfy the equations

$$\left.\begin{aligned}
\sum_{k=1}^{n} B_{i,k}(1)\xi_k(1) &= G_i(1), \\
\sum_{k=1}^{n} B_{i,k}(1)E_{k,\ell}(1) &= -D_{i,\ell}(1), \\
\sum_{k=1}^{n} B_{i,k}(1)x_{k,\ell} &= -X_{i,\ell},
\end{aligned}\right\} \quad \text{(C-15)}$$

which all have the same matrix of coefficients $B_{i,k}$ and which can be readily solved.

For the remaining points, except $j = j_{\max}$, the unknowns $C_k$ assume the form

$$C_k(j) = \xi_k(j) + \sum_{\ell=1}^{n} E_{k,\ell}(j)C_\ell(j+1). \quad \text{(C-16)}$$

Substitution of equation 16 into equation 4 to eliminate first $C_k(j-1)$ and then $C_k(j)$ and setting the remaining cofficient of each $C_k(j+1)$ equal to zero yield a set of equations for the determination of $\xi_k$ and $E_{k,\ell}$:

$$\sum_{k=1}^{n} b_{i,k}(j)\xi_k(j) = G_i(j) - \sum_{\ell=1}^{n} A_{i,\ell}(j)\xi_\ell(j-1), \qquad \text{(C-17)}$$

$$\sum_{k=1}^{n} b_{i,k}(j)E_{k,m}(j) = -D_{i,m}(j), \qquad \text{(C-18)}$$

where $\qquad b_{i,k}(j) = B_{i,k}(j) + \sum_{\ell=1}^{n} A_{i,\ell}(j)E_{\ell,k}(j-1)$ \qquad (C-19)

and is not to be confused with $b_{i,k}(x)$ in equation 1. The solution of these linear equations at each point $j$ is again straightforward, but the point at $j-1$ must be calculated first, since $\xi_k(j-1)$ appears on the right side of equation 17 and $E_{k,\ell}(j-1)$ appears in the matrix of coefficients $b_{i,k}$. The equations for $j=2$ actually take a slightly different form since equation 14 should be used instead of equation 16 to eliminate $C_k(1)$ from equation 4.

Finally, one has equation 13 for $j=j_{max}$. If $C_k(j-2)$ and $C_k(j-1)$ are eliminated by means of equation 16, then the values of $C_k(j_{max})$ can be determined from the resulting equations:

$$\sum_{k=1}^{n} b_{i,k}(j)C_k(j) = G_i(j) - \sum_{\ell=1}^{n} Y_{i,\ell}\xi_\ell(j-2) - \sum_{\ell=1}^{n} a_{i,\ell}(j)\xi_\ell(j-1),$$
$$\text{(C-20)}$$

where $\qquad a_{i,\ell}(j) = A_{i,\ell}(j) + \sum_{m=1}^{n} Y_{i,m}E_{m,\ell}(j-2)$

and $\qquad b_{i,k}(j) = B_{i,k}(j) + \sum_{\ell=1}^{n} a_{i,\ell}(j)E_{\ell,k}(j-1).$ \qquad (C-21)

Having in hand values for $C_k(j)$ for $j=j_{max}$, one is now in a position to determine $C_k(j)$ in reverse order in $j$ from equation 16 and finally to determine $C_k(1)$ from equation 14. Such repetitive calculations as are involved in the solution of the matrix equations and in the back substitution are, of course, ideally suited for a large, fast, digital computer.

Because the boundary-value problem involves boundary conditions at both $x=0$ and $x=L$, it is not possible to start at either end and obtain final values of the unknowns. This is possible only for initial value problems. Instead, one makes two passes through the domain of interest, in opposite directions. $E_{k,\ell}$ in equation 16 allows the effect of the boundary conditions at $x=L$ to be reflected back through this domain, the effect not being realized until the back substitution is carried out.

### Program for coupled, linear difference equations

All the steps for solving coupled, linear difference equations have been given above. However, to program these complicated steps is a bit tricky, and it is easy to make a mistake. Consequently, we give here subroutines for implementing the solution method. To solve a problem, one then only needs to write a main program which supplies values of $A$, $B$, $D$, and $G$

appropriate to that problem. An example of a main program which uses these subroutines is given after the discussion of the linearization of nonlinear problems.

To save storage space, the arrays $A$, $B$, $D$, and $G$ are to be supplied by the main program for each value of $j$, and the subroutine BAND(J) is to be called for each values of $j$. The values of $X$ are to be supplied for $j = 1$, and the values of $Y$ for $j = \text{NJ} = j_{\text{max}}$. The values of $X$ are not to be disturbed for any intermediate calculations between $j = 1$ and $j = j_{\text{max}}$. The dimensions have been selected for $n = 6$, the number of unknown variables at each mesh point, and $j_{\text{max}} = 103$, the number of mesh points including image points, if any. These can be changed appropriately for a particular problem. The second dimension of the $D$ array is to be $2n + 1$, although values need to be supplied only for the original $n$ by $n$ array. The second dimension of the $E$ array is $n + 1$ since $\xi_k$ is stored here.

```
      SUBROUTINE BAND(J)
      DIMENSION C(6,103),G(6),A(6,6),B(6,6),D(6,13),E(6,7,103),X(6,6),
     1Y(6,6)
      COMMON A,B,C,D,G,X,Y,N,NJ
  101 FORMAT (15HODETERM=0 AT J=,I4)
      IF (J-2)  1,6,8
    1 NP1= N + 1
      DO 2 I=1,N
      D(I,2*N+1)= G(I)
      DO 2 L=1,N
      LPN= L + N
    2 D(I,LPN)= X(I,L)
      CALL MATINV(N,2*N+1,DETERM)
      IF (DETERM)  4,3,4
    3 PRINT 101, J
    4 DO 5 K=1,N
      E(K,NP1,1)= D(K,2*N+1)
      DO 5 L=1,N
      E(K,L,1)= - D(K,L)
      LPN= L + N
    5 X(K,L)= - D(K,LPN)
      RETURN
    6 DO 7 I=1,N
      DO 7 K=1,N
      DO 7 L=1,N
    7 D(I,K)= D(I,K) + A(I,L)*X(L,K)
    8 IF (J-NJ)  11,9,9
    9 DO 10 I=1,N
      DO 10 L=1,N
      G(I)= G(I) - Y(I,L)*E(L,NP1,J-2)
      DO 10 M=1,N
   10 A(I,L)= A(I,L) + Y(I,M)*E(M,L,J-2)
   11 DO 12 I=1,N
      D(I,NP1)= - G(I)
      DO 12 L=1,N
      D(I,NP1)= D(I,NP1) + A(I,L)*E(L,NP1,J-1)
      DO 12 K=1,N
   12 B(I,K)= B(I,K) + A(I,L)*E(L,K,J-1)
      CALL MATINV(N,NP1,DETERM)
      IF (DETERM)  14,13,14
```

```
13 PRINT 101, J
14 DO 15 K=1,N
   DO 15 M=1,NP1
15 E(K,M,J)= - D(K,M)
   IF (J-NJ)  20,16,16
16 DO 17 K=1,N
17 C(K,J)= E(K,NP1,J)
   DO 18 JJ=2,NJ
   M= NJ - JJ + 1
   DO 18 K=1,N
   C(K,M)= E(K,NP1,M)
   DO 18 L=1,N
18 C(K,M)= C(K,M) + E(K,L,M)*C(L,M+1)
   DO 19 L=1,N
   DO 19 K=1,N
19 C(K,1)= C(K,1) + X(K,L)*C(L,3)
20 RETURN
   END
```

The subroutine MATINV is used to solve the linear equations 15, 17, 18, and 20 which arise at each value of $j$. If, at any value of $j$, the determinant of the matrix of these equations is found to be zero, this fact is reported in the output. This usually indicates that all the equations have not been programmed or that they are not all independent. It can also indicate that the equations for $j = 1$ are not sufficient to determine the image points although the equations for $j = 1$ and $j = 2$ would be sufficient to determine both the boundary point and the image point. In rare instances, it may indicate that the trial solution is inadequate and gives a zero determinant.

```
SUBROUTINE MATINV(N,M,DETERM)
DIMENSION ID(6),B(6,6),D(6,13),A(6,6),C(6,103)
COMMON A,B,C,D
DETERM= 1.0
DO 1 I=1,N
1 ID(I)= 0.0
DO 18 NN=1,N
BMAX= 0.0
DO 6 I=1,N
IF (ID(I))  2,2,6
2 DO 5 J=1,N
IF (ID(J))  3,3,5
3 IF (ABSF(B(I,J)) - BMAX)  5,5,4
4 BMAX= ABSF(B(I,J))
IROW= I
JCOL= J
5 CONTINUE
6 CONTINUE
IF (BMAX)  7,7,8
7 DETERM= 0.0
RETURN
8 ID(JCOL)= 1
IF (JCOL-IROW)  9,12,9
9 DO 10 J=1,N
```

```
      SAVE= B(IROW,J)
      B(IROW,J)= B(JCOL,J)
   10 B(JCOL,J)= SAVE
      DO 11 K=1,M
      SAVE= D(IROW,K)
      D(IROW,K)= D(JCOL,K)
   11 D(JCOL,K)= SAVE
   12 F= 1.0/B(JCOL,JCOL)
      DO 13 J=1,N
   13 B(JCOL,J)= B(JCOL,J)*F
      DO 14 K=1,M
   14 D(JCOL,K)= D(JCOL,K)*F
      DO 18 I=1,N
      IF (I-JCOL)  15,18,15
   15 F= B(I,JCOL)
      DO 16 J=1,N
   16 B(I,J)= B(I,J) - F*B(JCOL,J)
      DO 17 K=1,M
   17 D(I,K)= D(I,K) - F*D(JCOL,K)
   18 CONTINUE
      RETURN
      END
```

## Linearization of nonlinear problems

We have shown how to solve coupled, linear, difference equations. Often, however, one is faced with a set of coupled, nonlinear, differential equations. A variety of experience shows that repetitive calculation (or *iteration*) with a linearized form of the equations frequently converges to the correct result. That is to say, if the equations are linearized on the basis of a trial solution, the solution of the linearized equations is closer to the correct solution. The new solution is then used as the trial solution to obtain a second approximation, and the process is repeated until the desired accuracy is achieved. In the present context, experience shows that whether or not the method converges is not particularly sensitive to the choice of the first trial solution.

The equations

$$D_i c_i'' + z_i u_i F(c_i \Phi'' + c_i' \Phi') = 0 \qquad \text{(C-22)}$$

represent Fick's second law for one-dimensional, steady migration and diffusion of ionic species in a stagnant medium (see chapter 19). Here $c_i$ is the concentration, $\Phi$ is the electric potential, $D_i$ is the diffusion coefficient, $u_i$ is the mobility, $z_i$ is the charge number, and $F$ is Faraday's constant. The diffusion term is already linear, but the migration terms are nonlinear. Equations 22 can be linearized by assuming that one has nearly correct values of $c_i$ and $\Phi$, say $c_i^0$ and $\Phi^0$, and that the change in these quantities during one iteration is relatively small. Then we can write, for example,

$$c_i \Phi'' = (c_i^0 + \Delta c_i)(\Phi^{0''} + \Delta \Phi'') \approx c_i^0 \Phi^{0''} + \Phi^{0''} \Delta c_i + c_i^0 \Delta \Phi'',$$
(C-23)

where the term quadratic in the small quantities $\Delta c_i$ and $\Delta \Phi''$ has been neglected. Replacing $\Delta c_i$ by $c_i - c_i^0$ and $\Delta \Phi''$ by $\Phi'' - \Phi^{0''}$, we can then write

$$c_i \Phi'' \approx c_i \Phi^{0''} + c_i^0 \Phi'' - c_i^0 \Phi^{0''}.$$
(C-24)

The linearized form of equations 22 can now be written as

$$D_i c_i'' + z_i u_i F(c_i \Phi^{0''} + c_i^0 \Phi'' + c' \Phi^{0'} + c_i^{0'} \Phi')$$
$$= z_i u_i F(c_i^0 \Phi^{0''} + c_i^{0'} \Phi^{0'}). \quad \text{(C-25)}$$

These form a set of coupled, linear differential equations. The finite difference form is

$$c_i(j-1)(D_i - \tfrac{1}{2} z_i u_i F h \Phi^{0'}) + c_i(j)(-2D_i + z_i u_i F h^2 \Phi^{0''})$$
$$+ c_i(j+1)(D_i + \tfrac{1}{2} z_i u_i F h \Phi^{0'}) + \Phi(j-1)z_i u_i F(c_i^0 - \tfrac{1}{2} h c_i^{0'})$$
$$+ \Phi(j)(-2z_i u_i F c_i^0) + \Phi(j+1)z_i u_i F(c_i^0 + \tfrac{1}{2} h c_i^{0'})$$
$$= z_i u_i F h^2 (c_i^0 \Phi^{0''} + c_i^{0'} \Phi^{0'}). \quad \text{(C-26)}$$

The coefficients $A_{i,k}$ in equation 4 then become the coefficients of $c_i$ and $\Phi$ at $j-1$, etc., and $\Phi$ becomes one of the unknowns $C_k$. The term on the right in equation 26 becomes $G_i(j)$.

The one remaining equation for this problem would be the electroneutrality relation

$$\sum_{k=1}^{n-1} z_k c_k = 0, \quad \text{(C-27)}$$

which is already linear and does not involve the unknowns at $j-1$ or $j+1$.

For porous electrodes, the equation might involve a reaction term involving exponentials, such as

$$J = e^{-\phi} - c_3 e^{\phi}, \quad \text{(C-28)}$$

where $\phi$ is a dimensionless potential and $c_3$ is a reactant concentration. The linearized form of this equation would be

$$J \approx -\phi(e^{-\phi^0} + c_3^0 e^{\phi^0}) - c_3 e^{\phi^0} + (1 + \phi^0)e^{-\phi^0} + c_3^0 \phi^0 e^{\phi^0}. \quad \text{(C-29)}$$

### Program for the effect of ionic migration
### on limiting currents

This program gives an example of the use of the subroutines presented earlier to solve a particular problem, that of the effect of ionic migration on limiting currents (see chapter 19). Each iteration begins at statement 8 and involves setting up the coefficients $A$, $B$, $D$, and $G$ for each value of $j$ followed by calling subroutine BAND(J) for each value of $j$. In the program, U(I) is proportional to $z_i u_i$, and the electric potential is the $n$th unknown variable, the other unknowns being the $n-1$ species concentrations. MODE is 1 for a Nernst stagnant diffusion layer, 2 for a growing mercury drop or unsteady diffusion into a stagnant fluid, and 3 for a rotating disk. H is the mesh size, and CRO is the concentration of the reactant at the electrode (equal to zero at the limiting current).

```
      PROGRAM MIGR(INPUT,OUTPUT)
C     PROGRAM FOR EFFECT OF MIGRATION ON LIMITING CURRENT
      DIMENSION A(6,6),B(6,6),C(6,103),D(6,13),G(6),X(6,6),Y(6,6),U(6),
     1V(103),DIF(6),Z(6),S(6),CIN(6),REF(6)
      COMMON A,B,C,D,G,X,Y,N,NJ
  101 FORMAT (2I4,E8.4)
  102 FORMAT (4E8.4,A6)
  103 FORMAT (4H0NJ=,I4,5H, H=,F6.4/34H0SPECIES        U        DIF       Z
     1 S/(3X,A6,2F8.5,2F5.1))
  104 FORMAT (32H1NERNST STAGNANT DIFFUSION LAYER)
  105 FORMAT (32H1GROWING DROP OR PLANE ELECTRODE)
  106 FORMAT (14H1ROTATING DISK)
  107 FORMAT (5E8.4)
  108 FORMAT (30H0THE NEXT RUN DID NOT CONVERGE)
  109 FORMAT (1H0,26X,F10.6/(3X,A6,2F9.5))
  110 FORMAT (35H0SPECIES        CINF       CZERO       AMP)
      READ 101, MODE,NJ,H
      GO TO (1,2,3),MODE
    1 H= 1.0/(NJ-2)
      CONST= 0.0
      PRINT 104
      GO TO 99
    2 CONST= 2.0
      PRINT 105
      GO TO 99
    3 CONST= 3.0
      PRINT 106
   99 READ 101, N,J,CRO
      IF (N)  4,4,5
    4 STOP
    5 NM1= N - 1
      NM2= N - 2
      READ 102, (U(I),DIF(I),Z(I),S(I),REF(I),I=1,NM1)
      PRINT 103, NJ,H,(REF(I),U(I),DIF(I),Z(I),S(I),I=1,NM1)
      PRINT 110
   98 READ 107, (CIN(I),I=1,NM1)
      CIN(N)= 0.0
      IF (CIN(1))  99,6,6
    6 DO 7 J=1,NJ
      V(J)= CONST*(H*(NJ-J-1))**(MODE-1)*DIF(NM1)
      C(NM1,J)= CRO + (CIN(NM1) - CRO)*(NJ-J-1)/(NJ-2)
      C(N,J)= 0.0
      DO 7 I=1,NM2
    7 C(I,J)= CIN(I)
      JCOUNT = 0
      AMP= 0.0
    8 JCOUNT= JCOUNT + 1
      J= 0
      DO 9 I=1,N
      DO 9 K=1,N
      Y(I,K)= 0.0
    9 X(I,K)= 0.0
   10 J= J + 1
      DO 11 I=1,N
```

```
      G(I)= 0.0
      DO 11 K=1,N
      A(I,K)= 0.0
      B(I,K)= 0.0
   11 D(I,K)= 0.0
      IF (J-1)  12,12,14
   12 DO 13 I=1,N
      B(I,I)= 1.0
   13 G(I)= CIN(I)
      CALL BAND(J)
      GO TO 10
   14 DO 15 K=1,NM1
   15 B(N,K)= Z(K)
      IF (J-NJ)  16,18,18
   16 DO 17 I=1,NM1
      PP= U(I)/DIF(I)*(C(N,J+1)-C(N,J-1))/2.0
      PPP= U(I)/DIF(I)*(C(N,J+1)+C(N,J-1)-2.0*C(N,J))
      CP= (C(I,J+1) - C(I,J-1))/2.0
      A(I,I)= - 1.0 + PP/2.0 - H*V(J)/2.0/DIF(I)
      B(I,I)= 2.0 - PPP
      D(I,I)= - 1.0 - PP/2.0 + H*V(J)/2.0/DIF(I)
      A(I,N)= U(I)/DIF(I)*(CP/2.0 - C(I,J))
      B(I,N)= 2.0*U(I)/DIF(I)*C(I,J)
      D(I,N)= - U(I)/DIF(I)*(CP/2.0 + C(I,J))
   17 G(I)= - PPP*C(I,J) - PP*CP
      CALL BAND(J)
      GO TO 10
   18 DO 19 I=1,NM2
      PP= U(I)/DIF(I)*(C(N,NJ)-C(N,NJ-2))
      Y(I,I)= - 1.0
      A(I,I)= PP
      B(I,I)= 1.0
      Y(I,NM1)= S(I)*DIF(NM1)/S(NM1)/DIF(I)
      B(I,NM1)= - Y(I,NM1)
      Y(I,N)= (S(I)/S(NM1)*U(NM1)*CRO-U(I)*C(I,NJ-1))/DIF(I)
      B(I,N)= - Y(I,N)
   19 G(I)= PP*C(I,NJ-1)
      G(NM1)= CRO
      A(NM1,NM1)= 1.0
      CALL BAND(J)
      AMPO= AMP
      AMP= (U(NM1)*CRO*(C(N,NJ-2)-C(N,NJ))+DIF(NM1)*(C(NM1,NJ-2)-C(NM1,
     1NJ)))/2.0/H/(CIN(NM1)-CRO)/DIF(NM1)
      IF (ABSF(AMP-AMPO)-0.00001 *ABSF(AMP))  22,22,20
   20 IF (JCOUNT-10)  8,8,21
   21 PRINT 108
   22 PRINT 109, AMP,(REF(I),C(I,1),C(I,NJ-1),I=1,NM1)
      GO TO 98
      END
```

## Discussion and conclusions

The procedure outlined here for solving coupled, nonlinear, difference equations by linearization and subsequent iteration is quite general and flexible and has proved useful in a number of problems. For special problems it may be possible to devise more efficient methods, but with a loss of generality and an expense of effort.

Two other methods might occur to one faced with a problem of the type treated here. One is to linearize and decouple the equations by taking the coefficients of the derivatives to be given by a trial solution, for example, approximate $c_1 \, dc_2/dx$ by $c_1^0 \, dc_2/dx$. Then the decoupled equations are solved one after another in a cyclic process producing new functions to be used as a trial solution. In general, the convergence behavior is poorer than for the present method, although there are special problems where the coupling is not strong and the method works.

A second method would be to treat the problem as an initial-value problem and to fabricate the needed initial conditions. This method requires little storage space, but the adjustment of the added initial conditions so as to satisfy the boundary conditions at $x = L$ can be tricky or impossible.

The errors in the present method arise from three sources: convergence errors for the nonlinear problem (which can be made negligibly small here), errors in the difference approximation to the differential equations (which decrease with the mesh distance $h$), and round-off errors in the computer (which increase as the mesh distance is decreased). Convergence may not be possible if there are sharp variations of the unknowns in some region of $x$; in such a case a singular-perturbation method may be appropriate.

A remark is in order for first- and third-order differential equations. For the purpose of computation, a third-order equation involving, say, $d^3f/dx^3$ could be replaced by two second order equations (with $f = dc_1/dx$ and $c_2 = df/dx$) or a first- and a second-order equation (with $c_1 = f$ and $c_2 = df/dx$). In this way, the finite-difference forms still involve only the points at $j - 1, j$, and $j + 1$. For a first-order equation, it is probably better to use a backward difference rather than a central difference. The order of the approximation will still be $h^2$ if the coefficient takes on its average value:

$$K\frac{dc_1}{dx} = \frac{K(j) + K(j-1)}{2} \frac{C_1(j) - C_1(j-1)}{h} + 0(h^2). \qquad \text{(C-30)}$$

# *Index*

## A

Activation energy, 173–74
Activity
 absolute, 33
 relative, 53–54
Activity coefficients, 13, 22–23,
 27–28, 33–39, 45–46, 49–55, 59–60,
 76–104, 119, 121–22, 124–28, 133–38,
 152, 236–37, 242, 267, 383–84
  Debye-Hückel limiting law, 49, 84, 85,
  86, 88
  effect of dissociation, 37–38, 61, 78,
  98–101, 103–04
  electrostatic theory, 72, 78–88
  pressure dependence, 34, 59–60
  single ionic, 76, 89, 94, 124–25, 215,
  220, 235, 240, 242
Adsorption, 123, 141, 145, 146–48, 149–51,
 161–62, 164–65, 191–92
Amalgam, 30, 40–41, 48, 50, 53–54, 61–62,
 109, 136
Annulus, 272, 316–20, 322–24, 334
Anode, 3–5, 10, 14–15, 17, 23
Anodic protection, 185
Arsenic, 112–13

## B

Binary electrolyte, 12, 35–38, 44–47, 88–92,
 223, 241–46, 250–51, 255–57, 266–69,
 284, 286, 331–34, 336–37, 366, 373–79,
 380, 384–85, 394, 397, 398
Bisulfate ions, 98–101, 104, 364–68, 374
Boltzmann distribution, 79, 86, 153, 180,
 246

## C

Calomel electrode, 58, 114–16, 119–23,
 151–52, 153, 157–59, 163–64
Carbon dioxide, 109, 112
Catalysis, 109, 111–13, 331
Cathode, 3–5, 10, 12, 15–18, 20–22, 24
Centrifugal field, 254, 262
Charge density, 67, 79, 141, 192, 193–95,
 232, 284–85
Charging processes, 82–83, 87–88, 101–02
Chemical potential, 30–31
Colligative properties, 35
Complexes, 109, 115, 119
Concentrated solutions, transport in,
 239–52, 330–33